HIGH PERFORMANCE

CLOCK DISTRIBUTION

NETWORKS

edited by

Eby G. Friedman
University of Rochester

Reprinted from
a Special Issue of
JOURNAL OF VLSI SIGNAL PROCESSING SYSTEMS
for Signal, Image, and Video Technology
Vol. 16, Nos. 2 & 3
June/July 1997

KLUWER ACADEMIC PUBLISHERS
Boston / Dordrecht / London

Journal of VLSI SIGNAL PROCESSING SYSTEMS for Signal, Image, and Video Technology

Volume 16—1997

Special Issue on High Performance Clock Distribution Networks

Distributors for North America:
Kluwer Academic Publishers
101 Philip Drive
Assinippi Park
Norwell, Massachusetts 02061 USA

Distributors for all other countries:
Kluwer Academic Publishers Group
Distribution Centre
Post Office Box 322
3300 AH Dordrecht, THE NETHERLANDS

Library of Congress Cataloging-in-Publication Data

A C.I.P. Catalogue record for this book is available
from the Library of Congress.

ISBN 978-1-4684-8442-7 ISBN 978-1-4684-8440-3 (eBook)
DOI 10.1007/978-1-4684-8440-3

Softcover reprint of the hardcover 1st edition 1997

Printed on acid-free paper.

Journal of VLSI Signal Processing 16, 113–116 (1997)

High Performance Clock Distribution Networks

As semiconductor technologies operate at increasingly higher speeds, system performance has become limited not by the delays of the individual logic elements and interconnect but by the ability to synchronize the flow of the data signals. Different synchronization strategies have been considered, ranging from completely asynchronous to fully synchronous. However, the dominant synchronization strategy within industry will continue to be fully synchronous clocked systems. Systems ranging in size from medium scale circuits to large multimillion transistor microprocessors and ultra-high speed supercomputers utilize fully synchronous operation which require high speed and highly reliable clock distribution networks. Distributing the clock signals within these high complexity, high speed processors is one of the primary limitations to building high performance synchronous digital systems. Greater attention is therefore being placed on the design of clock distribution networks for large VLSI-based systems.

In a synchronous digital system, the clock signal is used to define the time reference for the movement of data within that system. Since this function is vital to the operation of a synchronous system, much attention has been given to the characteristics of these clock signals and the networks used in their distribution. Clock signals are often regarded as simple control signals; however, these signals have some very special characteristics and attributes. Clock signals are typically loaded with the greatest fanout, travel over the greatest distances, and operate at the highest speeds of any signal, either control or data, within the entire system. Since the data signals are provided with a temporal reference by the clock signal, the clock waveforms must be particularly clean and sharp. Furthermore, these clock signals are strongly affected by technology scaling in that long global interconnect lines become highly resistive as line dimensions are decreased. This increased line resistance is one of the primary reasons for the increasing significance of clock distribution networks on synchronous performance. The control of any differences in the delay of the clock signals can also severely limit the maximum performance of the entire system and create catastrophic race conditions in which an incorrect data signal may latch within a register.

In a synchronous system, each data signal is typically stored in a latched state within a bistable register awaiting the incoming clock signal, which determines when the data signal leaves the register. Once the enabling clock signal reaches the register, the data signal leaves the bistable register and propagates through the combinatorial network, and for a properly working system, enters the next register and is fully latched into that register before the next clock signal appears. Thus, the delay components that make up a general synchronous system are composed of the following three subsystems: 1) the memory storage elements, 2) the logic elements, and 3) the clocking circuitry and distribution network. Interrelationships among these three subsystems of a synchronous digital system are critical to achieving maximum levels of performance and reliability.

A number of fundamental topics in the field of high performance clock distribution networks are covered in this special issue. This special issue is composed of ten papers from a variety of academic and industrial institutions. Topically, these papers can be grouped within three primary areas. The first topic area deals with exploiting the localized nature of clock skew. The second topic area deals with the implementation of these clock distribution networks while the third topic area considers more longer range aspects of next generation clock distribution networks.

Until very recently, clock skew was considered to behave more as a global parameter rather than a local parameter. Clock skew was budgeted across a system, permitting a particular value of clock skew to be subtracted from the minimum clock period. This design perspective misunderstood the nature of clock skew, not recognizing that clock skew is local in nature and is specific to a particular local data path. Furthermore, if the data and clock signals flow in the same direction with respect to each other (i.e., negative clock skew), race conditions are created in which quite possibly the race could be lost (i.e., the clock signal would arrive at the register and shift the previous data signal out of the register before the current data signal arrives and is successfully latched). Thus strategies have only recently been developed to not only ensure that these race conditions do not exist, but to also exploit localized clock skew in order to provide additional time for the signals in the worst case paths to reach and set-up in the final register of that local data path, effectively permitting the synchronous system to operate at a higher maximum clock frequency. Thus, the localized clock skew of each local data path is chosen so as to minimize the system-wide clock period while ensuring that no race conditions exist. This process of determining a set of local clock skews for each local data path is called *clock skew scheduling* or clock skew optimization and is used to extract what has been called *useful clock skew*. Other names have been mentioned in the literature to describe different aspects of this behavior of clock distribution networks such as negative clock skew, double-clocking, deskewing data pulses, cycle stealing, and prescribed skew.

Four papers are included in this special issue that present different approaches and criteria for determining an optimal clock skew schedule and designing and building a clock distribution network that satisfies this target clock skew schedule. Little material has been published in the literature describing this evolving performance optimization methodology in which localized clock skew is used to enhance circuit performance while removing any race conditions. These performance improvements come in different flavors, such as increased clock frequency, decreased power dissipation, and quite recently, decreased $L\ di/dt$ voltage drops.

P. Vuillod, L. Benini, A. Bogliolo, and G. DeMicheli describe a new criterion for choosing the local clock skews. In their paper, "Clock Skew Optimization for Peak Current Reduction," the local clock skews are chosen so as to shift the relative transition time within the data registers, thereby decreasing the maximum peak current drawn from the power supply, minimizing the $L\ di/dt$ voltage drops within the power/ground lines. A related clock skew scheduling algorithm is described and demonstrated on benchmark circuits. This paper represents a completely new technique for minimizing inductive switching noise as well as describing an additional advantage to applying clock skew scheduling techniques.

Hong-Yean Hsieh, Wentai Lu, Paul Franzon, and Ralph Cavin III present a new approach for scheduling and implementing the clock skews. In their paper, "Clocking Optimization and Distribution of Digital Systems with Scheduled Skews," the authors describe a two step process for implementing a system that exploits non-zero clock skew. The initial step is to choose the proper values of the clock skews, while the final step is to build a system that is tolerant to process and environmental variations. The authors present an innovative self-calibrating all digital phase lock loop implementation to accomplish this latter task. Experimental results describing a manufactured circuit are also presented.

Jose Neves and Eby G. Friedman present a strategy for choosing a set of local clock skews while minimizing the sensitivity of these target clock skew values to variations in process parameters. Their paper, "Buffered Clock Tree Synthesis with Non-Zero Clock Skew Scheduling for Increased Tolerance to Process Parameter Variations," describes a theoretical framework for evaluating clock skew in synchronous digital systems and introduces the concept of a *permissible range* of clock skew for each local data path. Algorithms are presented for determining a clock skew schedule tolerant to process variations. These algorithms are demonstrated on benchmark circuits.

Joe Gufeng Xi and Wayne Wei-Ming Dai describe a related approach to implementing the physical layout of the clock tree so as to satisfy a non-zero clock skew schedule. In their paper, "Useful-Skew Clock Routing with Gate Sizing for Low Power Design," the authors present a new formulation and related algorithms of the clock routing problem while also including gate sizing to minimize the power dissipated within both the logic and the clock tree. A combination of simulated annealing and heuristics is used to attain power reductions of approximately 12% to 20% as compared with previous methods of clock routing targeting zero (or negligible) clock skew with no sacrifice in maximum clock frequency.

Another area of central importance to the design of high speed clock distribution networks is the capability for efficiently and effectively implementing these high performance networks. This implementation process is composed of two types: synthesis and layout. Four papers are included in this special issue that discuss this primary topic area of design techniques for physically implementing the clock distribution network.

Shatanu Ganguly, Daksh Lehther, and Satyamurthy Pullela describe the clock distribution design methodology used in the development of the PowerPC Microprocessor. In their paper, "Clock Distribution Methodology for PowerPC™ Microprocessors," the authors review specific characteristics and related constraints pertaining to the

PowerPC clock distribution network. The architecture of the clock distribution network is presented, and the clock design flow is discussed. Each step of the design process, synthesis, partitioning, optimization, and verification are reviewed and statistical data are presented. This paper represents an interesting overview of many issues and considerations related to timing and synchronization that are encountered when designing high performance microprocessors.

David J. Hathaway, Rafik R. Habra, Erich C. Schanzenbach, and Sara J. Rothman describe in their paper, "Placement, Chip Optimization, and Routing for IBM IC Technology," an industrial approach for physically optimizing the clock distribution network in high performance circuits. Iterative placement algorithms are applied to refine the timing behavior of the circuit. Optimization tools are used to minimize clock skew while improving wireability. Manual intervention is permitted during clock routing to control local layout constraints and restrictions. This tool has been successfully demonstrated on a number of IBM circuits.

Andrew Kahng and C.-W. Albert Tsao present new research in the development of practical automated clock routers. Specifically, in their paper, "Practical Bounded-Skew Clock Routing," the authors present problem formulations and related algorithms for addressing clock routing with multi-layer parasitic impedances, non-zero via resistances and capacitances, obstacle avoidance within the metal routing layers, and hierarchical buffered tree synthesis. A theoretical framework and new heuristics are presented and the resulting algorithms are validated against benchmark circuits.

Keith M. Carrig, Albert M. Chu, Frank D. Ferraiolo, John G. Petrovick, P. Andrew Scott, and Richard J. Weiss report in their paper, "A Clock Methodology for High Performance Microprocessors," on an efficient clock generation and distribution methodology that has been applied to the design of a high performance microprocessor (a single-chip 0.35 μm PowerPC microprocessor). Key attributes of this methodology include clustering and balancing of clock loads, variable wire widths within the clock router to minimize skew, hierarchical clock wiring, automated verification, an interface to commercial CAD tools, and a complete circuit model of the clock distribution network for simulation purposes. The microprocessor circuit technology is described in detail, providing good insight into how the physical characteristics of a deep submicrometer CMOS technology affect the design of a high performance clock distribution network.

A third topic area of investigation in high performance clock distribution networks deals with next generation strategies for designing and implementing the clock distribution network. One subject that has periodically been discussed over the past ten years is the use of electro-optical techniques to distribute the clock signal. This subject is discussed in great detail in the first paper in this topic area. The second paper offers new strategies for dealing with multi-gigahertz frequency systems built in superconductive technologies.

Stuart K. Tewksbury and L. A. Hornak provide a broad review of the many approaches for integrating optical signal distribution techniques within electronic systems with a specific focus on clock distribution networks. In their paper, "Optical Clock Distribution in Electronic Systems," the authors first present chip level connection schemes followed by board level connection strategies. Common optical strategies applied to both of these circuit structures are diffractive optical elements, waveguide structures, and free-space paths to provide the interconnection elements. General strategies for optical clock distribution are presented using single-mode and multi-mode waveguides, planar diffractive optics, and holographic distribution. Interfacing the electro-optical circuitry to VLSI-based systems is also discussed.

Kris Gaj, Eby G. Friedman, and Marc J. Feldman present new methodologies for designing clock distribution networks that operate at multi-gigahertz frequencies. In their paper, "Timing of Multi-Gigahertz Rapid Single Flux Quantum Digital Circuits," different strategies for distributing the clock signal based on a recently developed digital superconductive technology is presented. This technology, Rapid Single Flux Quantum (RSFQ) logic, provides a new opportunity for building digital systems of moderate complexity that can operate well into the gigahertz regime. Non-zero clock skew timing strategies, multi-phase clocking, and asynchronous timing are some of the synchronization paradigms that are reviewed in the context of ultra-high speed digital systems.

This special issue presents a number of interesting strategies for designing and building high performance clock distribution networks. Many aspects of the ideas presented in these articles are being developed and applied today in next generation high performance microprocessors. As the microelectronics community approaches and quickly exceeds the one gigahertz clock frequency barrier for silicon CMOS, aggressive strategies will be required to provide the necessary levels of circuit reliability, power dissipation density, chip die area, design productivity, and circuit testability. The design of the clock distribution network is one of the primary concerns at the center of each of these technical goals.

The guest editor would like to thank the Editor, S.Y. Kung, for suggesting and supporting the development of this special issue, Carl Harris for his continued interest and friendship while developing important publications for the microelectronics community, Lorraine M. Ruderman, Julie Smalley, and the staff at Kluwer Academic Press for their support in producing this special issue, and Ruth Ann Williams at the University of Rochester for her dependable and cheerful assistance throughout the entire review and evaluation process. It is my sincere hope that this special issue will help augment and enhance the currently scarce material describing the design, synthesis, and analysis of high performance clock distribution networks.

Eby G. Friedman
University of Rochester

Eby G. Friedman was born in Jersey City, New Jersey in 1957. He received the B.S. degree from Lafayette College, Easton, PA, in 1979, and the M.S. and Ph.D. degrees from the University of California, Irvine, in 1981 and 1989, respectively, all in electrical engineering.

He was with Philips Gloeilampen Fabrieken, Eindhoven, The Netherlands, in 1978 where he worked on the design of bipolar differential amplifiers. From 1979 to 1991, he was with Hughes Aircraft Company, rising to the position of manager of the Signal Processing Design and Test Department, responsible for the design and test of high performance digital and analog IC's. He has been with the Department of Electrical Engineering at the University of Rochester, Rochester, NY, since 1991, where he is an Associate Professor and Director of the High Performance VLSI/IC Design and Analysis Laboratory. His current research and teaching interests are in high performance microelectronic design and analysis with application to high speed portable processors and low power wireless communications.

He has authored many papers and book chapters in the fields of high speed and low power CMOS design techniques, pipelining and retiming, and the theory and application of synchronous clock distribution networks, and has edited one book, *Clock Distribution Networks in VLSI Circuits and Systems* (IEEE Press, 1995). Dr. Friedman is a Senior Member of the IEEE, a Member of the editorial board of *Analog Integrated Circuits and Signal Processing*, Chair of the VLSI track for ISCAS '96 and '97, Technical Co-Chair of the International Workshop on Clock Distribution Networks, and a Member of the technical program committee of a number of conferences. He was a Member of the editorial board of the *IEEE Transactions on Circuits and Systems II: Analog and Digital Signal Processing*, Chair of the VLSI Systems and Applications CAS Technical Committee, Chair of the Electron Devices Chapter of the IEEE Rochester Section, and a recipient of the Howard Hughes Masters and Doctoral Fellowships, an NSF Research Initiation Award, an Outstanding IEEE Chapter Chairman Award, and a University of Rochester College of Engineering Teaching Excellence Award.

Journal of VLSI Signal Processing 16, 117–130 (1997)

Clock Skew Optimization for Peak Current Reduction

L. BENINI, P. VUILLOD,* A. BOGLIOLO,† AND G. DE MICHELI
Computer Systems Laboratory, Stanford University, Stanford, CA 94305-9030

Received August 1, 1996; Revised October 21, 1996

Abstract. The presence of large current peaks on the power and ground lines is a serious concern for designers of synchronous digital circuits. Current peaks are caused by the simultaneous switching of highly loaded clock lines and by the signal propagation through the sequential logic elements. In this work we propose a methodology for reducing the amplitude of the current peaks. This result is obtained by clock skew optimization. We propose an algorithm that, for a given clock cycle time, determines the clock arrival time at each flip-flop in order to minimize the current peaks while respecting timing constraint. Our results on benchmark circuits show that current peaks can be reduced without penalty on cycle time and average power dissipation. Our methodology is therefore well-suited for low-power systems with reduced supply voltage, where low noise margins are a primary concern.

1. Introduction

Clock skew is usually described as an undesirable phenomenon occurring in synchronous circuits. If clock skew is not properly controlled, unexpected timing violations and system failures are possible. Mainly for this reason, research and engineering effort has been devoted to tightly control the misalignment in the arrival times of the clock [1]. Although clock-skew control is still an open issue for extremely large chip-level and board-level designs, recently proposed algorithms for skew minimization have reported satisfying results [1–4]. For a large class of systems skew control can therefore be achieved with sufficient confidence margin.

Conservative design styles (such as those adopted for FPGAs) explicitly discourage "tampering with the clock" [5]. Nevertheless, the arrival time of the clock is often purposely skewed to achieve high performance in more aggressive design styles. In the past, several algorithms for *cycle-time minimization* have been proposed [6–10]. The common purpose of these methods was to find an *optimum clock-skewing strategy* that allows the circuit to run globally faster. Average power dissipation can also be reduced by clock skewing coupled with gate resizing [11].

In this work, we discuss the productive use of clock skew in a radically new context. We target the minimization of the peak power supply current. Peak current is a primary concern in the design of power distribution networks. In state-of-the-art VLSI systems, power and ground lines must be over-dimensioned in order to account for large current peaks. Such peaks determine the maximum voltage drop and the probability of failure due to electromigration [12]. In synchronous systems, this problem is particularly serious. Since *all sequential elements* are clocked, huge current peaks are observed in correspondence of the clock edges. These peaks are caused not only by the large clock capacitance, but also by the switching activity in the sequential elements and by the propagation of the signals to the first levels of combinational logic.

In this paper, we focus application specific integrated circuits implemented with semi-custom technology. We do not address the complex issues arising in custom-designed chips with clock frequencies over 150 MHz. For such high-end circuits, achieving adequate skew control is already a challenging task. We assume a single-clock edge-triggered clocking style, because it represents the worst case condition for current peaks. We propose an algorithm that determines the clock

*On leave from INPG—CSI, Grenoble, France.
†Also with DEIS, Università di Bologna, Italy.

arrival times at the flip-flops in order to minimize the maximum current on the power supply lines, while satisfying timing constraints for correct operation.

In addition, we propose *a clustering* technique that groups flip-flops so that they can be driven by the same clock driver. Since the number of sequential elements is generally large, it would not be practically feasible to specify a skew value for each one of them. In our tool, the user can specify the maximum number of clock drivers, and the algorithm will find a clustering that always satisfies the timing constraints while minimizing the peak current.

Any optimization technique based on clock control cannot neglect the structure and the performance of the clock distribution network and clock buffers [13]. Implementing skewed clocks with traditional buffer architectures imposes sizable power costs that may swamp the advantages obtained by clock skew. Our clocking strategy is based on a customized driver that achieves good skew control with negligible cost in power, area and performance.

Our technique is particularly relevant for low-power systems with reduced supply voltage, where the noise margins on power and ground are extremely low. Experimental results show that our method not only reduces the current peaks, but it does not increase the average power consumption of the system. We tested our approach on several benchmark circuits. On average, current peak reduction of more than 30% has been observed. Average power dissipation is unchanged and timing constraints are satisfied.

The results were further validated by accurate *post-layout* electrical simulation of circuits of practical size (over 100 flip-flops). The power dissipation due to the clock network and buffers was taken into account. The post-layout results confirm the practical interest of our method and the effectiveness of our clustering heuristic.

2. Skew Optimization

It is known that clock skew can be productively exploited for obtaining faster circuits. *Cycle borrowing* is an example of such practice: if the critical path delay between two consecutive pipeline stages is not balanced, it is possible to skew the clock in such a way that the slower logic has more time to complete its computation, at the expense of the time available for the faster logic. For large and unstructured sequential networks, finding the best cycle borrowing strategy is a complex task that requires the aid of automatic tools.

2.1. Background

We will briefly review the basic concepts needed for the formal definition of the skew optimization problem. The interested reader can refer to [1, 7, 9] for further information. Clock-skew optimization is achieved by assigning an arrival time to the local clock signals of each sequential element in the circuit. We consider rising-edge-triggered flip-flops and single clock. The clock period is T_{clk}. For the generic flip-flop i ($i = 1, 2, \ldots, N$, where N is the number of flip-flops in the network) we define its *arrival time* T_i, $0 \leq T_i < T_{\text{clk}}$. The arrival time represents the amount of skew between the reference clock and the local clock signal of flip-flop i. A *clock schedule* is obtained by specifying all arrival times T_i. Obviously not all clock schedules are valid. The combinational logic between the flip-flops has finite delay. The presence of delays imposes constraints on the relative position of the arrival times.

The classical clock-skew optimization problem can be stated as follow: *find the optimal clock schedule* $\mathbf{T} = [T_1, T_2, \ldots, T_N]$ *such that no timing constraint is violated and the cycle time* T_{clk} *is minimized.* This problem has been analyzed in detail and many solutions have been proposed. Here we follow the approach presented in [7] where edge-triggered flip-flops are considered.

We assume for simplicity that all flip-flops have the same setup and hold times, respectively called T_{SU} and T_{HO}. If there is at least one combinational path from the output of flip-flop i to the input of flip-flop j, we call the maximum delay on these paths $\delta_{i,j}^{\max}$. The minimum delay $\delta_{i,j}^{\min}$ is similarly defined. If no combinational path exists between the two flip-flops, $\delta_{i,j}^{\max} = -\infty$ and $\delta_{i,j}^{\min} = +\infty$. For each pair of flip-flops i and j, two constraints must be satisfied.

First, if a signal propagating from the output of i reaches the input of j before the clock signal for j is arrived, the data will propagate through two consecutive sequential elements in the same clock cycle. This problem is called *double clocking* and causes failure. The first kind of constraints prevents double clocking:

$$T_i + \delta_{i,j}^{\min} \geq T_j + T_{\text{HO}} \tag{1}$$

On the other hand, if a signal propagating from i to j arrives with a delay larger than the time difference between the next clock edge on j and the current clock edge on i, the circuit will fail as well. This phenomenon is called *zero clocking*. Zero clocking avoidance is enforced by the following constraint:

$$T_i + T_{SU} + \delta_{i,j}^{max} \leq T_j + T_{clk} \tag{2}$$

Input and output impose constraints as well. Input constraints have the same format as regular constraints, where the constant value of the input arrival time T_{in} replaces the variable T_i. For output constraints the variable T_j is replaced by the constant output required time T_{out}.

The total number of constraint inequalities constructed by this method is $O(N^2 + I + O)$, where, I and O are the number of inputs and outputs respectively. In practice, this number can be greatly reduced. Techniques for the reduction of the number of constraints are described in [6, 8] and are not discussed here for space reasons.

Example. We obtain the constraint equations for the circuit in Fig. 1. There are two variables T_1 and T_2, representing the skew of the clocks CLK1 and CLK2. The clock period is T_{clk}. We assume that $T_{SU} = T_{HO} = 0$. The constraints for variable T_1 are the following:

$$T_1 + \delta_{1,2}^{max} \leq T_2 + T_{clk}$$
$$T_1 + \delta_{1,2}^{min} \geq T_2$$
$$T_1 + \delta_{1,out}^{max} \leq T_{out} + T_{clk}$$
$$T_{in} + \delta_{in,1}^{min} \geq T_1$$

Moreover, $0 \leq T_1 \leq T_{clk}$. Similar constraints hold for T_2. We have eliminated one input constraint and one output constraint because we assume that skews are positive and that the circuit with no skews was originally satisfying all input and output constraints. Notice that all constraints are linear. The feasibility of a set of linear constraints can be checked in polynomial time by the Bellman-Ford algorithm [14].

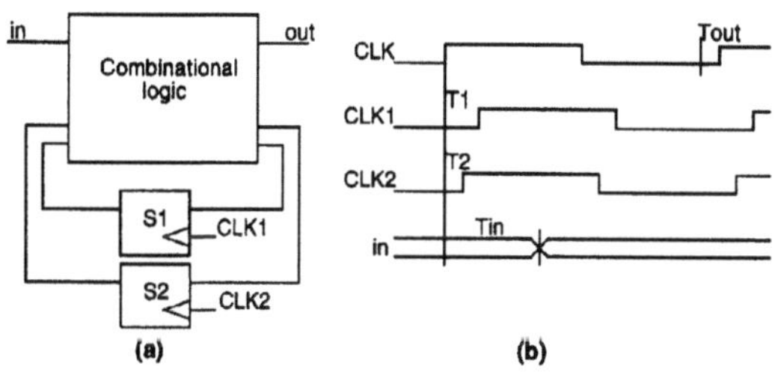

Figure 1. (a) Example circuit, with two flip-flops. (b) Timing waveform representing the skewed clocks.

An important practical consideration that is often overlooked in the literatures is the generation of the skewed clocks. Although generating delays is a relatively straightforward task, the cost (in power, area and signal quality degradation) of the delay elements is an important factor in the evaluation of optimization techniques based on clock skewing. We will first concentrate on the theory of clock skew optimization for the sake of simplicity. Circuits for the generation of skewed clocks will be discussed in a later section.

Cycle time minimization is an optimization problem targeting the minimization of a linear cost function (i.e., $F(T_1, T_2, \ldots, T_N, T_{clk}) = [0, 0, \ldots, 0, 1] \cdot [T_1, T_2, \ldots, T_N, T_{clk}]$) of linearly constrained variables. It is therefore an instance of the well-known *linear programming* (LP) problem. Several efficient algorithms for the solution of LP have been proposed in the past [15]. Our problem is radically different and substantially harder. It can be stated as follows: *find a clock schedule such that the peak current of the circuit is minimum.* The cost function that we want to minimize is *not* linear in the variables T_i. In the following subsection, we discuss this issue in greater detail.

2.2. *Cost Function*

In peak current minimization, the constraints are exactly the same as for the traditional cycle time minimization, the only difference being that we consider T_{clk} as a constant. Unfortunately, our cost function is much more complex. Ideally, we would like to minimize the maximum current peak that the circuit can produce. This is however a formidable task, because such peak can be found by exhaustively simulating the system for all possible input sequences (and a circuit level simulation would be required, because traditional gate-level simulators do not give information on current waveforms). To simplify the problem, we make two important assumptions. First, we only minimize the current peak directly caused by clock edges (i.e., caused by the switching of clock lines and sequential elements' internal nodes and outputs). This approximation is justified by experimental evidence. In all circuits we have tested, the largest current peaks are observed in proximity of the clock edges. The current profile produced by the propagation of signals through the combinational logic is usually spread out and its maximum value is sensibly smaller.

Notice that we are *not* neglecting the combinational logic, but we consider its current as a phenomenon on which we have no control. Again, this choice is motivated by experimental evidence: our tests show that in most cases, the current profile of the combinational logic is not very sensitive to the clock schedule. For some circuits, the combinational logic may be dominant and strongly influenced by the clock schedule. We will discuss this case in a later section.

The second approximation regards the shape of the current waveform. Each sequential element produces two peaks, one related to the rising edge of the clock, and the other to the falling edge. For a given flip-flop, the shape of the current peaks is weakly pattern dependent. We approximate the current peaks produced by each sequential element (or group of sequential elements) with two triangular shapes, that are fully characterized by four parameters: starting time t_s, maximum time t_m, maximum value current I_m and final time t_f.

To compute these parameters we run several current simulations [16] (see Section 4) and we obtain current waveform envelopes $I_{av}(t)$ ($I_{av}(t)$ is obtained by averaging the current at t on different input patterns). For each peak of the curve I_{av}, we define the four parameters as shown in Fig. 2: t_s is the time at which the current first reaches 1% of the maximum value, t_f is the time at which the current decreases below 1% of the maximum value, I_m and t_m are respectively the maximum current value and the time when it is reached. Experimentally we observed that the triangular approximation is satisfactory for the current profiles of the sequential elements. For combinational logic, this approximation is generally inaccurate. The current profile of combinational logic is more adequately modeled by a *piecewise linear* approximation. Fortunately, any piecewise linear function can be decomposed in the sum of one or more triangular functions.

The total current is the sum of the current contributions represented as triangular shapes. Every flip-flop i has two associated contributions $\Delta_i^r(t, T_i)$ and $\Delta_i^f(t, T_i)$, representing respectively the current drawn on the raising and falling edge of the clock. Notice that such contributions are functions of time t and of the clock arrival time T_i. In fact, the curve translates rigidly with T_i. The current drawn by the combinational logic is approximated with a sum of triangles (i.e., a piecewise linear waveform) $\Delta_C(t)$. Note that $\Delta_C(t)$ is *not* a function of the arrival time of any clock. The total current is the sum of the contributions due to flip-flops and combinational logic:

$$I_{tot}(t, \mathbf{T}) = \Delta_C(t) + \sum_{i=1}^{N} \Delta_i^r(t, T_i) + \sum_{i=1}^{N} \Delta_i^f(t, T_i) \tag{3}$$

We clarify this equation through an example.

Example. The current profiles for the flip-flops of the circuit in Fig. 1 are shown in Fig. 3 for one assignment of T_1 and T_2. The current profile of the combinational logic for this example is shown in Fig. 4 with its approximation.

The contribution of a flip-flop is approximated by two triangular shapes. The first corresponds to the

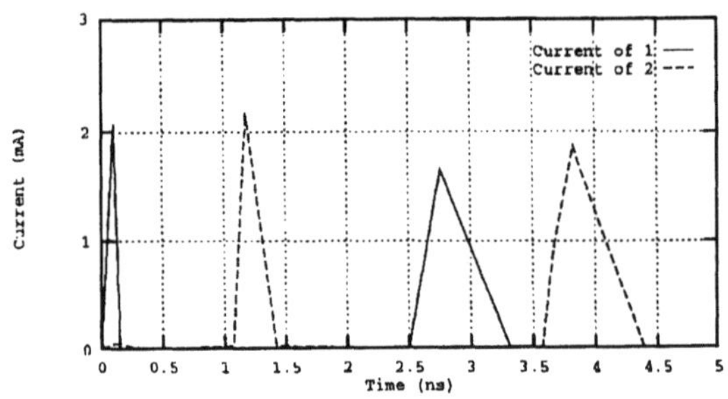

Figure 3. Current profiles for the two flip-flops 1 and 2 from simulation of our example circuit.

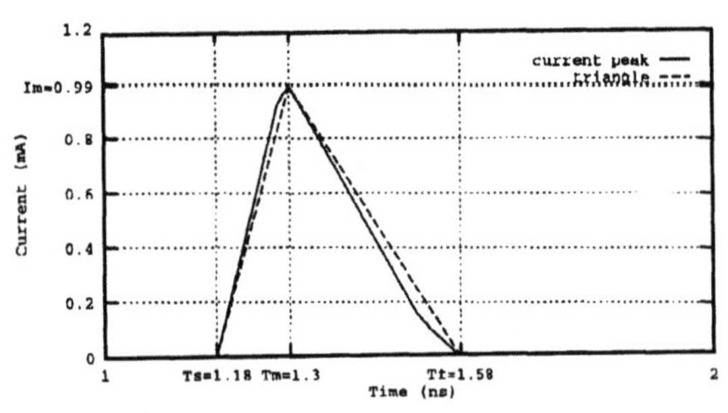

Figure 2. The four parameters characterizing the triangular approximation of the average current profile. t_s and t_e are the times at which the current reaches 1% of its maximum value.

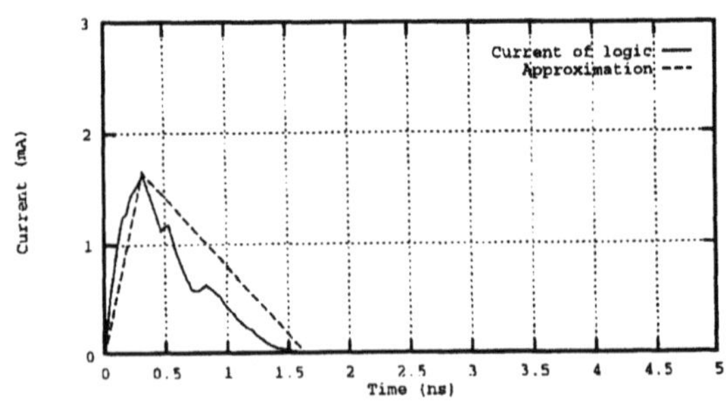

Figure 4. Current profile corresponding to the combinational logic from simulation of our example circuit. The dashed line is its piecewise linear approximation.

rising edge of the clock, the second to the falling edge. Here we have $T_1 = 0$ ns and $T_2 = 1.07$ ns. Notice that the current profile of flip-flop 2 is shifted to the right. The profiles for the two flip-flop do not have exactly the same shape because they are differently loaded. Notice that when $T_1 = T_2$ the two current profiles of the flip-flops are perfectly overlapped. When $T_1 \neq T_2$, the two contributions are skewed.

The cost function F that approximates the peak current is the maximum value of the (approximate) current waveform over the clock period T_{clk}:

$$F(\mathbf{T}) = \max_{t \in [0, T_{\mathrm{clk}}]} \{I_{\mathrm{tot}}(t, \mathbf{T})\} \tag{4}$$

For the above example, the value of the cost function $F(T_1, T_2)$ is the maximum value of the sum of the five triangles over the clock period T. In this case $F(0, 1.07) = 2.7$, whereas initially $F(0, 0) = 4.2$. Our target is to find the optimum clock schedule $\mathbf{T}_{\mathrm{opt}}$ which minimizes the cost function F, while satisfying the timing constraints for correct operation of the circuit.

3. Peak Current Minimization

We now describe our approach to the minimization of the cost function described in the previous section. The first key result of this section is summarized in the following proposition.

Theorem 1. *The cost function F of Eq.* (4) *can be evaluated in quadratic time* (*in the number of triangular contributions*).

Proof: The proof of this theorem is given in a constructive fashion, by describing a $O(N_\Delta^2)$ algorithm (N_Δ is the number of triangular current contributions) for the evaluation of the cost function. The algorithm is based on the observation that the maximum of the cost function can be attained in a finite number of points, namely the points of maximum of the triangles that compose it. In order to evaluate the value of F in one of such points, we must check if the corresponding triangle is overlapping with any of the other contributions. The quadratic complexity stems from this check: for each maximum value v_i (`val` in the pseudo-code), we check if its corresponding triangle Δ_i is overlapping with any other triangle. In case there is overlap, v_i is incremented by the value of the overlapping waveform at the maximum point. Thus, we have two nested loops with iteration bound N_Δ. The pseudo-code of the algorithm is shown in Fig. 5. □

```
/* Let T[i] (i..N) be the variable vector */
/* Delta_orig[i] (i..2N+1) are the 2N+1 contributions when T[i]=0 */
float evaluate (T)
  /* computes the contributions for the vector T */
  Delta = translate_triangles (Delta_orig, T);
  max = 0;
  foreach (c1 in [0..2N])
    val = max(Delta([c1]));
    foreach (c2 in [0..2N])
      if (c2 != c1) then
        if (overlap (Delta[c1], Delta[c2])) then
          /* we look if the 2 triangles overlap and add the value */
          /* of c2 at the maximum point of c1 */
          val += get_value (Delta[c2], time_max (Delta[c1]));
        endif;
      endif;
    endfor;
    if (val ≥ max) then max = val;
  endfor;
  return (max);
end evaluate;
```

Figure 5. $O(N^2)$ algorithm for the computation of the cost function F.

The second key result is summarized by the following theorem:

Theorem 2. *The peak current minimization problem is an instance of the* **constrained DC optimization problem** (*DC optimization problems are those where the cost function can be expressed as the difference of two concave functions* [17]).

Proof: The proof of the theorem is straightforward. The cost function $F(\mathbf{T})$ is the maximum over a finite interval of I_{tot} which is obtained by summing triangular current contributions. Hence, I_{tot} is piecewise-linear. The maximum of a piecewise-linear function is piecewise-linear [17]. The Theorem is therefore proven, because piecewise-linear functions are DC [17]. □

An important consequence of Theorem 2 is the NP-completeness of the current minimization problem (since DC optimization is NP-complete). Our solution strategy is heuristic and it is based on a genetic algorithm (GA) [18]. We will briefly discuss the application of the genetic algorithm for the solution of the problem at hand. Refer to [18] for a more in-depth treatment of genetic search and optimization techniques.

3.1. Heuristic Peak Current Minimization

The minimization of a multi-modal cost function such as the one representing the current peak is a difficult task. Gradient-based techniques [17] are fast and well-established, but they tend to rapidly converge to a local minimum. The *genetic algorithm* is a global optimization technique that mimics the dynamics of natural evolution and survival of the fittest.

A set of initial random solutions (a *population*) is generated. For each solution (an *individual* of the population) the cost function is evaluated. From the initial population a new population is created. The best individuals in the old population have a high probability of either becoming member of the new population or participating in the generation of new solution points. New solutions are created by combining couples of good solutions belonging to the old population. This process is called *crossover*. Weak individuals (i.e., points with a high value of the cost function) have a low probability of being selected for crossover or replication.

The creation and cost evaluation of new sets of solutions is carried on until no improvement is obtained on the best individuals over several successive generations. Alternatively, a maximum number of cost function evaluations is specified as a stopping rule. The basic genetic algorithm and many advanced variations have been applied to a number of hard optimization problems for which local search techniques are not successful. The interested reader can refer to [18] for several examples and theoretical background.

The GA approach is attractive in our case because we have an efficient way to compute the cost function (with low-order polynomial complexity). GA-based functional optimization requires a very large number of function evaluations (proportional to the number of generations multiplied by the size of the population). Since F can be efficiently evaluated, large instances of the problem can be (heuristically) solved.

Notice two important facts. First, our algorithm heavily relies on the triangular approximation. If we relax this assumption, the evaluation of F becomes an extremely complex problem (finding the maximum of a multi-modal function), and the GA approach would not be practical. Second, we consider the contribution of the combinational logic as function of time only (independent from the clock schedule). As a consequence, if the maximum current is produced by the combinational logic, $F(T_1, \ldots, T_N)$ is a constant, and no optimization is achievable.

Although the experimental results seem to confirm that the GA is an effective optimization algorithm for peak current minimization, there are margins of improvement. First, the GA does not provide any insight on how far is the best individual from the absolute minimum of the cost function over the feasible region. Moreover, the quality of the results can be improved if the GA is coupled with gradient techniques that are applied starting from the GA-generated solutions and lead to convergence towards local minima.

3.2. Clustering

Up to now, we have assumed that the arrival time T_i of each individual flip-flop can be independently controlled. This is an unrealistic assumption. In VLSI circuits the clock is distributed using regular structures such as *clock trees* [1, 19]. Usually, sub-units of a complex system have local clocks, connected with buffers (drivers) to the main clock tree. The buffers are the ideal insertion points for the delays needed for skew optimization (a practical implementation of such delays will be discussed later). In general it would not be feasible to provide each flip-flop with its own buffer and delay element, for obvious reasons of layout complexity, routability and power dissipation.

Since clock-skew optimization is practical only if applied at a coarser level of granularity, we have developed a strategy that allows the user to specify the number of clusters (i.e., the number of available clock buffers with adjustable delay), and heuristically finds flip-flops that can be clustered without large penalty on the cost function. Here we assume that no constraints on the grouping of flip-flops have been previously specified. This is often the case for circuits generated by automatic synthesis. Structured circuits (data-path, pipelined systems) with pre-existing clustering constraints are discussed later.

Our clustering algorithm can be summarized as follows. The user specifies the number of clusters N_P. First, we solve the peak current minimization problem without any clustering (every flip-flop may have a different arrival time). We then insert the flip-flops in a list ordered by clock arrival times. The list is partitioned in N_P equal blocks. New constraint equations and new current profiles are obtained for the blocks of the partition. A new peak current minimization is solved where the variables are the arrival times T_j^P, $j = 1, 2, \ldots, N_P$, one for each cluster. We also recompute the delays from cluster i to cluster j. The number of equations reduces to $O(N_P^2 + I + O)$. The pseudocode of the clustering algorithm is shown in Fig. 7.

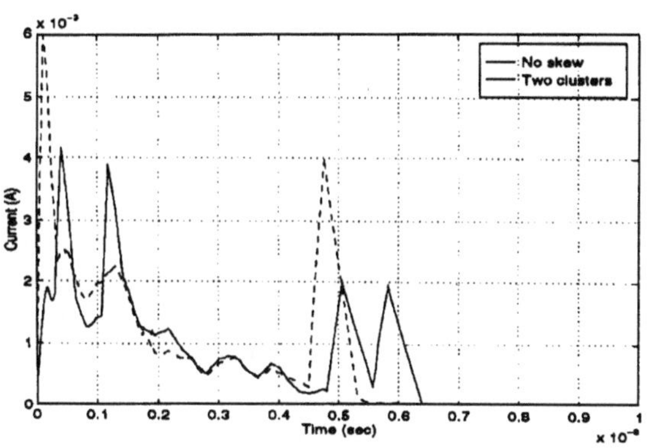

Figure 6. Current profile for benchmark s208 before and after skew optimization with two clusters. The current profiles are obtained by accurate current simulation.

```
/* Let F[i] (i..N) be the instances of the flip-flops */
/* Let T[i] (i..N) be the values given by the MA for instance i */
/* Let N_p be the number of clusters to obtain */
F_sort[i] = sort_by_skew (F[i], T[i]);
size_cluster = N / N_p;
tum_cluster = 0;
foreach (i in F_sort[i])
  if (size (Cluster [num_cluster] == size_cluster)) then
    num_cluster++;
  endif;
  add_in_cluster (Cluster [num_cluster], F_sort[i]);
endfor;
return (Cluster);
```

Figure 7. Clustering algorithm.

The complexity of the clustering algorithm is dominated by the complexity of the ordering of the clock arrival times. Thus, the overall complexity is $O(N \log N)$. Clearly, the overall computational cost of our procedure is not dominated by the clustering step.

Using clustering, we can control the granularity of the clock distribution. The first step of our partitioning strategy is based on the optimal clock schedule found without constraints on the number of partitions. Clustering implies loss in optimality, because some degrees of freedom in the assignment of the arrival times are lost. Our clustering strategy reduces the loss by trying to enforce a natural partitioning. The second iteration of current peak optimization guarantees correctness and further reduces the optimality loss.

Example. Consider the small benchmark s208. It consists of 84 combinational gates and 8 flip-flops. The cycle time is 10 ns, the clock has 50% duty cycle. The current profile for the circuit is shown in Fig. 6 with the dashed line. Observe the two current peaks synchronized with the raising and falling edge of the clock. The irregular shape that follows the first peak shows the current drawn by the combinational logic.

The skew is then optimized with the constraint of 2 partition blocks (i.e., two separate clock drivers allowed). The current profile after skew optimization is shown in Fig. 6 with continuous line. The beneficial effect of our transformation is evident. The two current peaks due to the two skewed clusters of switching flip-flops have approximatively one half of the value of the original peaks. The irregular current profile between peaks is due to the propagation of the switching activity through the combinational logic. Notice that skewing the clock does not have a remarkable impact on the overall current drawn by the combinational logic.

Several different clustering heuristics could be tried. In our experiments we observed that our heuristic produced consistently good results, and did not excessively degrade the quality of the solution with no clustering. However, notice that our heuristic can be applied only if an optimal clock schedule with fine granularity has already been found. For large circuits this preliminary step may become very computationally intensive. In these cases, the user can specify clusters using a different heuristic. In the following sub-section a clustering technique is discussed for dealing with large and structured data-path circuits.

3.3. *Clustering for Staged Circuits*

In the previous discussion, we have solved the current peak optimization problem assuming that we cannot control the current profile of the combinational logic. For many practical circuits this is an overly pessimistic assumption, because the data path of large synchronous systems is often *staged.* In a staged structure, a set of flip-flops A feeds the inputs of a combinational logic block. The outputs of the block are connected to the inputs of a second set of flip-flops B. The sets A and B are disjoint. The flip-flops in A and the block of combinational logic are called a *stage*. Pipelined circuits are staged, and most data paths have this structure, that makes the design easier and the layout much more compact.

If the circuit has a staged structure, the behavior of the combinational logic is much more predictable. If we cluster the flip-flops at the input of each stage, by imposing the same arrival time (i.e., assigning the same clock driver) to their clock signal, we can guarantee that all inputs of the combinational logic of the stage are synchronized. As a consequence, the current profile of the combinational logic translates rigidly with the arrival time of the clock of the flip-flops at its inputs.

For staged circuits our algorithm is more effective, because the clock schedule controls the current profile of the combinational logic as well. The current peak can therefore be reduced even if it is entirely dependent on the combinational logic. Interestingly, the application of clock skew to pipelined circuits has been investigated in [20], where the authors describe a high-performance design style called *counter-flow clocked pipelining* based on multiple skewed clocks. Although the methodology in [20] was not developed to reduce current peaks, the authors observe that clock skewing has beneficial effects on peaks for practical chip level designs.

4. Layout and Clock Distribution

To make our methodology useful in practice, several issues arising in the final steps of the design process need to be addressed. First, pre-layout power and delay estimates are inaccurate and constraints met before layout may be violated in the final circuit. Second, and more importantly, the impact of the clock distribution scheme is not adequately considered when performing pre-layout estimation. Any optimization exploiting clock skew is not practical if the skew cannot be controlled with sufficient accuracy or the cost of generating skewed clocks swamps the reductions that can be obtained.

In the following discussion we assume that the layout of the circuit is automatically generated by placement and routing tools starting from structural gate-level specification. Clusters are specified by providing different names for clock wires coming from different buffers. Flip-flops connected to the same buffer will have the same clock wire name.

To overcome the uncertainty in pre-layout power and delay estimation, two different approaches can be envisioned. We can apply our methodology as a post-processing step after layout. In this case, the constraints can be formulated with high accuracy, and the clock schedule computed with small uncertainty. After finding the optimal clock scheduling and clustering, we need to iterate placement and routing, specifying the new clock clusters and their skews. Alternatively, we can find the clock schedule using pre-layout estimates and allowing a safety margin on the constraint equations. This can be done by increasing the length of the longest paths estimates and decreasing that of the shortest paths, and considering some delay inaccuracy on the computed skews. The effect of the margins is to potentially decrease the effectiveness of the optimization, but in this approach the lay out has to be generated only once.

We chose the second approach for efficiency reasons. For large circuits, the automatic layout generation step dominates the total computation time. The first approach was disregarded because it requires the iteration of the layout step, with an unacceptable computational cost. Notice that this is not always the best choice: if an advanced and efficient layout system is available, which allows incremental modifications (local re-wiring of the clock lines) at low computational cost, the first approach becomes preferable. Moreover, if clustering is user-specified and consistent with the partitioning of the clock distribution implemented in the layout, there would be no need of re-wiring at all, and the first approach would always lead to better results.

4.1. Clock Distribution

After placement and routing, we have complete and accurate information on the load that must be driven by the clock buffer of each cluster. Although many algorithms have been developed for the design of topologically balanced clock trees considering wire lengths and tree structure, for the technology targeted by this work such algorithms are an overkill. Algorithms based on wire length and width balancing become necessary for clock frequencies and die sizes much larger than the ones we deal with [19]. In our case, clock distribution design is simply a buffer design problem.

We assume that we have no control on how the clock tree will be routed, once we specify the clock clusters (i.e., the flip-flops to be connected to the same buffer). From layout we extract the equivalent passive network representing the clock tree for each cluster. We need to design a clock buffer that drives the load with satisfactory clock waveform and skew. The clock waveform must have fast and sharp edges (to avoid short circuit power dissipation on the flip-flops and possible timing violations), and the skew must be as close as possible to the one specified by our algorithm.

Numerous techniques for buffer sizing have been proposed [1, 21] and empirical formulas are available. We used computer-aided optimization methods based on iterative electrical simulation (such as those implemented in HSPICE [22]) that have widespread usage in real-life designs. The main advantage of this approach

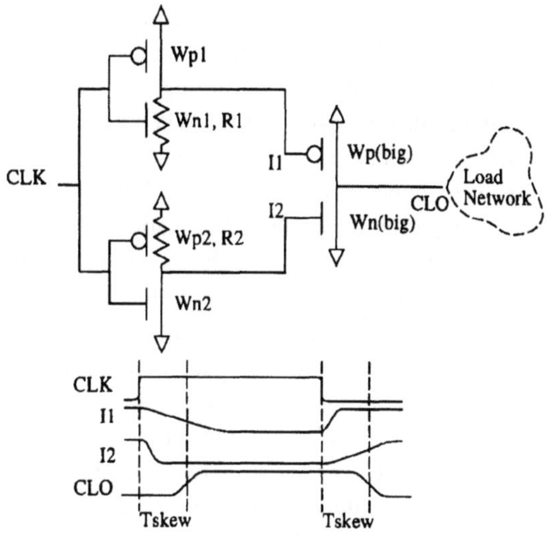

Figure 8. Buffer for generation of skewed clock and signal waveforms.

is that no simplifying assumptions are made on the transistor models and on the buffer architecture. Although the basic clock buffer architecture (a chain of scaled inverters) is well-suited for driving large loads with satisfactory clock waveform, its performance for generating controlled clock skew is poor. There are two standard ways to generate clock skews using the basic buffer: i) add an even number of suitably scaled inverters ii) add capacitance and/or resistance between stages to slow down the output.

Both methods have considerable area and power dissipation overhead. The first method adds stages that dissipate additional power (and use additional area), the second method is probably even worse for both cost measures, because it produces slow transitions inside the buffer, that imply a large amount of short circuit power dissipation. We briefly discuss a clock buffer architecture that has a limited overhead in area and almost no penalty in power dissipation. Our architecture is shown in Fig. 8 for a simple two-stage buffer. The key intuition in this design is that the two large transistors in the output stage are never on at the same time, thus eliminating the short circuit dissipation. The clock skew is obtained by dimensioning the resistances of the two inverters in the first stage.

The transition that controls the output edge is always produced by the transistor in series with the resistance and it can be slowed down using large values $R1$ and $R2$. The penalty is in less sharp output edges (although the gain of the output inverter mitigates this effect) and in the presence of a period when both output transistors are off (the clock line is prone to the damaging effect of cross-talk). Both these effects are greatly reduced by adding another output stage (i.e., two inverters). The complete discussion of this buffer, its dimensioning and its comparison with standard implementation is outside the scope of this paper. However, our HSPICE simulations show that the power overhead of this buffer is negligible and the area overhead is very small.

5. Implementation and Results

The implementation of a program for peak current minimization depends on the availability of a tool that provides accurate current waveforms for circuits of sufficiently large size. Electrical simulators such as SPICE are simply too slow to provide the needed information. In our tool, pre-layout current waveforms are estimated by an enhanced version of PPP [16], a multi-level simulator specifically designed for power and current estimation [23] of digital CMOS circuits. PPP has performance similar to logic level simulators, it is fully compatible with Verilog XL and provides power and current data with accuracy comparable to electrical simulators. Input signal and transition probabilities for all the simulations are set to 50%.

The starting point for our tool is a mapped sequential network (we accept Verilog, SLIF and BLIF netlists). First, the sequential elements are isolated and current profiles are obtained. Alternatively, pre-characterized current models of all flip-flops in the library can be provided. The combinational logic between flip-flops is then simulated and its average current profile is obtained. The first simulation step assumes no skews.

Timing information is extracted from the network. Maximum and minimum delays are estimated with safe approximations (i.e., topological paths). Input arrival times and output required times are provided by the user. The uncertainties in pre-layout estimates are accounted for by specifying a safety margin of 15% on the delay values. The constraint inequalities are generated taking the margin into account. In this step several optimizations, such as those described in [6, 8], are applied to reduce the number of constraint inequalities. Data needed for the evaluation of the cost function are produced: the triangular approximations are extracted from the current profiles and passed to the GA solver [24].

The GA solver is then run to find the optimal schedule that minimizes the peak current. The initial population is generated by perturbing an initial feasible solution (zero skew). The GA execution terminates after an user-specified number of generations. The resulting optimal clock schedule is then applied in a last simulation pass, where the effect on current peaks and average power dissipation is evaluated.

If the maximum number of clock drivers has been specified, the tool first clusters the solution with the algorithm described in Fig. 7, then it runs another simulation to obtain the new current profiles for the clusters (which are now regarded as atomic blocks). A second GA run is performed to re-optimize the clustered solution. Finally, simulation is repeated to check the quality of the result.

The results on a set of benchmark circuits (from the MCNC91 suite [25]) are reported in Table 1. The first two columns represent the name of the circuit and the number of flip-flops. For each of the following columns, two rows are reported for each benchmark. The first row refers to the results obtained with no clustering (i.e., clusters of size 1), the second lists the results obtained with the number of partitions reported in column three. Columns four, five and six describe the effect of clock-skew optimization on average power dissipation. The last three columns describe the effect on current peaks. Without clustering, we reduced on average the current peak by 39%. When we constraint the number of clock drivers, we reduce it by 27%.

We were concerned about a possible increase in power dissipation inside the combinational logic due to unequal arrival times of the clocks controlling flip-flops at its inputs (i.e., increased glitching). From the analysis of the results it appears that skew optimization does not have a sizable impact on average power dissipation. The area of the circuits is unchanged. On the other hand, the effect on current peaks is always positive, and often very remarkable. For some circuits, current peaks are reduced to less than a half the original values. The range in quality of the results is due to the relative importance of the current in the combinational logic. For circuit where the current peak produced by the combinational logic is close to that produced on the clock edges, only marginal improvements are possible. Notice however that some improvements have always

Table 1. Results of our procedure applied to MCNC91 benchmarks.

			AvgPower (μW)			Current peak (mA)		
Bench	FF	P	Before	After	Ratio	Before	After	Ratio
s15850	550	90	46732	46342	0.992	320	176	0.550
		20	46731	46717	1.000	320	219	0.680
s13207	490	80	52094	48476	0.931	267	165	0.619
		20	52094	48856	0.938	267	196	0.733
dsip	224	224	70081	70038	0.999	270	230	0.852
		20	70081	69720	0.995	270	240	0.889
s5378	163	163	71565	72813	1.017	99.3	60.6	0.610
		15	71587	72420	1.012	99.5	75.0	0.754
s9234	135	135	19364	20154	1.041	49.7	12.0	0.241
		20	19364	20619	1.065	49.7	18.0	0.362
mm30a	90	90	14141	14040	0.993	163	138	0.846
		9	14141	14239	1.007	163	131	0.807
s1423	74	74	8043	7276	0.905	50.0	22.2	0.444
		10	8043	7965	0.990	50.0	35.1	0.702
mult32b	61	61	71970	72462	1.007	40.7	24.5	0.602
		7	71970	71944	1.000	40.7	28.2	0.693
sbc	27	27	34285	34178	0.997	47.4	40.8	0.861
		4	34285	34619	1.010	47.4	43.2	0.911
s400	21	21	9773	9751	0.998	12.6	7.94	0.630
		3	9777	9854	1.008	12.6	10.6	0.844
s208	8	8	4207	4095	0.973	5.98	3.19	0.533
		2	4207	4077	0.969	5.98	3.86	0.645

Table 2. Results of our procedure after layout.

Bench	Area	FF	Clustering Type	Clustering nb	rms current	Peak current Estimated	Peak current P&R
s15850	75620	515	Auto	10	0.91	0.72	0.731
s15850_random	75620	515	Random	10	0.95	0.893	0.847
s13207	65995	490	Auto	10	0.968	0.8	0.85
s5378	26431	163	Auto	4	0.915	0.742	0.711
s5378_random	26272	163	Random	4	0.936	0.748	0.767

been obtained even for small circuits with few flip-flops. This result may seem surprising and warrants further explanation. In the combinational logic, signals propagate through cascade connections of gates, therefore only a relatively small number of logic gates is switching at any given time. In contrast, on a clock transition (with zero skew) all flip-flips switch and all gates directly connected to them draw current approximatively at the same time.

The running time of the algorithm is dominated by the first skew optimization step and it ranges from a few minutes to one hour (on a DEC station 5000/240). On average, the simulation time is approximatively 40% of the total. A larger fraction (55% in average) is spent in the GA solver. The remaining 5% is spent in generating the constraints and parsing the files. When the clustered solution is simulated and optimized, the speedup is almost linear in the size of the clusters.

5.1. *Layout Results*

To further validate our method and prove its applicability in real-life circuits, we run placement and routing for some of the largest benchmarks. Since our method targets relatively large circuits with many flip-flops we present the results for three benchmarks with more than 100 flip-flops. The size of the clusters (i.e., the number of flip-flops connected to each clock driver) was set to 50 flip-flops per clock driver (reasonable loads for local clock drivers usually range between 50 and 100).

We used LAGER IV [26] for automatic placement and routing on a gate array. The technology used was SCMOS 1.2 μm. The complete flattened transistor-level netlist of the circuits was extracted using Magic [26], and the circuits were simulated with PowerMill [27]. The time spent in layout completely swamps the total time spent in optimization and simulation (pre and post layout). As mentioned in Section 4, a safety margin was needed on pre-layout delay estimates: our simple delay model and the absence of wiring capacitance information caused sizable errors in the estimates. With the margin set to 0, two of the circuits had timing violations after layout. However, with a 15% margin, all circuits performed correctly. To further increase accuracy, the clock buffers were simulated with HSPICE, and their load network was extracted from layout as well. The power dissipated by the clock buffers was taken into account in the final power estimation. Every step was taken to obtain the level of confidence on the results that is required in real-life design environments.

The results are shown in Table 2. The average power dissipation and area are virtually unchanged (1–4% variations). Each line of the table reports the name of the benchmark, the area in terms of transistors, the number of flip-flops, the clustering technique used, the number of drivers used, the rms current reduction achieved and the peak current reduction achieved. We report as *estimated* the peak reduction estimated by PPP at gate level, and as P&R the reduction given by electrical simulation with PowerMill after placement and routing. The error estimating peak reduction before layout does not go beyond 10%. This validates the results obtained in Table 1.

For the benchmarks, we carried out five layout processes, using two different partitioning techniques. We achieved an average peak reduction of 26% after layout using the automatic clustering algorithm (`auto` in the table) discussed in previous sections. The average rms current is also reduced for all the experiments after layout. In a second set of experiments (`random` in the table) we created random clusters, in order to have feeling on the impact of our clustering heuristic and emulate a worst-case scenario for the applicability of our method. If clustering is externally imposed, the peak current reduction is generally less marked.

The results on the two benchmarks with random clustering give a gain of 20% compared to a gain of 28% with automatic clustering, confirming the effectiveness of the automatic clustering technique. On the other hand, good reductions in peak current are achieved even when the clusters are user-specified. This is an encouraging result, because it extends the applicability of our method to design environments where the clustering of flip-flops is decided by factors such as clock routability or global floorplanning, that may have higher priority than peak current.

6. Conclusions and Future Work

We proposed a new approach for minimizing the peak current caused by the switching of the flip-flops in a sequential circuit using clock scheduling. The peak current was reduced by 30% on average, without any increase of power consumption. Moreover, the initial clock frequency of the circuit was preserved. Our results were fully validated for practical size circuits using post-layout electrical simulation. The impact of clock distribution and buffering was also taken into account and a buffer architecture for generation of skewed clocks with low power overhead was introduced.

We showed that linear programming approaches traditionally used for clock scheduling are not suitable for solving the current minimization problem, and we proposed a heuristic solution strategy based on a genetic algorithm. Clustering techniques have been introduced to account for constraints on the maximum number of available clock drivers.

Although we conservatively assumed that we have no control on the current profiles of the combinational logic, this assumption can be relaxed for staged circuits. In such circuits, the combinational logic can be clustered with the sequential elements. In this case the peak current of the combinational logic plays a role in the cost function of the peak reduction algorithm: the waveform of this combinational logic would be shifted if the clock schedule changes. Clock skewing in this case would also reduce the current peak caused by combinational logic, therefore allowing more effective minimization.

Our technique can be combined with behavioral peak power optimization approaches based on unit selection [28] to achieve even more sizable peak current reductions at the chip level. In this case, however, accurate analysis of current profiles for chip I/O pads would be required, since pads are important contributors to the overall chip-level current profiles.

Acknowledgments

This research is partially supported by NSF under contract MIP-9421129. We would like to thank Enrico Macii for reviewing the manuscript and for many useful suggestions.

References

1. E. Friedman (Ed.), *Clock Distribution Networks in VLSI Circuits and Systems*, IEEE Press, 1995.
2. R. Tsay, "An exact zero-skew clock routing algorithm," *IEEE Transactions on CAD of Integrated Circuits and Systems*, Vol. 12, No. 2, pp. 242–249, Feb. 1993.
3. J.-D. Cho and M. Sarrafzadeh, "A buffer distribution algorithm for high performance clock net optimization," *IEEE Transactions on VLSI Systems*, Vol. 3, No. 1, pp. 84–97, March 1995.
4. N.-C. Chou et al., "On general zero-skew clock net construction," *IEEE Transactions on VLSI Systems*, Vol. 3, No. 1, pp. 141–146, March 1995.
5. Actel, *FPGA Databook and Design Guide*, 1994.
6. T. Szymanski, "Computing optimal clock schedules," *Proceedings of the Design Automation Conference*, pp. 399–404, 1992.
7. J. Fishburn, "Clock skew optimization," *IEEE Transactions on Computers*, Vol. 39, No. 7, pp. 945–951, July 1990.
8. N. Shenoy, R. Brayton, and A. Sangiovanni-Vincentelli "Graph algorithms for clock schedule optimization," *Proceedings of the International Conference on Computer-Aided Design*, pp. 132–136, 1992.
9. K. Sakallah, T. Mudge, and O. Olukotun, "Analysis and design of latch-controlled synchronous digital circuits," *IEEE Transactions on CAD of Integrated Circuits and Systems*, Vol. 11, No. 3, pp. 322–333, March 1992.
10. T. Burks and K. Sakallah, "Min-max linear programming and the timing analysis of digital circuits," *Proceedings of the International Conference on Computer-Aided Design*, pp. 152–155, 1993.
11. J. Xi and W. Dai, "Useful-skew clock routing with gate sizing for low power design," *Proceedings of the Design Automation Conference*, pp. 383–388, 1996.
12. S. Chowdury and J. Barkatullah, "Estimation of maximum currents in MOS IC logic circuits," *IEEE Transaction on CAD of Integrated Circuits and Systems*, Vol. 9, No. 6, pp. 642–654, 1990.
13. J. Neves and E. Friedman, "Design methodology for synthesizing clock distribution networks exploiting nonzero localized clock skew," *IEEE Transactions on VLSI systems*, Vol. 4, No. 2, pp. 286–291, June 1996.
14. E. Lawler, *Combinatorial Optimization: Networks and Matroids*, Holt, Rinehard and Winston, 1976.
15. K. Murty, *Linear Programming*, Wiley, 1983.

16. A. Bogliolo, L. Benini, and B. Riccò, "Power estimation of cell-based CMOS circuits," *Proceedings of the Design Automation Conference*, pp. 433–438, 1996.
17. R. Horst and P. Pardalos (Ed.), *Handbook of Global Optimization*, Kluwer, 1995.
18. D. Goldberg, "Genetic algorithms in search," *Optimization and Machine Learning*, Addison-Wesley, 1989.
19. M. Horowitz, "Clocking strategies in high performance processors," *Symposium on VLSI Circuits Digest of Technical Papers*, pp. 50–53, 1996.
20. J. Yoo and G. Gopalakrishnan et al., "High speed counterflow-clocked pipelining illustrated on the design of HDTV sub-band vector quantizer chips," *Advanced Research on VLSI*, Chapel Hill, 1995, pp. 112–118.
21. J. Xi and W. Dai, "Buffer insertion and sizing under process variations for low power clock distribution," *Proceedings of the Design Automation Conference*, pp. 491–496, 1995.
22. Meta-Software Inc., *Hspice User Manual*, v. H9001, 1990.
23. A. Bogliolo, L. Benini, G. De Micheli, and B. Riccò, "Gate-level current waveform simulation," *International Symposium on Low Power Electronics and Design*, pp. 109–112, 1996.
24. J. Grefenstette, *A User's Guide to GENESIS*, 1990.
25. S. Yang, "Logic synthesis and optimization benchmarks user guide. Version 3.0," MCNC Technical Report, 1991.
26. R. Brodersen (Ed.), *Anatomy of a Silicon Compiler*, Kluwer, 1992.
27. Epic Design Technology, Inc., PowerMill, v. 3.3, 1995.
28. R. San Martin and J. Knight, "Power-profiler: Optimizing ASICs power consumption at the behavioral level," *Proceedings of the Design Automation Conference*, pp. 42–47, 1995.
29. T. Szymanski and N. Shenoy, "Verifying clock schedules," *Proceedings of the International Conference on Computer-Aided Design*, pp. 124–131, 1992.
30. T. Burd, "Low-power CMOS library design methodology," M.S. Report, University of California, Berkeley, UCB/ERL M94/89, 1994.

Luca Benini received a Ph.D. degree in electrical engineering at Stanford University in 1997. Previously he was a research assistant at the Department of Electronics and Computer Science, University of Bologna, Italy. His research interests are in synthesis and simulation techniques for low-power systems. He is also interested in logic synthesis, behavioral synthesis and design for testability. Mr. Benini received an M.S. degree in 1994 in electrical engineering from Stanford University, and a Laurea degree (summa cum laude) in 1991 from University of Bologna. He is a student member of the IEEE.
luca@pampulha.stanford.edu

Patrick Vuillod was a visiting scholar at Stanford University in 1996, while on leave from INPG-CSI, France. Previously he worked in Grenoble in research and development for IST in cooperation with INPG-CSI. His current research interests are in logic synthesis and synthesis for low-power systems. His previous works were on high level description languages and synthesis for FPGAs. Mr. Vuillod received the computer science engineering degree of Ingénieur ENSIMAG, Grenoble, France in 1993, and a master of computer science (DEA) at INPG, Grenbole France in 1994.
vuillod@pampulha.stanford.edu

Alessandro Bogliolo graduated in Electrical Engineering from the University of Bologna, Italy, in 1992. In the same year he joined the Department of Electronics and Computer Science (DEIS), University of Bologna, where he is presently a Ph.D. candidate in Electrical Engineering and Computer Science. From September 1995 to September 1996 he was a visiting scholar at the Computer Systems Laboratory (CSL), Stanford University. His research interests are in the area of power modeling and simulation of digital ICs. He is also interested in reliability, fault-tolerance and computer-aided design of low-power systems.
alex@pampulha.stanford.edu

Giovanni De Micheli is Professor of Electrical Engineering, and by courtesy, of Computer Science at Stanford University. His research interests include several aspects of the computer-aided design of integrated circuits and systems, with particular emphasis on automated synthesis, optimization and validation. He is author of: Synthesis and Optimization of Digital Circuits, McGraw-Hill, 1994, and co-author or co-editor of three other books. He was co-director of the NATO Advanced Study Institutes on Hardware/Software Co-design, held in Tremezzo, Italy, 1995 and on Logic Synthesis and Silicon Compilation, held in L'Aquila, Italy, 1986.

Dr. De Micheli is a Fellow of IEEE. He was granted a Presidential Young Investigator award in 1988. He received the 1987 IEEE Transactions on CAD/ICAS Best Paper Award and two Best Paper Awards at the Design Automation Conference, in 1983 and in 1993. He is the Program Chair (for Design Tools) of the 1996/97 Design Automation Conference. He was Program and General Chair of International Conference on Computer Design (ICCD) in 1988 and 1989 respectively.

nanni@stanford.edu

Journal of VLSI Signal Processing 16, 131–147 (1997)

Clocking Optimization and Distribution in Digital Systems with Scheduled Skews*

HONG-YEAN HSIEH, WENTAI LIU† AND PAUL FRANZON‡
Department of Electrical and Computer Engineering, North Carolina State University, Raleigh, NC 27695-7911

RALPH CAVIN III
Semiconductor Research Center, Research Triangle Park, NC 27709

Received September 30, 1996; Revised November 15, 1996

Abstract. System performance can be improved by employing scheduled skews at flip-flops. This optimization technique is called skewed-clock optimization and has been successfully used in memory designs to achieve high operating frequencies. There are two important issues in developing this optimization technique. The first is the selection of appropriate clock skews to improve system performance. The second is to reliably distribute skewed clocks in the presence of manufacturing and environmental variations. Without the careful selection of clocking times and control of unintentional clock skews, potential system performance might not be achieved. In this paper a theoretical framework is first presented for solving the problem of optimally scheduling skews. A novel self-calibrating clock distribution scheme is then developed which can automatically track variations and minimize unintentional skews. Clocks with proper skews can be reliably delivered by such a scheme.

1. Introduction

For single-phase clocking, circuit designers ordinarily try to deliver a skew-free clock to each flip-flop. As chip size increases, the resistance and capacitance of global interconnections increase linearly with the chip dimension [1]. As a result, the clock network produces a large RC load. The large loads greatly increase the unintentional skews, originated from process and environmental variations. These skews may constitute a significant portion of the cycle time and limit the clocking rate. At the same time, the cycle time of advanced VLSI designs is being reduced rapidly with the reduction of feature size. For a high-speed VLSI design, these factors make the design of clock distribution networks with relatively low unintentional skew more and more challenging. Synchronization will become more difficult in the future due to the unintentional skews.

However, clock skew is not always useless. System cycle time or its latency can be reduced by employing scheduled (intentional) skews at flip-flops. This design technique is called *skewed-clock optimization* [2–5] and has been used in memory designs [6, 7] to achieve high operating frequencies. As an example, Fig. 1(a) shows a two-stage pipelined design. Numbers shown inside the circle are the longest/shortest propagation delays of each combinational logic block. For single-phase clocking, the minimum cycle time is 15 ns if the effects of the setup/hold times and propagation delays of the flip-flops are neglected. However, with the insertion of a scheduled skew of 5 ns as shown in Fig. 1(b), the minimum cycle time can be reduced to 10 ns.

In this paper, a theoretical framework is developed for optimally scheduling skews into single-phase designs using edge-triggered flip-flops to increase system performance. At first, the temporal behavior of a

*This work is supported by NSF Grant MIP-92-12346 and MIP-95-31729.
†Wentai Liu is partially supported by NSF Grant MIP-92-12346 and MIP-95-31729.
‡Paul Franzon is supported by NSF Young Investigator's Award.

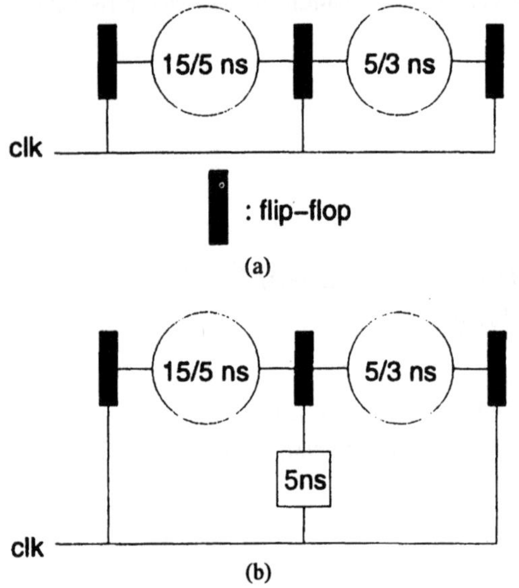

Figure 1. A pipelined design.

single-phase design is analyzed. Based on these investigations, a succinct, yet complete, formulation of the timing constraints is presented to minimize system cycle time. The solution of the resulting skewed-clock optimization problem is then achieved to within the required accuracy by a fully polynomial-time approximation scheme.

After obtaining a set of scheduled skews, it is natural to ask how to deliver them. In delivering skewed clocks for high-speed digital systems, the primary challenge is to minimize unintentional skews. In previous work a passive interconnect tree [8] or an active buffered clock tree [9] have been proposed for skewed-clock distribution. However, process and temperature variations, line loading, and supply voltage changes can cause delays along the clock tree to range from 0.4 to 1.4 times their nominal values [10]. These variations make the previously proposed schemes unreliable in delivering skewed clocks to improve system performance. In this paper, a self-calibrating clock distribution scheme is provided which generates multiple phases based on a reference clock. The scheme dynamically adjusts its phase across manufacturing and environmental variations to minimize unintentional skews. The tracking process is implemented with an all-digital pseudo phase-locked loop [11]. It is theoretically shown that the absolute value of unintentional skew, originating from the quantization error, is limited to Δ, where Δ is the resolution of the sampling and compensation circuitry. This tracking scheme has been verified through the implementation of a demonstration chip. Test results are consistent with the theoretical predictions and show that unintentional skews can be well controlled with such a scheme.

This paper is organized as follows: Sections 2–6 present a theoretical framework for optimally scheduling skews, and Sections 8–13 show a self-calibrating clock distribution scheme for reliably delivering skewed clock. Section 2 defines the temporal behavior of single-phase designs. The timing and graph models for sequential circuits are then defined in Section 3. In Section 4, the mathematical formulation and relaxed linear constraints are derived to specify the temporal behavior and guarantee correct operation when the skewed-clock optimization technique is applied. Section 5 explains how to obtain a set of skews for an unbounded feasible clock period. Section 6 shows a fully polynomial-time approximation scheme for solving the skewed-clock optimization problem. In Section 7, skewed-clock optimization is applied to a set of sequential circuits to demonstrate the performance improvements. Section 8 gives an overview of the presented clocking scheme and its basic operation principle. The algorithm and circuitry used to implement the all-digital phase-locked loop are presented in Sections 9 and 10, respectively. Section 11 shows the quantization error generated by this scheme. Simulation and test results are then presented in Section 12. Section 13 proposes two improved structures to reduce quantization error. Finally, we conclude the paper in Section 14.

2. Temporal Behavior of Sequential Circuits

The functional behavior of sequential circuits has been well investigated. However, two functionally equivalent circuits may not have identical temporal behavior. For example, a ripple adder and a carry look ahead adder perform the same function, but they may require different cycle times. Also, edge-triggered flip-flops and level sensitive transparent latches both are used to latch data and function as storage elements, but they have distinct temporal behavior. An edge-triggered flip-flop transfers the value at its data input to the output at one predetermined edge transition of the clock signal, while a transparent latch transfers the contents at its data input unimpeded to the output when the clock signal is in one predetermined logic level.

In this Section, the temporal 2 behavior of single-phase designs is examined. Figure 2(a) shows an

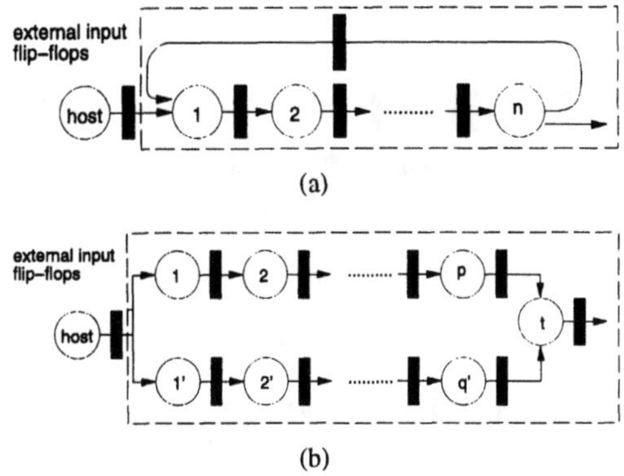

Figure 2. (a) Feedback loop. (b) Re-convergent fanout paths.

example in which there are n flip-flops along the feedback loop. The number shown inside a circle is the node name. A host machine applies data, d_i, to the design through *external input flip-flops* at time $i \cdot T_c$, where T_c is the system cycle time. In a single-phase design, both d_{i-n} and d_i are simultaneously available at node 1 [2]. Figure 2(b) shows the case of a system with two re-convergent fanout paths, in which there are $p+1$ and $q+1$ flip-flops along the top and bottom paths, respectively. Data d_i arrives at node t via the top route after p cycles. At the same time, data $d_{i-(q-p)}$ arrives at node t via the bottom route. In the process of introducing scheduled skews, this property should be taken into account in order to guarantee correct operation.

Given a path p from the host to node v, *temporality*, $\varphi(v, p)$, is defined as follows [12]:

Definition. *Temporality*, $\varphi(v, p)$, is defined as the number of clock cycles for data originating from the host to reach node v along path p.

For the single-phase design shown in Fig. 2(b), temporality $\varphi(t, top\_path)$ is $p+1$ and temporality $\varphi(t, bottom\_path)$ is $q+1$. Thus d_i, arriving at node t via the top route, should meet with $d_{i-(\varphi(t,bottom\_path)-\varphi(t,top\_path))}$ via the bottom route. This amount, $(\varphi(t, bottom\_path) - \varphi(t, top\_path))$, is defined as the *temporal shift* for the bottom path with respect to the top path.

Definition. With re-convergent fanout paths of $v_{pi} \overset{p}{\rightsquigarrow} v$ and $v_{pi} \overset{q}{\rightsquigarrow} v$, the *temporal shift* for path p with respect to path q is $(\varphi(t, p) - \varphi(t, q))$ and for path q with respect to path p is $(\varphi(t, q) - \varphi(t, p))$.

Table 1. Notations in graph model G.

V	Set of functional nodes in the system
V_{fo1}	Set of dummy fanout nodes (explained below)
V_{fo2}	Set of dummy fanout nodes (explained below)
V_l	Set of dummy loop nodes (explained below)
V_{pi}	Vertex set as a host driving primary inputs
V_{po}	Vertex set as a host which is driven by primary outputs
E	Set of directed edges
w	Number of flip-flops along each edge
$t_{max}(v)$	Longest propagation delay at each node v of $V \cup V_{fo1} \cup V_{fo2} \cup V_{po}$
$t_{min}(v)$	Shortest propagation delay at each node v of $V \cup V_{fo1} \cup V_{fo2} \cup V_{po}$
$\tau(v)$	Temporal shift at each node v of V_l

3. Timing and Graph Models

A general sequential circuit can be modeled as a directed graph $G = (V, V_{fo1}, V_{fo2}, V_l, V_{pi}, V_{po}, E, w, t_{max}, t_{min}, \tau)$. Table 1 gives the notations used. A functional node refers to either a gate or a complex combinational module. The longest propagation delay, t_{max}, and the shortest propagation delay, t_{min}, are defined for each node $v \in V$. All the other nodes $v \in V_{fo1} \cup V_{fo2} \cup V_{pi} \cup V_{po}$ are dummy nodes and have zero delay. Delay values are assumed to be measured under worst case conditions. Although the timing model assumes constant longest and shortest propagation delays for all input-output pairs of an individual node, it can be easily generalized to the case in which the propagation delays for each input-output pair of a node are not necessarily equal.

For modeling purposes, special vertex sets V_{pi} and V_{po} are included as shown in Fig. 3. All primary inputs are clocked in through *external input flip-flops* from a single host node, $v \in V_{pi}$, while all primary outputs are sent to a single host node, $v \in V_{po}$, which is clocked out by *external output flip-flops*. All external input flip-flops are clocked synchronously, and so are all external output flip-flops.

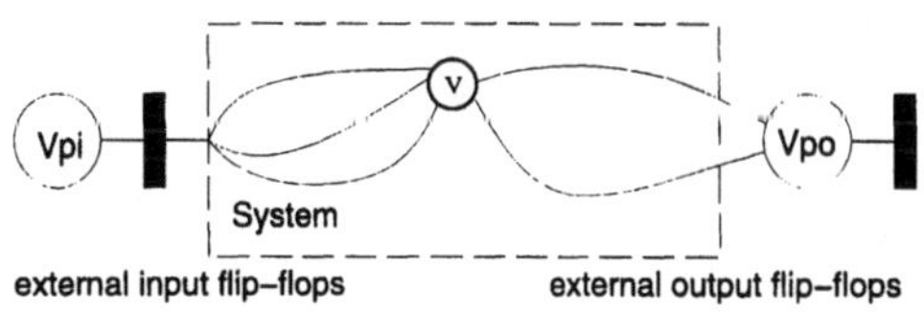

Figure 3. System view.

A directed edge, $e(u, v)$ or $u \xrightarrow{e} v$, is pointed from node u to node v if the output of the gate at u is an input of the gate at v. In this case, u is called a *fan-in node* of v, and v is a *fanout node* of u. Edge $e(u, v)$ is a *fan-in edge* of node v and a *fanout edge* of u. A path $v_0 \overset{p}{\rightsquigarrow} v_n$ is a set of alternating nodes and edges $v_0 \xrightarrow{e_0} v_1 \xrightarrow{e_1} \cdots \xrightarrow{e_{n-1}} v_n$. The edge weight, $w(e)$, indicates if a flip-flop exists along edge e. If a flip-flop is present, $w(e) = 1$, otherwise, $w(e) = 0$.

Three kinds of dummy nodes could be used in the model as described below. If the output of one flip-flop is directly connected to another flip-flop, we introduce a dummy node $v_d \in V_{\text{fo1}}$ between the flip-flops with zero propagation delay. Also, a dummy node $v_d \in V_{\text{fo2}}$ with zero delay is inserted along an edge $e(u, v)$ if flip-flops exist along more than one fanout edge of node u and $w(e) = 1$. As described later the clocking time of the flip-flop along edge $e(u, v)$ will be defined at node u. Introducing the dummy node, $v_d \in V_{\text{fo2}}$, removes the restriction of having the same clocking time for all the flip-flops along the fanout edges of node u. In this procedure, edge weight $w(e(u, v_d))$ is set to zero and $w(e(v_d, v))$ to $w(e)$.

The third kind of dummy node, $v_l \in V_l$, is used to satisfy the temporal relationship required by the original design. Initially a spanning tree is chosen in the graph. The timing information, such as the latest and earliest arrival times at the output of each node, is then calculated for the data d_i propagating along the spanning tree. Next, each remaining edge, called a chord, defines a semi-loop or a loop with respect to the chosen spanning tree. The temporal shift, τ, associated with each chord $e(u, v)$ with respect to the chosen spanning tree can then be calculated. If $\tau \neq 0$, it means that data d_i propagating along the spanning tree should meet with data $d_{i-\tau}$ from the chord instead of d_i. This information is back-annotated into the graph. A dummy node $v_l \in V_l$ is then inserted along the chord, and the vertex weight $\tau(v_l)$ is set to the temporal shift, τ. In the insertion process, the edge weight $w(e(u, v_l))$ is set to zero and $w(e(v_l, v))$ to $w(e)$.

A sequential circuit and its corresponding graph model are shown in Figs. 4(a) and (b). The number shown inside a circle is the node name. The solid lines and all the vertices form a spanning tree, and the dashed lines are chords. For this example, V is the set of {1, 2, 3, 4, 5, 6, 7}, $V_{\text{pi}} = \{0\}$, and $V_{\text{po}} = \{100\}$. Vertex sets V_{fo1} is {8, 9}, V_{fo2} is {10, 11, 12}, and V_l is {14, 15}. The temporal shift for node $14 \in V_l$ is 2, and for node $15 \in V_l$ is 1.

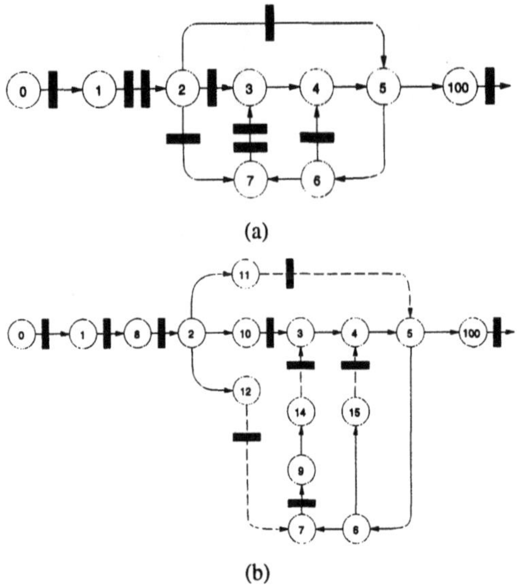

Figure 4. (a) Circuit. (b) Graph model.

4. Mathematical Formulation and Relaxed Linear Constraints

In this section, a succinct, yet complete, formulation of the timing constraints for designs with scheduled skews is presented. The constraints in this formulation are easily constructed for any circuit topology. The notations used in this formulation are listed in Table 2.

For a general system, its graph model, G, is first defined according to the procedure described in the previous section. Mathematical formulation for modeling the temporal behavior is given in Table 3. It should be pointed out that variables $l(v)$, $s(v)$, and $b(v)$ are defined with reference to a specific data d_i and a chosen spanning tree. And $t_{\text{skew}}(v)$ is the scheduled skew for the clocking time $l(v)$.

Table 2. List of symbols.

T_c	Clock period
T_{opt}	Optimum clock period
$l(v)$	Clocking time of flip-flop at the output node v
$s(v)$	Latest arrival time at the output of node v
$b(v)$	Earliest arrival time at the output of node v
$t_{\text{skew}}(v)$	Scheduled skew of flip-flop at the output node v
t_s	Setup time of flip-flops
t_h	Hold time of flip-flops
t_{cl}	Longest delay of flip-flops
t_{cs}	Shortest delay of flip-flops

Table 3. Mathematical formulation.

Delay and synchronization constraints:

$v \in V \cup V_{fo1} \cup V_{fo2} \cup V_{po}$

$$s(v) = \max_{u \xrightarrow{e} v} \begin{cases} s(u) + t_{max}(v) & \text{if } w(e) = 0, \\ l(u) + t_{cl} + t_{max}(v) & \text{if } w(e) = 1 \end{cases}$$

$$b(v) = \min_{u \xrightarrow{e} v} \begin{cases} b(u) + t_{min}(v) & \text{if } w(e) = 0, \\ l(u) + t_{cs} + t_{min}(v) & \text{if } w(e) = 1 \end{cases}$$

Zero and double clocking constraints: $u \xrightarrow{e} v$ & $w(e) = 1$

$s(u) + t_s \leq l(u)$

$l(u) + t_h \leq b(u) + T_c$

Loop/semi-loop constraints: $u \xrightarrow{e} v$ & $v \in V_l$

$s(v) = s(u) - \tau(v) \cdot T_c$

$b(v) = b(u) - \tau(v) \cdot T_c$

External constraints (optional): $u \in V_{pi}$ & $v \in V_{po}$

$l(v) - l(u) = k \cdot T_c$

Delay and synchronization constraints define the timing relationship between node v and its fan-in node u. The signal at the output of node v for data d_i becomes valid if *all* of its input signals have had sufficient time to propagate through the combinational circuit of node v. The latest arrival time, $s(v)$, is the maximum of the time, $s(u) + t_{max}(v)$, if no flip-flop is present along the edge and the time, $l(u)+t_{cl}+t_{max}(v)$, if a flip-flop is present. However, the signal at the output of node v for data d_i becomes invalid if *any* of data d_{i+1} input signals have had enough time to propagate through the combinational circuit of node v. The earliest arrival time, $b(v)$, must be the minimum of the time, $b(u) + t_{min}(v)$, if no flip-flop exists along the edge and the time, $l(u) + t_{cs} + t_{min}(v)$, if a flip-flop exists. Zero and double clocking constraints are required for each flip-flop in order to prevent setup and hold time violations, respectively.

Assuming the chord connecting node v to the chosen spanning tree is $e(v, w)$, data d_i propagating along the spanning tree should meet data $d_{i-\tau(v)}$ from the chord, $e(v, w)$. On the basis of this reasoning, the value of variables $s(v)$ and $b(v)$ should be reduced by the amount $\tau(v) \cdot T_c$. Loop/semi-loop constraints incorporate these effects to ensure that the right data will meet. If designers require the optimized circuit to preserve the original latency, defined as the number of the clock cycles for completing a job, an external constraint must be satisfied. In the formulation, k is the temporality along the path in the spanning tree from node $u \in V_{pi}$ to node $v \in V_{po}$.

Table 4. Relaxed linear constraints.

Relaxed delay constraints:

$u \xrightarrow{e} v$ & $v \in V \cup V_{fo1} \cup V_{fo2} \cup V_{po}$ & $w(e) = 0$

$s(u) + t_{max}(v) \leq s(v)$

$b(v) \leq b(u) + t_{min}(v)$

Relaxed synchronization constraints:

$u \xrightarrow{e} v$ & $v \in V \cup V_{fo1} \cup V_{po}$ & $w(e) = 1$

$l(u) + t_{cl} + t_{max}(v) \leq s(v)$

$b(v) \leq l(u) + t_{cs} + t_{min}(v)$

Relaxed loop/semi-loop constraints: $u \xrightarrow{e} v$ & $v \in V_l$

$s(v) \geq s(u) - \tau(v) \cdot T_c$

$b(v) \leq b(u) - \tau(v) \cdot T_c$

For skewed-clock optimization, the minimum clock cycle time can be obtained by solving the following optimization problem:

Problem	*Opt*1
Minimize	T_c
Subject to	delay constraints
	synchronization constraints
	zero and double constraints
	loop/semi-loop constraints
	external constraints

Problem *Opt*1 is a nonlinear optimization problem since max and min functions exist in the delay and synchronization constraints. However a linear optimization problem can be formulated if these constraints are relaxed as shown in Table 4. In the relaxation process, operator max is replaced by $\geq$ and operator min is replaced by $\leq$. Then the relaxed linear optimization can be expressed as follows:

Problem	*Opt*2
Minimize	T_c
Subject to	relaxed delay constraints
	relaxed synchronization constraints
	zero and double constraints
	relaxed loop/semi-loop constraints
	external constraints

It can be shown that the minimum cycle time obtained by solving the nonlinear optimization problem *Opt*1 is the same as the minimum cycle time obtained by solving the linear problem *Opt*2. The following theorem establishes the equivalence.

1. **Procedure** *Update(Opt2)*
2. while ($s(v)$ and $b(v)$ are not at their minimum and maximum values, respectively)
3. if ($v \in V \cup V_{fo1} \cup V_{fo2} \cup V_{po}$)
4. update $s(v)$ and $b(v)$ by delay and synchronization constraints as defined in Table 3
5. if ($v \in V_l$)
6. update $s(v)$ and $b(v)$ by loop/semi-loop constraints as defined in Table 3

Theorem 1. *If the optimal cycle times for Problems Opt1 and Opt2 are denoted by T_{c1} and T_{c2}, respectively, then $T_{c1} = T_{c2}$.*

Proof: Since Problem *Opt2* is a relaxed version of Problem *Opt1*, the solution obtained by solving Problem *Opt1* is also a solution of Problem *Opt2*. In other words, $T_{c1} \geq T_{c2}$.

This theorem is proved if we can show that $T_{c1} \leq T_{c2}$. The proof involves showing that the solution obtained by *Opt2* can be iteratively refined by Procedure *Update* until it becomes a solution of Problem *Opt1*.

It is not difficult to argue that this procedure will terminate in a finite number of steps. In this process only the values of variables $s(v)$ or $b(u)$ are settled to their minimum and maximum values, respectively, so that the delay, synchronization, and loop/semi-loop constraints in Problem *Opt1* are satisfied. Since the values of variables, $s(v)$, are decreasing and the values of variables, $b(u)$, are increasing, this refinement procedure will not result in violating any other constraint such as zero/double clocking and external constraints. The terminated result is thus a solution of Problem *Opt1*. □

5. Skew Scheduling

After obtaining a feasible solution for the clock period T_c, the next step is to determine the value of implemented skews, $t_{skew}(v)$. This step is called *skew scheduling*. For each flip-flop, its clocking time, $l(v)$, can be implemented in various ways. It is a feasible value if the value of the scheduled skew satisfies $l(v) = i \cdot T_c + t_{skew}(v)$, where i is an integer. Nevertheless different values of the skew might give different operating frequency ranges. As an example, the system shown in Fig. 5(a) consists of a single combinational logic block with edge-triggered flip-flops. The longest and shortest propagation delays of this combinational logic are 15 ns and 10 ns, respectively. Neglecting the

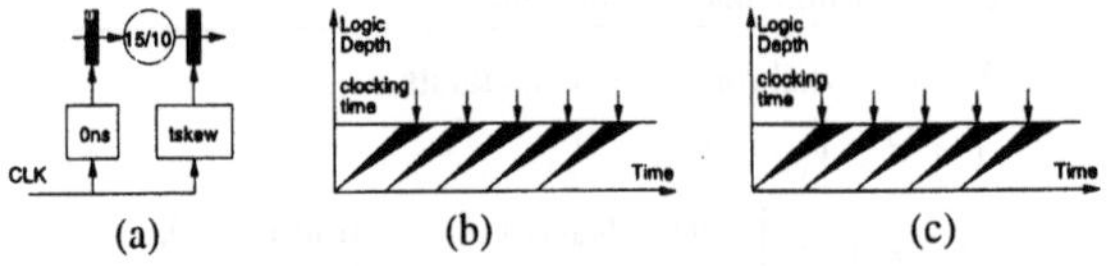

Figure 5. (a) Example, (b) $t_{skew} = 5$ ns, (c) $t_{skew} = 10$ ns.

setup/hold times and propagation delay of each flip-flop, the optimization problem *Opt2* gives the clocking time of input and output flip-flops at 0 ns and 15 ns. This results in the optimal clock period of 5 ns. Due to the cyclic nature of a clock, the clocking time 15 ns can be implemented by a delay of t_{skew} of 0 ns, 5 ns, 10 ns, 15 ns, etc. Consider the cases of $t_{skew} = 5\ ns$ and $t_{skew} = 10$ ns. Shown in Figs. 5(b) and (c) are graphical representations of the data flow through combinational logic, which are called logic-depth timing diagrams. The shaded regions, bounded by the longest and shortest delays through the logic, depict the flow of data through the combinational logic. The unshaded areas depict time at which the logic is stable. If the clock period is increased to 8 ns, the system no longer works with $t_{skew} = 5$ ns. As shown in Fig. 5(b), the output flip-flops sample data at the unstable (shaded) regions. In contrast, as shown in Fig. 5(c), the system works correctly if the clock period is 8 ns and $t_{skew} =$ 10 ns.

From the infinitely possible solutions, proper skews could be scheduled such that the system can be operated in a clock period ranging from infinity to its optimum value. In other words, the feasible system cycle time is unbounded above. Algorithm *Skew_scheduling* generates the set of proper skews. The proof is given in Theorem 2. Steps 2–6 calculate the implemented skews by subtracting $\varphi(v_n, p) \cdot T_c$ from $l(v_n)$ where p is along the chosen spanning tree. If $v_n \in V_l$, the additional amount $\tau(v_n) \cdot T_c$ is being added. Steps 7–9 set the minimum skew to be 0 ns by subtracting the value of *shift_to_zero* from all the skews.

1. **Algorithm** *Skew_scheduling*(T_c)
2. for each variable $l(v_n)$
3. calculate $\varphi(v_n, p)$
4. $t_{skew}(v_n) \leftarrow l(v_n) - \varphi(v_n, p) \cdot T_c$
5. If ($v_n \in V_l$)
6. $t_{skew}(v_n) \leftarrow t_{skew}(v_n) + \tau(v_n) \cdot T_c$
7. *shift_to_zero* $\leftarrow \min_{\forall v}\{t_{skew}(v)\}$
8. for each variable $t_{skew}(v)$
9. $t_{skew}(v) \leftarrow t_{skew}(v) -$ *shift_to_zero*

Theorem 2. *If the clock period T_{opt} is the solution of Problem Opt2, Algorithm* Skew_scheduling *generates a set of skews such that the feasible operating period is in the range from T_{opt} to ∞.*

Proof: Path p, $v_{pi} \overset{p}{\rightsquigarrow} v$, is defined as the path from node $v_{pi} \in V_{pi}$ to node v along the spanning tree. Assume that $v_{pi} \overset{p}{\rightsquigarrow} v \equiv v_{pi} \overset{p_1}{\rightsquigarrow} v_1 \overset{p_2}{\rightsquigarrow} v$ and $v_1 \overset{p_2}{\rightsquigarrow} v \equiv v_1 \overset{e_1}{\rightarrow} v_2 \overset{e_3}{\rightarrow} v_3 \cdots v_{n-1} \overset{e_{n-1}}{\rightarrow} v_n \overset{e_n}{\rightarrow} v$ where $w(e_1) = w(e_n) = 1$ and $w(e_2) = w(e_3) = \cdots = w(e_{n-1}) = 0$. Two cases should be investigated: $v_n \in V_l$ and $v_n \ni V_l$. Due to space limitations, we only analyze the second case. However this analysis can be extended to the first case.

A feasible solution of Problem *Opt2* satisfies the constraints shown in Table 5. These constraints depict that the data launched from the flip-flop at the output of node v_1 can be correctly captured by the flip-flop at the output of node v_n. The longest delay and synchronization constraints give the latest arrival time at the output of node v_n, which is $l(v_1) + t_{cl} + \sum_{i=2}^{n} t_{max}(v_i)$. Also the shortest delay and synchronization constraints give the earliest arrival time at the output of node v_n, which is $l(v_1) + t_{cs} + \sum_{i=2}^{n} t_{min}(v_i)$. Added to the zero and doubleclocking constraints, the equivalent constraints are generated in Table 6. It can be proved that the constraints in Table 5 are satisfied if and only if constraints in Table 6 are satisfied.

Table 5. Feasible constraints.

Longest delay constraints:

$$s(v_i) + t_{max}(v_{i+1}) \le s(v_{i+1}) \quad \text{where } 2 \le i \le n-1$$

Shortest delay constraints:

$$b(v_{i+1}) \le b(v_i) + t_{min}(v_{i+1}) \quad \text{where } 2 \le i \le n-1$$

Longest synchronization constraint:

$$l(v_1) + t_{cl} + t_{max}(v_2) \le s(v_2)$$

Shortest synchronization constraint:

$$b(v_2) \le l(v_1) + t_{cs} + t_{min}(v_2)$$

Zero clocking constraint:

$$s(v_n) + t_s \le l(v_n)$$

Double clocking constraint:

$$l(v_n) + t_h \le b(v_n) + T_{opt}$$

Table 6. Equivalent constraints.

Setup constraint:

$$l(v_1) + t_{cl} + \sum_{i=2}^{n} t_{max}(v_i) + t_s \le l(v_n)$$

Hold constraint:

$$l(v_n) + t_h \le l(v_1) + t_{cs} + \sum_{i=2}^{n} t_{min}(v_i) + T_{opt}$$

Table 7. Constraints expressed in terms of scheduled skews.

Setup constraint:

$$t_{skew}(v_1) + \varphi(v_1, p) \cdot T_{opt} + \sum_{i=2}^{n} t_{max}(v_i) + t_{cl} + t_s \le t_{skew}(v_n) + \varphi(v_n, p) \cdot T_{opt}$$

Hold constraint:

$$t_{skew}(v_n) + \varphi(v_n, p) \cdot T_{opt} + t_h \le t_{skew}(v_1) + \varphi(v_1, p) \cdot T_{opt} + t_{cs} + \sum_{i=2}^{n} t_{min}(v_i) + T_{opt}$$

Table 8. Alternative expressions of constraints.

Setup constraint:

$$t_{skew}(v_1) + t_{cl} + \sum_{i=2}^{n} t_{max}(v_i) + t_s \le t_{skew}(v_n) + T_c$$

Hold constraint:

$$t_{skew}(v_n) + t_h \le t_{skew}(v_1) + t_{cs} + \sum_{i=2}^{n} t_{min}(v_i)$$

Algorithm *Skew_scheduling* generates a set of skews such that $t_{skew}(v_1) = l(v_1) - \varphi(v_1, p) \cdot T_{opt}$ and $t_{skew}(v_n) = l(v_n) - \varphi(v_n, p) \cdot T_{opt}$. Substituting these scheduled skews into the constraints in Table 6, the resulting constraints are shown in Table 7. The object is to prove that if the constraints in Table 7 are satisfied at $T_c = T_{opt}$, then they are also satisfied at $T_c = \infty$. If so, from the linear programming theory, we know that the feasible operation period is unbounded above. Based on the discussion in Section 2, it can be shown that $\varphi(v_n, p) = \varphi(v_1, p) + 1$. The constraints in Table 7 can be rewritten as shown in Table 8 for the clock period T_c. It is noted that this only gives T_c a lower bound such that the feasible clock period is unbounded above. □

6. A Fully Polynomial-Time Approximation Scheme

The optimization problem *Opt2* is linear and can be solved by the simplex algorithm. However, the time complexity of the simplex algorithm is exponential in the worst case. In this section, we first present a fully polynomial-time approximation scheme that achieves the optimal clock period to within any given error bound ϵ for the optimization problem *Opt2* [13]. It provides a theoretical proof that the optimization problem *Opt2* can be solved in polynomial time instead of exponential time with the required accuracy.

As stated in Section 5, the feasible clock period T_c can be unbounded above if skews are appropriately scheduled. Using this property, Algorithm *Search_*T_{opt} performs a binary search to achieve the optimum clock period T_{opt} to within any given error bound ϵ. The running time of *Search_*T_{opt} is in $O(\log((T_{\text{upper}} - T_{\text{lower}})/\epsilon)|V_t||E|)$, where the number of variables and constraints in Problem *Opt*2 is in $O(|V_t|)$ and $O(|E|)$, respectively, and $V_t \equiv V \cup V_{\text{fo1}} \cup V_{\text{fo2}} \cup V_l \cup V_{\text{pi}} \cup V_{\text{po}}$. T_{upper} and T_{lower} are the given upper and lower bounds of the clock period, respectively. The term $|V_t||E|$ is contributed by Algorithm *Feas*, which is used to check the feasibility of the constraints in Problem *Opt*2 for a given clock period. If the clock period T_c is given, the constraints in the optimization problem *Opt*2 would become a system of difference constraints. For a system of difference constraints, Algorithm *Feas*, which is a variant of the Bellman-Ford algorithm, takes the advantage of this special structure of the constraint set to test the feasibility in polynomial time [14].

```
Algorithm Search_T_opt()
  T_u ← T_upper; T_l ← T_lower
  while T_u − T_l > ε
    T_c ← 1/2(T_u + T_l)
    if Feas(T_c) T_u = T_c
    else T_l = T_c
  T_opt ← T_u
```

```
Algorithm Feas(T_c)
  /* Each constraint con is written as follows:
     x_l ≥ x_r + c_lr.
  x_l and x_r are the variables in constraint con.
  c_lr is the constant in constraint con. */
  for each variable s(v), b(v), and l(v)
    s(v), b(v), and l(v) ← −∞
  l(v) ← 0 where v ∈ V_pi
  for i = 1 to |var| − 1
    for each constraint con
      if (x_l < x_r + c_lr) x_l ← x_r + c_lr
    for each constraint con
      if (x_l < x_r + c_lr)
        return(FALSE)
    return(TRUE)
```

7. Experimental Results

In this section, we have applied our formulation to design examples in order to demonstrate the performance improvements from skewed-clock optimization. The examples include a set of sequential circuits from ISCAS89 and the experimental results are shown in Table 9. For each circuit, the table provides data that describes its size in terms of the number of flip-flops and gates in the second column. All gates are assumed to have unit delays, and the setup/hold times and propagation delays of flip-flops are arbitrarily set to zero. T_{c1} is the minimum clock period for single-phase clocking, and the optimal clock period T_{c2} is for skewed clocking. The corresponding percentage improvement over the initial clock period T_{c1} is calculated as follows: $gain\% = \frac{T_{c1}-T_{c2}}{T_{c1}} \times 100\%$. In this experiment, T_{upper} is set to the minimum clock period obtained by single-phase clocking, and T_{lower} is set to the maximal value of $t_{\max} - t_{\min}$ for any pair of flip-flops. A binary search is then performed to achieve the optimal clock period to within the given error bound of 0.1 ns. The total CPU times for running Algorithm *Search_*T_{opt} on

Table 9. Performance improvements with scheduled skews.

ckt name	# of ff/gate	T_{c1}	T_{c2}	CPU(s)
s27	3/10	6	4.03	0.26
s208.1	8/104	11	8.00	0.43
s298	14/119	9	6.00	0.60
s344	15/160	20	14.04	0.73
s382	21/158	9	6.00	0.86
s420.1	16/218	13	10.00	0.99
s444	21/181	11	7.02	0.96
s499	22/174	4	1.00	0.90
s526	21/193	9	6.00	1.02
s635	32/286	127	124.00	2.27
s938	32/446	17	14.00	3.37
s991	19/535	59	55.02	2.37
s1269	37/570	35	30.03	4.33
s1423	74/657	59	54.03	11.73
s1488	6/653	17	15.03	1.66
s1494	6/647	17	14.54	1.75
s1512	57/798	30	24.04	9.08
prolog	136/1602	26	14.02	45.45
s3271	116/1572	28	19.02	35.10
s3330	132/1789	29	14.00	48.46
s3384	183/1685	60	51.02	62.98
s4863	104/2342	58	53.03	44.93
s5378	179/2779	25	16.39	92.66
s6669	239/3080	93	81.02	162.26
s9234	228/5597	58	38.07	260.42

a DECstation 5000 are shown in the last column. It is worth mentioning that the average speedup of 25 circuits is over 26.27%.

8. Overview of the Clocking Scheme

In this section, a self-calibrating clock scheme for skewed clocking is presented. Figure 6 shows a skewed-clock design which is partitioned into several different regions. In each region i, the clock is delivered with a scheduled skew $t_{skew}(i)$. The synchronization within the region can be maintained with either an H-tree [1] or a zero-skew tree [8].

For each region i, the clock is delivered from the clock generator to the root of the tree through path $p_1(i)$ and then to its leaves through path $p_4(i)$. Assume that the clock takes time $d_1(i)$ from the clock generator to reach the root of the clock tree and time $d_4(i)$ from its root to reach its leaves. The sum of $d_1(i)$ and $d_4(i)$ is a portion of the skew at the flip-flops in region i. To have the scheduled skew $t_{skew}(i)$ at these flip-flops, the clock generator needs to calculate the difference between these two amounts, $t_{skew}(i) - (d_1(i) + d_4(i))$, and dynamically compensate an amount of $d_c(i)$ for it. The following constraint must be satisfied.

$$d_c + d_1 + d_4 - \left\lfloor \frac{d_c + d_1 + d_4}{T_c} \right\rfloor T_c = t_{skew}$$

In the above constraint, t_{skew} is smaller than the system clock period T_c due to the periodic nature. The amount of d_c is provided by tapping the closest phase from the clock generator. Thus it is clear that the accuracy depends on the phase resolution. As an example, d_c is 3 ns if $T_c = 24$ ns, $d_1 = 6$ ns, $d_4 = 9$ ns, and $t_{skew} = 18$ ns.

Two feedback paths, path $p_2(i)$ from the root and path $p_3(i)$ from the leaf of the clock tree, provide the timing information of paths $p_1(i)$ and $p_4(i)$ to the clock generator. The clock signal, originating from the clock generator, takes the amounts of time of $d_1 + d_2$ and $d_1 + d_4 + d_3$ to traverse back to the clock generator via paths p_2 and p_3, respectively. In our design, the nominal delays of these two paths are equalized with that of path $p_1(i)$ by carefully designing p_1, p_2, and p_3 [15]. Thus d_1 is equal to half of $d_1 + d_2$, and d_4 is equal to the difference of $d_1 + d_4 + d_3$ and $d_1 + d_2$.

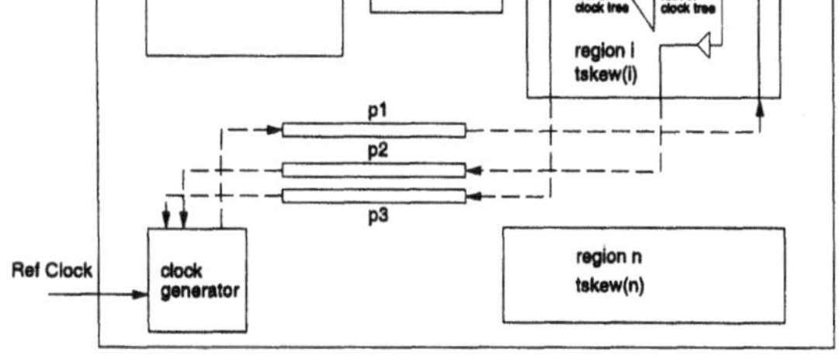

Figure 6. A skewed-clock design.

9. A Dynamically Tracking Scheme

To determine both d_1 and d_4, the elapsed time between the rising edges of the reference and the feedback clock signals needs to be calculated. This duration is measured in a digital format by a time-to-digital converter. For illustrative purposes, the transfer characteristics of a 3-bit time-to-digital converter are shown in Fig. 7(a). An encoding example is shown in Fig. 7(b). The transfer function is not continuous and the output is

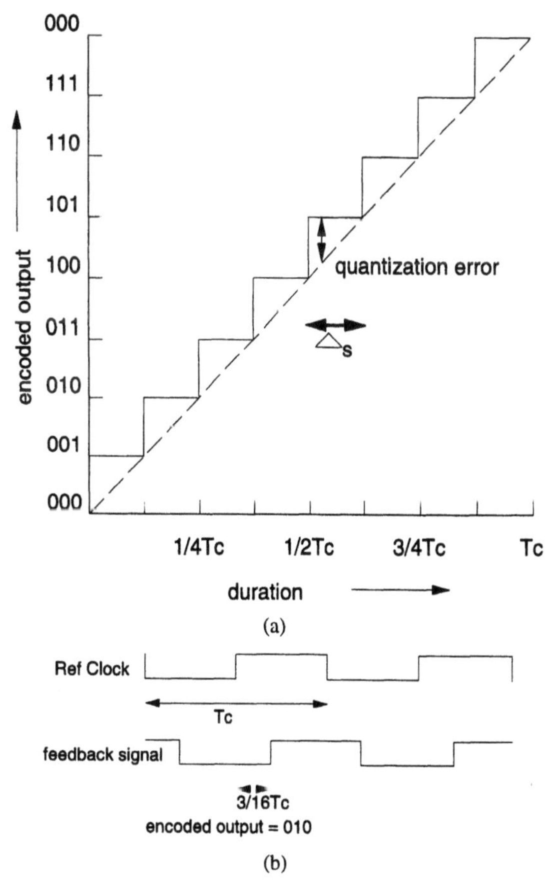

Figure 7. Transfer characteristics.

"quantized". As a result, each output code corresponds to a small range Δ_s of duration. This conversion process results in an irreducible quantization error, which is equal to the difference between the transfer curve and the straight dotted line. This effect will be discussed in Section 11.

Relative to the reference clock, the time intervals for the feedback signals via p_2 and p_3 are then digitally encoded to E_{p2} and E_{p3}, respectively, by a time-to-digital converter. Assume that the desired scheduled skew is $E_i \cdot \Delta_p$, and the current tapped phase is E_c. Algorithm *Adjust_tap* is used to dynamically adjust the tapped phase in order to provide the proper compensation amount.

1. **Algorithm** *Adjust_tap*()
2. for each region i
3. $m_1(i) \leftarrow \frac{1}{2}(E_{p2}(i) - E_c(i))$
4. $m_4(i) \leftarrow E_{p3}(i) - E_{p2}(i)$
5. $m_5(i) \leftarrow m_1(i) + m_4(i)$
6. $m_6(i) \leftarrow E_i(i) - m_5(i)$
7. $E_c(i) \leftarrow m_6(i)$

Since E_{p2} is the digitally encoded value of $2d_1$ plus E_c, Step 3 gives the digitally encoded value of d_1. The division by 2 is performed by a shift right operation. Step 4 calculates the digitally encoded value of d_4. Step 5 produces the digitally encoded value of the internal delay $d_1 + d_4$. The chosen tap for the proper compensation amount is then obtained in Step 6 by calculating the difference between $E_i(i)$ and $m_5(i)$. It should be pointed out that the results obtained in all the steps are computed modulo 2^n.

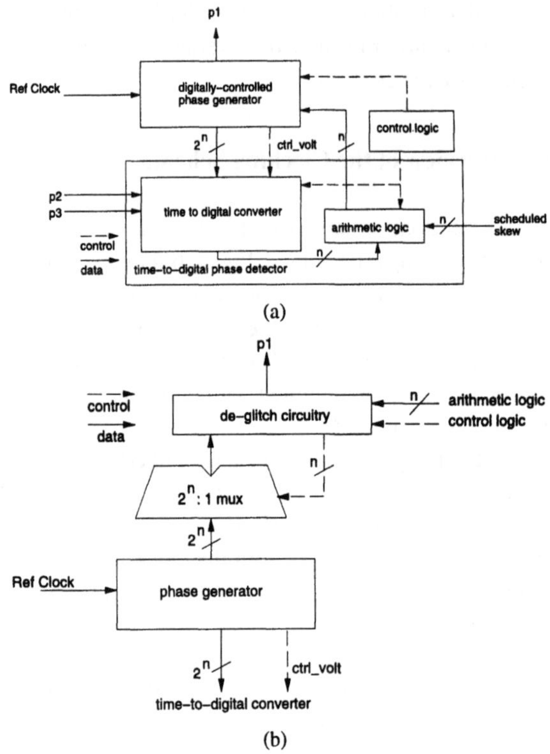

Figure 8. (a) Block diagram. (b) Digitally-controlled phase generator.

10. Circuit Design

As shown in Fig. 8(a), the clock generator consists of a time-to-digital phase detector, a digitally-controlled phase generator, and control logic. The time-to-digital phase detector calculates the difference of the times $d_1 + d_4$ and t_{skew}, and performs a similar function as a phase detector does in an analog PLL. Also, the digitally-controlled phase generator provides the compensation amount and functions as a VCO. Thus the clock generator is an all-digital pseudo PLL [11].

The time-to-digital phase detector consists of a time-to-digital converter and arithmetic logic. The time intervals between the rising edges of the reference clock and feedback signals are measured by the time-to-digital converter. The arithmetic logic calculates the difference. The next chosen phase is calculated according to the algorithm described in the Section 9. The calculated result is then sent to the digitally-controlled phase generator for selecting the right phase from 2^n equally spaced phases. As described in the next section, the digitally-controlled phase generator uses a delay-locked loop (DLL) to generate these 2^n equally spaced phases.

10.1. Digitally-Controlled Phase Generator

As shown in Fig. 8(b), the digitally-controlled phase generator consists of a phase generator, a $2^n : 1$ multiplexer, and de-glitch circuitry. The phase generator generates $N = 2^n$ equally spaced phases, which are then tapped out by the dedicated multiplexer for compensation. The de-glitch circuitry is used to ensure a clean clock waveform.

Figure 9(a) shows a delay-locked loop used in the phase generator. The reference clock passes through a 2^{n+1} stage delay chain to generate 2^{n+1} phases. Each delay element, which consists of two inverters, has the nominal delay value $\Delta = \frac{T_c}{2^n}$, which is the phase resolution Δ_p of the design. The first 2^n phases are sent to a $2^n : 1$ multiplexer to be used for compensation.

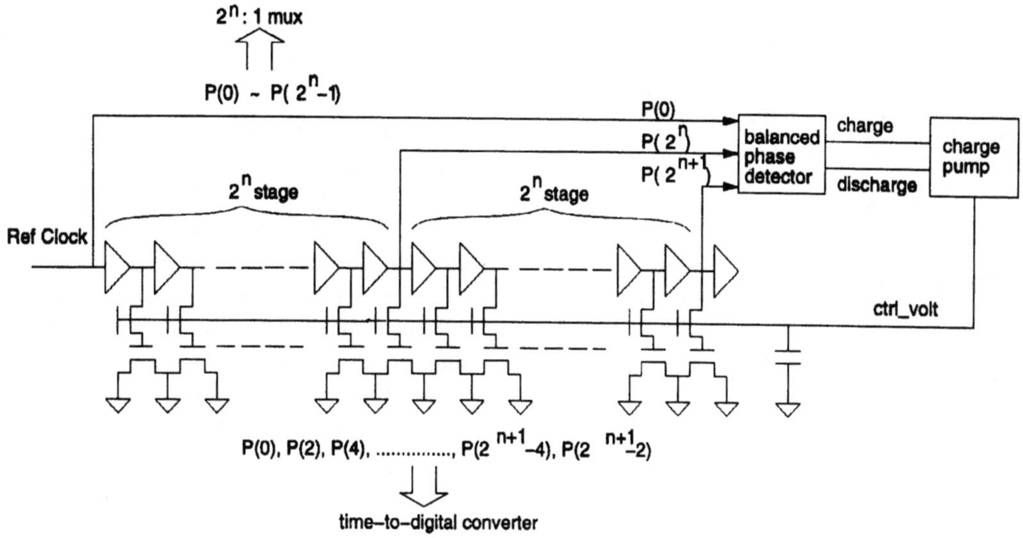

Figure 9a. Phase generator.

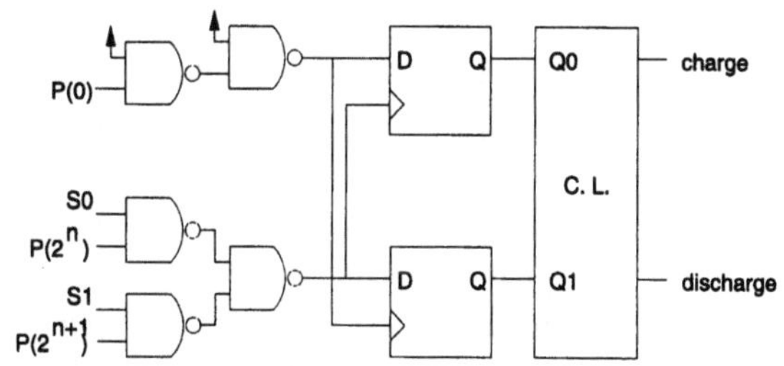

Figure 9b. Balanced phase detector.

Meanwhile, phases $p(0)$, $p(2)$, $p(4)$, ..., $p(2^{n+1}-4)$, $p(^{n+1}-2)$ are sent to the time-to-digital converter for sampling the feedback signals. The reason for generating 2^{n+1} instead of 2^n phases is described in Section 10.2.

Ideally, synchronization among the phases of $p(i)$, $p(i+2^n)$, and $p(i+2^{n+1})$ is maintained for each $i \in [0, 2^n - 1]$. This is achieved by a balanced phase detector [16, 17] as shown in Fig. 9(b). At the beginning, the phase of $p(2^n)$ is selected by setting $s0 = 1$ and $s1 = 0$ such that it is aligned with the phase of $p(0)$. The balanced phase detector decides to increase or decrease the bias voltage in the delay chain by charging or discharging the charge pump circuitry. Once locked, the phase of $p(2^{n+1})$ is selected by $s0 = 0$ and $s1 = 1$, and it is aligned with the phase of $p(0)$. Simulation shows that the skew among these three phases is less than $40ps$ for a chip implementation using 2 μm N-well CMOS technology available at MOSIS.

The multiplexer changes its state only when both the clocks of the present and next chosen taps are off. When the control inputs change their values, a pulse is generated at *net_deg* to ensure there is no spurious signal at *net_clock* as shown in Fig. 10(a). A timing circuit schedules these events shown in Fig. 10(b). Signals *pre_phase* and *next_phase* represent the clocks of the previous and next taps, respectively.

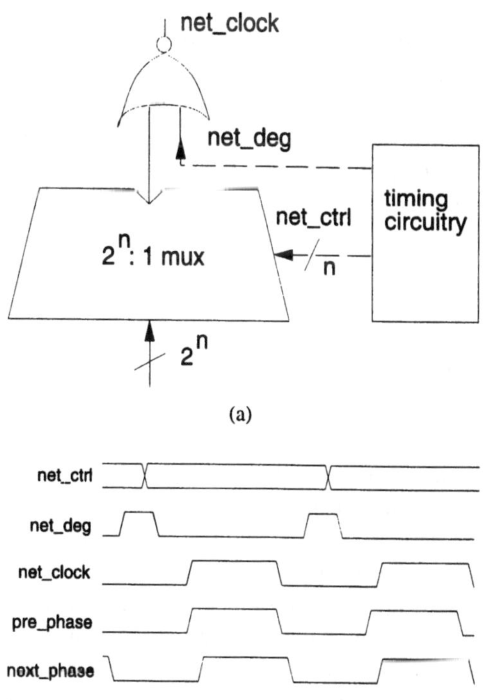

Figure 10. De-glitch circuitry.

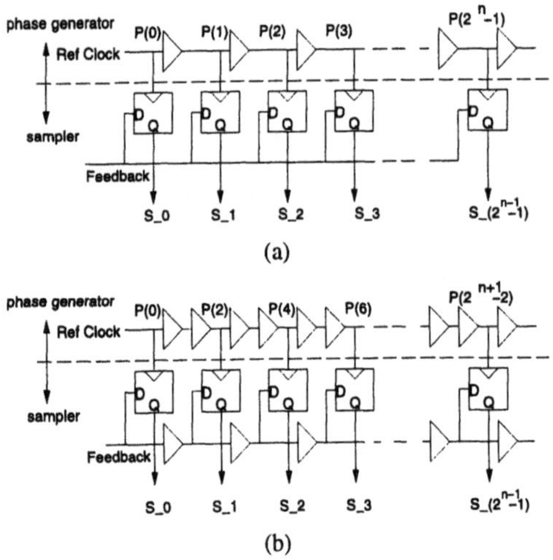

Figure 11. Sampler.

10.2. *Time-to-Digital Converter*

Relative to the reference clock, the elapsed time of the feedback signals are digitally encoded by a time-to-digital converter, which consists of a sampler and an encoder. Figure 11(a) shows that feedback signals can be sampled by the structure. However this structure creates a large load attached to the feedback signal. This possibly results in the distortion of the timing information.

Alternatively, sampling can be done with a matched delay structure [18] as shown in Fig. 11(b). This structure reduces the error due to the mismatch of input driving resistance and output loading. The reference clock is tapped to the flip-flops at every other delay element, while the feedback signal is tapped at every delay element. Consequently the sampling resolution, Δ_s, is equal to Δ. The outputs of the sampler are then encoded into n bits by an octal-to-binary encoder.

11. Quantization Error

In our implementation, $\Delta_s = \Delta_p$. Thus we analyze the quantization error for the special case of $\Delta = \Delta_s = \Delta_p$. This analysis can be easily extended to general cases where $\Delta_p = 2i\Delta_s$ and $i \in \mathbf{I}$. Basically, the total quantization error comes from two sources: one is from the quantization of d_1, the other is from the quantization of d_4. In the following, these errors, $d_1 - m_1 \cdot \Delta$ and $d_4 - m_4 \cdot \Delta$, are calculated. Algorithm *Adjust_tap* gives the values of m_1 and m_4, which are the quantized values of d_1 and d_4, respectively.

The feedback signal through the path p_2 takes $2d_1$ to come back to the clock generator. The amount of $2d_1$ can be expressed as $2q_1\Delta + r_1$, where $q_1 \in \mathbf{I}$ and $r_1 \in [-\Delta, \Delta]$, and is digitally encoded into $E_{p2} = 2q_1 + E_c$ (for $r_1 < 0$) or $2q_1 + 1 + E_c$ (for $r_1 \geq 0$). For either case, Algorithm *Adjust_tap* calculates the value of m_1 as q_1 in Step 2. Then the quantization error, err_1, for d_1 is equal to $\frac{r_1}{2} = d_1 - q_1\Delta$.

Also, the feedback signal through the path p_3 takes $2d_1 + d_4$. The amount of $2d_1 + d_4$ can be expressed as $q_3\Delta + r_3$, where $q_3 \in \mathbf{I}$ and $r_3 \in [-\Delta, 0]$, and is digitally encoded into $E_{p3} = q_3 + E_c$. Step 3 in *Adjust_tap* calculates the value of m_4 as $q_3 - (2q_1 + 1)$ if $r_1 \geq 0$ or $q_3 - 2q_1$ if $r_1 < 0$. Depending on the sign of r_1, the quantization error, err_4, for d_4 becomes

$$err_4 = \begin{cases} r_3 - r_1 + \Delta & \text{if } r_1 \geq 0 \\ r_3 - r_1 & \text{if } r_1 < 0. \end{cases}$$

The sum of err_1 and err_4 is thus the total quantization error as listed below.

$$err_1 + err_4 = \begin{cases} r_3 - \frac{1}{2}r_1 + \Delta & \text{if } r_1 \geq 0 \\ r_3 - \frac{1}{2}r_1 & \text{if } r_1 < 0. \end{cases}$$

For the first case, $err_1 + err_4$, is in $[-\frac{\Delta}{2}, \Delta]$, and $err_1 + err_4$ is in $[-\Delta, \frac{\Delta}{2}]$ for the second case. In summary, the total quantization error is in $[-\Delta, \Delta]$. In the general situation where $\Delta_p = 2i\Delta_s$, it can be shown [18] that the quantization error is in the range of $[-\frac{1}{2}(\Delta_p + \Delta_s), \frac{1}{2}(\Delta_p + \Delta_s)]$.

12. Simulation and Test Results

This clocking scheme has been implemented in a chip using 2 μm N-well CMOS technology available at MOSIS. Figure 12 shows a microphotograph of the chip at 2.22 × 2.25 mm^2. In this implementation, n is set to 4 and only one clock tree is driven by the clock generator. However, as stated before, this scheme can be extended to drive several different clock trees with a small increase of area. Design statistics are given in Table 10.

Table 10. Design statistics: Demonstration chip.

Transistors count	4215
Die size	2220 μm × 2250 μm
Total IO pins	15
Power/ground pins	18
Process	MOSIS 2 μm NWELL
Package	40 pin DIP
Power	100 mW

Figure 12. Microphotograph of the demonstration chip.

The physical design was done with the Magic layout system. The Cadence Spectre and CAzM simulators were used for circuit level simulation. In the simulation, the reference clock is set to 24 ns. Once the clock generator is powered on, it takes several cycles to calculate the compensation amount and settle the clock at each leaf of the clock tree to its scheduled skew. The capturing process of the clock generator is illustrated by Fig. 13(a). The dashed signal represents the skewed clock appearing at the leaf of the clock tree, and the solid signal is the desired clock. It takes 20 clock cycles for these two signals to be locked together.

In our chip, two microprobe pads were used to verify the clock alignment. One is connected to the reference clock, and the other is connected to the leaf of the clock tree. These two signals were measured using a Tektronix 11801A digital sampling oscilloscope. The locked waveforms are shown in Fig. 13(b).

Figure 14(a) shows the unintentional skew of the chip at different internal delays. In this experiment,

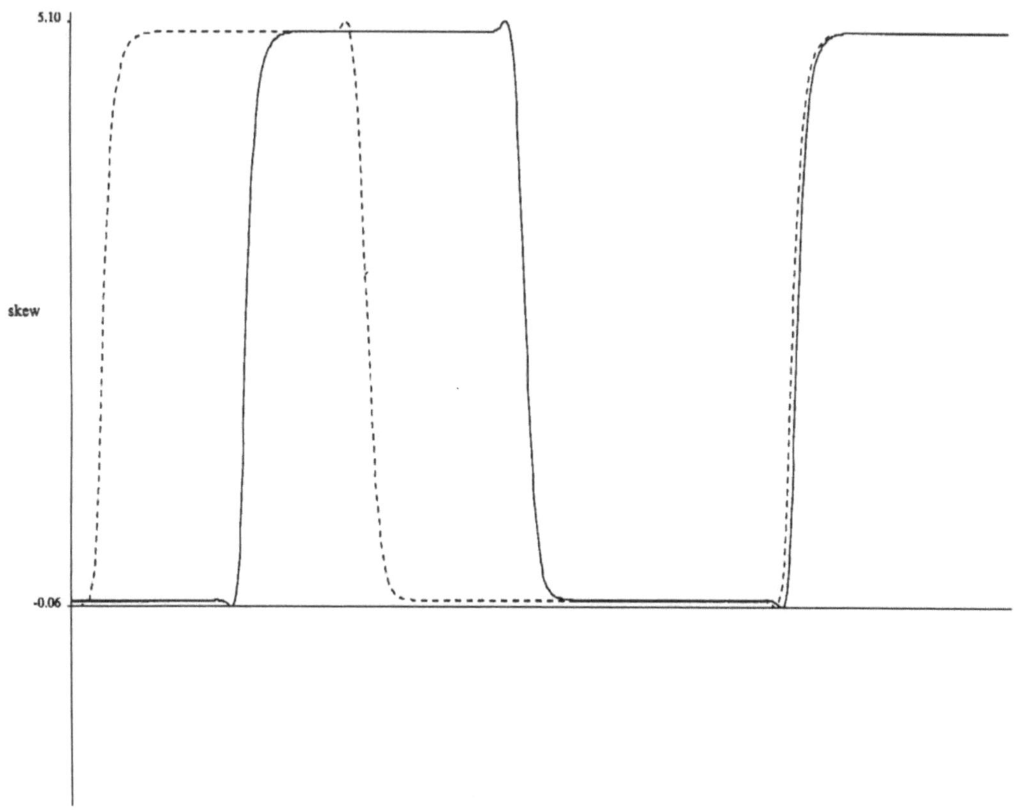

Figure 13a. Capturing process.

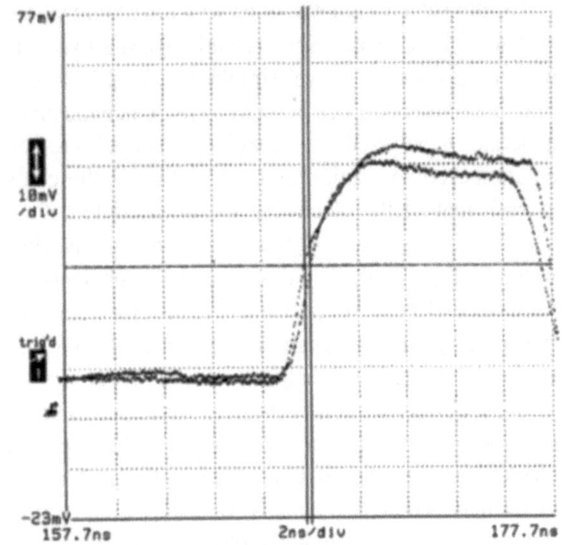

Figure 13b. Locked waveforms for $T_c = 24$ ns.

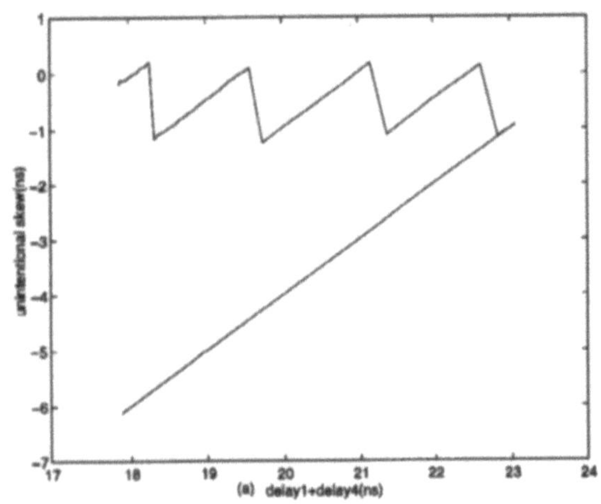

(a)

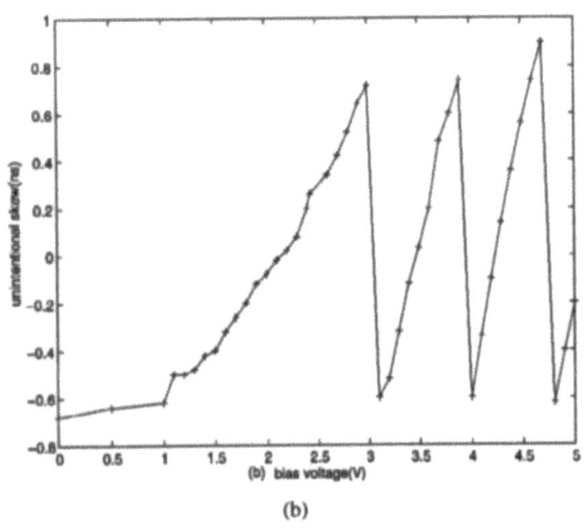

(b)

Figure 14. (a) Simulation results. (b) Test results.

the delay of clock buffer tree, d_4, is adjusted externally by a bias voltage and then the unintentional skew appearing at the leaves of the clock tree is monitored. Also the scheduled skewed-clock is set to the reference clock. Without the clock generator, the unintentional skew is represented by the straight line. With the clock generator, the unintentional skew is limited to the range of $[-\Delta, \Delta]$. The range is $[-1.5$ ns, 1.5 ns$]$ if the cycle time is 24 ns. The clock generator generates a sawtooth curve in the top of the figure. The granularity of adjacent phases gives the abrupt transitions in this curve. The corresponding test results at different bias voltages are shown in Fig. 14(b), which are consistent with the simulation results.

13. Improvements

As discussed in Section 11, the quantization error is in the range of $[-\frac{1}{2}(\Delta_p + \Delta_s), \frac{1}{2}(\Delta_p + \Delta_s)]$. For our demonstration chips, the sampling and phase resolutions are two inverter delays, which result in the quantization error, $[-\Delta, \Delta]$. Actually this error can be reduced if either a smaller phase resolution or a smaller sampling resolution is used. The structures in Fig. 15 can reduce the quantization error with a small increase of area. For the case shown in Fig. 15(a), the reference clock is tapped to the flip-flops at every delay element, and the feedback signal is tapped at every inverter. Then the sampling resolution is only one inverter delay, i.e., $\Delta_s = \frac{1}{2}\Delta_p = \frac{1}{2}\Delta$. Accordingly the

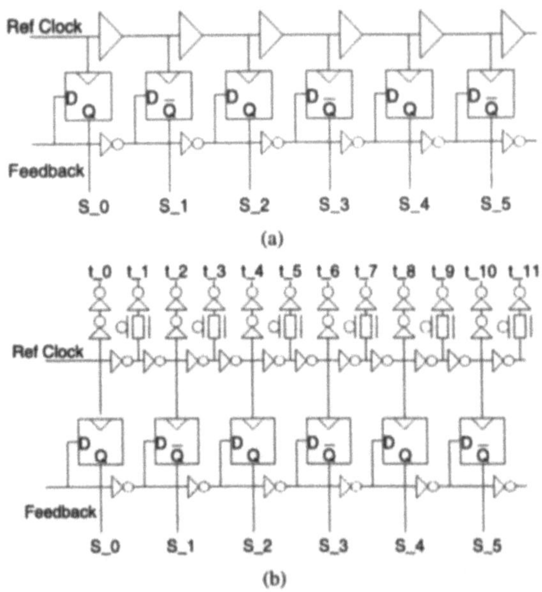

Figure 15. (a) Improved Δ_s (b) Improved Δ_s and Δ_p.

quantization error is in $[-\frac{3}{4}\Delta, \frac{3}{4}\Delta]$. Furthermore, the phase resolution of the clock generator can be halved ($\Delta_s = \Delta_p = \frac{1}{2}\Delta$) using a balancer circuit as shown in Fig. 15(b). The balancer circuit is used to tap out phase for compensation from the delay chain at every inverter. Thus the quantization error is in $[-\frac{1}{2}\Delta, \frac{1}{2}\Delta]$ for this design.

14. Conclusions

A skewed-clock optimization technique has been successfully used in the design of high performance systems. The feasibility of this technique depends on the solutions of two subproblems. First, optimal performance depends on intentional skews at the flip-flops. They must be carefully chosen. Second, skewed clocks must be reliably delivered. Thus for overcoming the process and environmental variations, a dynamically adjustable capability is required in the clock generator.

In this paper, a theoretical framework was developed for optimally scheduling skews. We concentrate on single-phase designs using edge-triggered flip-flops. However, this framework can also be extended to include multi-phase designs using either flip-flops or transparent latches or both. In addition, two other optimization techniques, retiming and resynchronization, can be incorporated into this framework for further optimization of systems with scheduled skews. Retiming is a technique for maximizing the speed of operation by relocating storage elements, while resynchronization allows optimal insertion of storage elements to remedy race-through conditions. These two techniques give designers additional ways to improve the system performance.

Another important issue in applying scheduled skews is the implementation of the desired amount of skew in real systems. In practice, designers are not allowed to choose arbitrary skews because of the difficulty in creating and distributing them. Thus scheduled skews can only be chosen from a set of predetermined values. This problem is called constrained-skew optimization and has not been reported. For further discussion, please refer to [19].

A self-calibrating clocking scheme was also presented in this paper. The scheme was implemented using a 2 μm N-well CMOS technology. With this scheme, unintentional skews can be limited to $[-\frac{1}{2}(\Delta_p+\Delta_s), \frac{1}{2}(\Delta_p+\Delta_s)]$ if paths p_1, p_2, and p_3 are well balanced. To maintain well balanced paths across all process variations, the design technique described in [15] was applied. This scheme was implemented digitally to enable the clock generator to be shared by several clock trees in different regions. This digital implementation effectively reduces the area for the clock generator. In this particular implementation, the die size of the core for the clock generator is only 1.8×1.8 mm^2 using 2 μm N-well CMOS technology. As feature size decreases and chip size increases, the area occupied by the core becomes an insignificant portion of the total system.

References

1. H.B. Bakoglu, *Circuits, Interconnection, and Packaging for VLSI*, Addison-Wesley, 1990.
2. H. Hsieh, W. Liu, C.T. Gray, and R. Cavin, "Concurrent timing optimization of latch-based digital systems," *International Conference on Computer Design*, pp. 680–685, 1995.
3. C.T. Gray, W. Liu, and R. Cavin III, *Wave Pipelining: Theory and CMOS Implementations*, Kluwer Academic Publisher, Oct. 1993.
4. J. Neves and E. Friedman, "Design methodology for synthesizing clock distribution networks exploiting non-zero localized clock skew," *IEEE Transactions on VLSI Systems*, Vol. 4, pp. 286–291, June 1996.
5. J.P. Fishburn, "Clock skew optimization," *IEEE Transactions on Computers*, Vol. 39, pp. 945–951, July 1990.
6. M. Heshami and B. Wooley, "A 250-MHz skewed-clock pipelined data buffer," *IEEE Journal of Solid-State Circuits*, Vol. 31, No. 3, pp. 376–383, March 1996.
7. H. Toyoshima, "A 300-MHz 4-Mb wave-pipeline CMOS SRAM using a multi-phase PLL," *IEEE Journal of Solid-State Circuits*, Vol. 30, No. 11, pp. 1189–1202, Nov. 1995.
8. R. Tsay, "An exact zero-skew clock routing algorithm," *IEEE Transactions on Computer Aided Design*, Vol. 12, No. 2, pp. 242–249, Feb. 1993.
9. S. Pullela, N. Menezes, and L.T. Pillage, "Reliable non-zero skew clock trees using wire width optimization," *30th Design Automation Conference*, pp. 165–170, 1993.
10. R. Watson and R. Iknaian, "Clock buffer chip with multiple target automatic skew compensation," *IEEE Journal of Solid-State Circuits*, Vol. 30, No. 11, pp. 1267–1276, Nov. 1995.
11. E. Best, *Phase Locked Loops: Theory, Design, and Applications*, McGraw-Hill, c1984.
12. N. Shenoy et al., "On the temporal equivalence of sequential circuits," *29th Design Automation Conference*, pp. 405–409, 1992.
13. M.R. Garey and D.S. Johnson, *Computers and Intractability*, W.H. Freeman and Company, 1979.
14. T.H. Cormen, C.E. Leiserson, and R.L. Rivest, *Introduction to Algorithms*, McGraw-Hill, 1990.

15. M. Shoji, "Elimination of process-dependent clock skew in CMOS VLSI," *IEEE Transactions on Computers*, Vol. C-39, No. 7, pp. 945–951, July 1990.
16. J. Kang, W. Liu, and R. Cavin III, "A monolithic 625Mb/s data recovery circuit in 1.2 μm CMOS," *Custom Integrated Circuits Conference*, pp. 463–465, March 1994.
17. M.G. Johnson and E.L. Hudson, "A variable delay line PLL for CPU-Coprocessor synchronization," *IEEE Journal of Solid-State Circuits*, pp. 1218–1223, 1988.
18. C.T. Gray, W. Liu, W. van Noije, T. Hughes, and R. Cavin III, "A sampling technique and its CMOS implementation with 1 Gb/s bandwidth and 25 ps resolution," *IEEE Journal of Solid-State Circuits*, Vol. 29, No. 3, pp. 340–349, March 1994.
19. H. Hsieh, Clocking Optimization and Distribution in Digital Systems with Scheduled Skews, Ph.D. Thesis, North Carolina State University, 1996.

Hong-yean Hsieh received his B.S. and M.S. degrees in Electrical Engineering from National Taiwan University, Taiwan in 1988 and 1990 respectively, and he received his Ph.D. degree in Computer Engineering from North Carolina State University in 1996. His research interests include VLSI designs and CAD for high-speed analog/digital systems.

Wentai Liu received his BSEE degree from National Chiao-Tung University, and MSEE degree from National Taiwan University, Taiwan, and his Ph.D. degree in computer engineering from the University of Michigan at Ann Arbor in 1983.

Since 1983, he has been on the faculty of North Carolina State University, where he is currently a Professor of Electrical and Computer Engineering. He has been a consultant and developed several VLSI CAD tools for microelectronic companies. He holds three U.S. patents. In 1986 he received an IEEE Outstanding Paper Award. His research interests include high speed VLSI design/CAD, microelectronic sensor design, high speed communication networks, parallel processing, and computer vision/image processing.

Dr. Liu has led a research group on wave pipelining and high speed digital circuit design at North Carolina State University. As a pioneer in the research area of CMOS wave pipelining and timing optimization, he has been invited to present research results in Germany, Brazil, and Taiwan. He has co-authored a book entitled "Wave Pipelining: Theory and CMOS Implementation" published by Kluwer Academic in 1994. His research results have been reported in news media such as CNN, EE Times, Electronic World News, and WRAL-TV. He is a council member of IEEE Solid-State Circuit Society.

Paul D. Franzon is currently an Associate Professor in the Department of Electrical and Computer Engineering at North Carolina State University. He has over eight years experience in electronic systems design and design methodology research and development. During that time, in addition to his current position, he has worked at AT&T Bell Laboratories in Holmdel, NJ, at the Australian Defense Science and Technology Organization, as a founding member of a successful Australian technology start up company, and as a consultant to industry, including technical advisory board positions.

Dr. Franzon's current research interests include design sciences/methodology for high speed packaging and interconnect, for high speed and low power chip design and the application of Micro Electro Mechanical Machines to electronic systems. In the past, he has worked on problems and projects in wafer-scale integration, IC yield modeling, VLSI chip design and communications systems design. He has published over 45 articles and reports. He is also the co-editor and author on a book about multichip module technologies to be published in October, 1992.

Dr. Franzon's teaching interests focuses on microelectronic systems building including package and interconnect design, circuit design, processor design and the gaining of hands-on systems experience for students.

Dr. Franzon is a member of the IEEE, ACM, and ISHM. He serves as the Chairman of the Education Committee for the National IEEE-CHMT Society. In 1993, he received an NSF Young Investigator's Award. In 1996, he was the Technical Program Chair at the IEEE MultiChip Module Conference and in 1997, the General Chair.

Ralph K. Cavin, III received his BSEE and MSEE degrees from Mississippi State University and his Ph.D. degree from Auburn University.

From 1962–65, he was a member of technical staff of the Martin Marietta Corporation in Orlando Florida working in the area of intermediate range and tactical missile guidance and control. From 1968 to 1983, he was a member of the faculty of Texas A&M University where he attained the rank of Professor of Electrical Engineering. He served as Director of the Design Science research program of the Semiconductor Research Corporation from 1983 to 1989. He joined North Carolina State University as Professor and Head of the Department of Electrical and Computer Engineering in 1989. Between 1994–1995, he was Dean of College of Engineering, North Carolina State University. Currently he is the Vice President of Research Operations at Semiconductor Research Corporation.

Dr. Cavin has authored over 100 reviewed papers. His research interests currently are in the areas of very high performance VLSI circuits and modeling and control of semiconductor processes.

He served as a member of the Board of Governors of the IEEE Circuits and Systems Society from 1990–1992. He is a member of the IEEE Strategic Planning Committee, chairs the New Technical Directions Committee of the IEEE Technical Activities Board, and serves as editor for the Emerging Technology series for the TAB/IEEE Press. He has served on numerous IEEE conference committees.

Journal of VLSI Signal Processing 16, 149–161 (1996)

Buffered Clock Tree Synthesis with Non-Zero Clock Skew Scheduling for Increased Tolerance to Process Parameter Variations*

JOSÉ LUIS NEVES AND EBY G. FRIEDMAN
Department of Electrical Engineering University of Rochester
Rochester, NY 14618

Received August 15, 1996; Revised November 20, 1996

Abstract. An integrated top-down design system is presented in this paper for synthesizing clock distribution networks for application to synchronous digital systems. The timing behavior of a synchronous digital circuit is obtained from the register transfer level description of the circuit, and used to determine a non-zero clock skew schedule which reduces the clock period as compared to zero skew-based approaches. Concurrently, the *permissible range* of clock skew for each local data path is calculated to determine the maximum allowed variation of the scheduled clock skew such that no synchronization failures occur. The choice of clock skew values considers several design objectives, such as minimizing the effects of process parameter variations, imposing a zero clock skew constraint among the input and output registers, and constraining the permissible range of each local data path to a minimum value.

The clock skew schedule and the worst case variation of the primary process parameters are used to determine the hierarchical topology of the clock distribution network, defining the number of levels and branches of the clock tree and the delay associated with each branch. The delay of each branch of the clock tree is physically implemented with distributed buffers targeted in CMOS technology using a circuit model that integrates short-channel devices with the signal waveform shape and the characteristics of the clock tree interconnect. A bottom-up approach for calculating the worst case variation of the clock skew due to process parameter variations is integrated with the top-down synthesis system. Thus, the local clock skews and a clock distribution network are obtained which are more tolerant to process parameter variations.

This methodology and related algorithms have been demonstrated on several MCNC/ISCAS-89 benchmark circuits. Increases in system-wide clock frequency of up to 43% as compared with zero clock skew implementations are shown. Furthermore, examples of clock distribution networks that exploit intentional localized clock skew are presented which are tolerant to process parameter variations with worst case clock skew variations of up to 30%.

1. Introduction

Most existing digital systems utilize fully synchronous timing, requiring a reference signal to control the temporal sequence of operations. Globally distributed signals, such as clock signals, are used to provide this synchronous time reference. These signals can dominate and limit the performance of VLSI-based digital systems. The importance of these global signals is, in part, due to the continuing reduction of feature size concurrent with increasing chip dimensions. Thus interconnect delay has become increasingly significant, perhaps of greater importance than active device delay. The increased global interconnect delay also leads to significant differences in clock signal propagation within the clock distribution network, called *clock skew*, which occurs when the clock signals arrive at the

*This research is based upon work supported by Grant 200484/89.3 from CNPq (Conselho Nacional de Desenvolvimento Científico e Tecnológico—Brasil), the National Science Foundation under Grant No. MIP-9208165 and Grant No. MIP-9423886, the Army Research Office under Grant No. DAAH04-93-G-0323, and by a grant from the Xerox Corporation.

storage elements at different times. The clock skew can be further increased by unintentional factors such as process parameter variations which may limit the maximum frequency of operation, as well as create race conditions independent of clock frequency, leading to circuit failure. Therefore, the design of high performance, process tolerant clock distribution networks is a critical phase in the synthesis of synchronous VLSI digital circuits. Furthermore, the design of the clock distribution network, particularly in high speed applications, requires significant amounts of time, inconsistent with the high turnaround in the design of the more common data flow elements of digital VLSI circuits.

Several techniques have been developed to improve the performance and design efficiency of clock distribution networks, such as placing distributed buffers within clock tree layouts [1] to control the propagation delay and power consumption characteristics of the clock distribution networks, resizing clock nets for speed optimization and clock path delay balancing [2, 3], perform simultaneous buffer and interconnect sizing to optimize for speed and reduce power dissipation [4], using symmetric distribution networks, such as H-tree structures [5], to minimize clock skew, and applying zero-skew clock routing algorithms [e.g., 6, 7] to the automated layout of high speed clock distribution networks in cell-based circuits. Effort has also been placed on reducing clock skew due to process variations [e.g., 8–10], and on designing clock distribution networks so as to ensure minimal variation in clock skew [1, 7]. Alternative approaches have been developed for using intentional non-zero clock skew to improve circuit performance and reliability by properly choosing the local clock skews [10–12]. Targeting non-zero local clock skew, a synthesis methodology has been developed for designing clock distribution networks capable of accurately producing specific clock path delays [13, 14]. These clock distribution networks exploit intentional localized clock skew while taking into account the effects of process parameter variations on the clock path delays.

A design environment is presented in this paper for efficiently synthesizing distributed buffer, tree-structured clock distribution networks. This methodology is illustrated in terms of the IC design process cycle in Fig. 1. The IC design cycle typically begins with

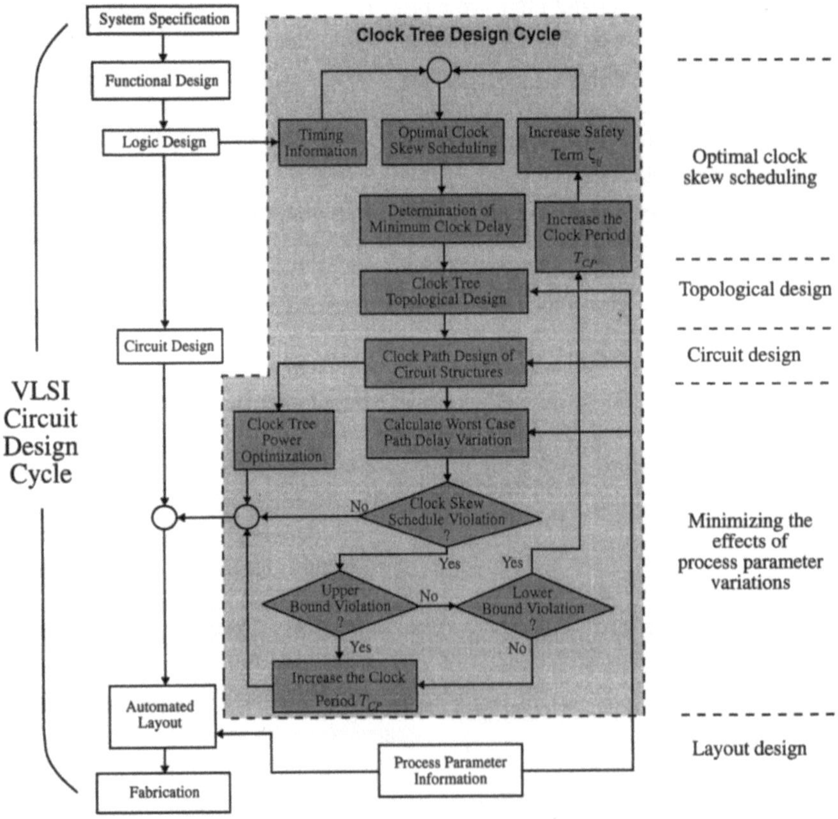

Figure 1. Block diagram of the clock tree design cycle integrated with standard IC design flow.

the System Specification phase. The Clock Tree Design Cycle utilizes timing information from the Logic Design phase, such as the minimum and maximum delay values of the logic blocks and the registers. The timing information is used to determine the maximum frequency of operation of the circuit, the non-zero clock skew schedule, the permissible range of clock skew between any pair of sequentially adjacent registers, and the minimum clock path delay to each register. The topology of the clock tree is designed to enforce the clock skew schedule. The delay of each clock path is accurately implemented using repeaters targeting CMOS technology. Finally, the clock tree is validated by ensuring that the worst case clock path delays caused by process parameter variations do not create clock skew values outside the allowed permissible range of each pair of sequentially adjacent registers. Process parameter information is extensively used in several stages of the design environment for ensuring the accuracy of the clock tree. The output of the Clock Tree Design Cycle is a detailed circuit description of the clock distribution network, including the number and geometric size of each buffer stage within each branch of the clock tree.

This paper is organized as follows: in Section 2, a localized clock skew schedule is derived from the effective permissible range of the clock skew for each local data path considering any global clock skew constraints and process parameter variations. In Section 3, a topology of the clock distribution network is obtained, producing a clock tree with specific delay values assigned to each branch. The design of circuit structures for implementing the individual branch delay values is summarized in Section 4. In Section 5, techniques for compensating the scheduled local clock skew values to process-dependent clock path delay variations are presented. In Section 6, these results are evaluated on a series of circuits, thereby demonstrating performance improvements and immunity to process parameter variations. Finally, some conclusions are drawn in Section 7.

2. Optimal Clock Skew Scheduling

A synchronous digital circuit C can be modeled as a finite directed multi-graph $G(V, E)$. Each vertex in the graph, $v_j \in V$, is associated with a register, circuit input, or circuit output. Each edge in the graph, $e_{ij} \in E$, represents a physical connection between vertices v_i and v_j, with an optional combinational logic path between the two vertices. An edge is a bi-weighted connection representing the maximum (minimum) propagation delay $T_{\mathrm{PD\,max}}$ ($T_{\mathrm{PD\,min}}$) between two sequentially adjacent storage elements. The propagation delay T_{PD} includes the register, logic, and interconnect delays of a local data path [13], as described in (1),

$$T_{\mathrm{PD}} = T_{C-Q} + T_{\mathrm{Logic}} + T_{\mathrm{Int}} + T_{\text{Set-up}}, \quad (1)$$

where T_{C-Q} is the time required for the data to leave R_i once it is triggered by a clock pulse C_i, T_{Logic} is the propagation delay through the logic block between registers R_i and R_j, T_{Int} accounts for the interconnect delay, and $T_{\text{Set-up}}$ is the time to successfully propagate to and latch the data within R_j [15].

A local data path L_{ij} is a set of two vertices connected by an edge, $L_{ij} = \{v_i, e_{ij}, v_j\}$ for any $v_i, v_j \in V$. A global data path, $P_{kl} = v_k \overset{p}{\rightarrow} v_l$, is a set of alternating edges and vertices $\{v_k, e_{k1}, v_1, e_{12}, \ldots, e_{n-1l}, v_l\}$, representing a physical connection between vertices v_k and v_l, respectively. A multi-input circuit can be modeled as a single input graph, where each input is connected to vertex v_0 by a zero- weighted edge. $Pl(L_{ij})$ is defined as the permissible range of a local data path and $Pg(P_{kl})$ is the permissible range of a global data path.

2.1. *Timing Constraints*

The timing behavior of a circuit C can be described in terms of two sets of timing constraints, local constraints and global constraints. The local constraints are designed to ensure the correct latching of data into the registers of a local data path. In particular, (2) prevents latching the incorrect data signal into R_j by the clock pulse that latched the same data into R_i (preventing double clocking [10, 11]),

$$T_{\mathrm{Skew}}(L_{ij}) \geq T_{\mathrm{Hold}\,j} - T_{\mathrm{PD(min)}} + \zeta_{ij}, \quad (2)$$

where ζ_{ij} is a safety term to provide some margin in a local data path against race conditions due to process parameter variations, and (3) guarantees that the data signal latched in R_i is latched into R_j by the following clock pulse (preventing zero clocking [10, 11]),

$$T_{\mathrm{Skew}}(L_{ij}) \leq T_{\mathrm{CP}} - T_{\mathrm{PD(max)}}. \quad (3)$$

Constraints (2) and (3) are similar to the synchronous constraints introduced in [11, 12, 16, 17], where

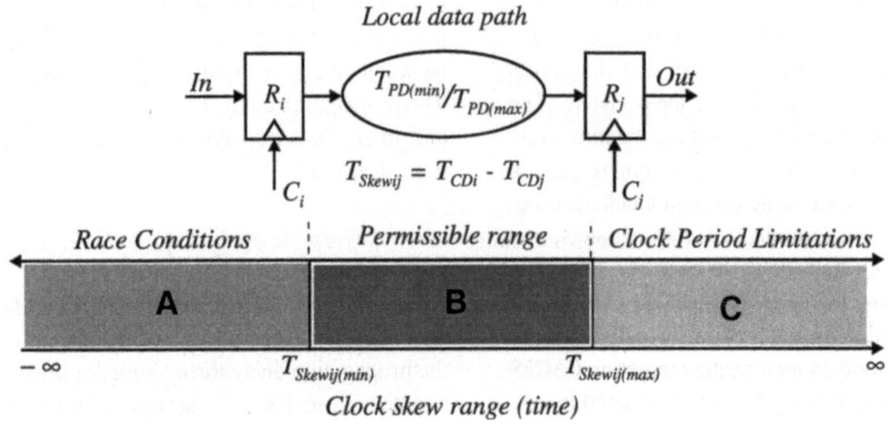

Figure 2. Permissible range of the clock skew of a local data path.

the clock skew $T_{\text{Skew}\,ij} = T_{\text{CD}i} - T_{\text{CD}j}$ and where $T_{\text{CD}i}(T_{\text{CD}j})$ is the delay of the Ith (Jth) clock path.

Assuming that the minimum and maximum delay of each combinational logic block and register are known, a region of valid clock skew is assigned to each local data path, called the permissible range $Pl(L_{ij})$ [13, 18], as shown in Fig. 2. The bounds of $Pl(L_{ij})$ are determined from the local constraints, (3) and (4), for a given clock period T_{CP}. Also, the *width* of a permissible range is defined as the difference between the maximum ($T_{\text{Skew}\,ij(\text{max})}$) and the minimum ($T_{\text{Skew}\,ij(\text{min})}$) clock skew.

Satisfying the clock skew constraints of each individual local data path does not guarantee that the clock skew between two vertices of a global data path P_{kl} is satisfied, particularly when there are multiple parallel and feedback paths between the two vertices. Since any two registers connected by more than one global data path are each driven by a single clock path, the clock skew between these two registers is unique and the permissible range of every path connecting the two registers must contain this clock skew value to ensure that the circuit will operate correctly. As an example to illustrate that the clock skew between registers must be contained within the permissible range of each global data path connecting both registers, consider the circuit illustrated in Fig. 3, where the numbers assigned to the edges are the maximum and minimum propagation delay of each local data path L_{ij}, and the register set-up and hold times are assumed to be zero. Furthermore, the pair of clock skew values associated with a vertex are the minimum and maximum clock skew calculated with respect to the origin vertex v_0 for a given clock period. The minimum bound of

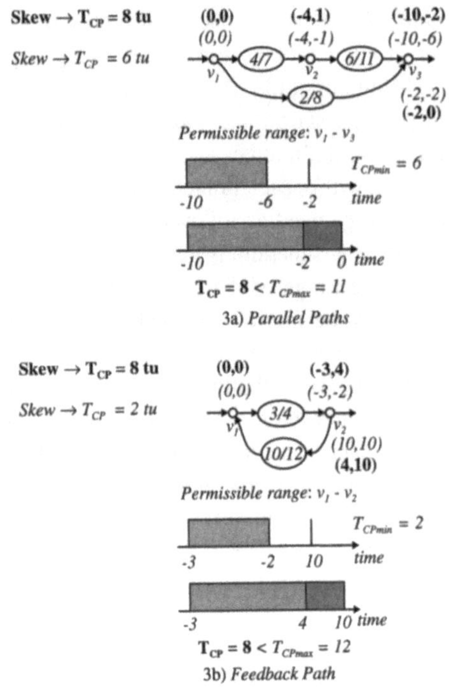

Figure 3. Matching permissible clock skew ranges by adjusting the clock period T_{CP}.

$Pl(L_{ij})$ is given by (3) and is $T_{\text{Skew}\,ij(\text{min})} = -T_{\text{PD}\,\text{min}}$ and the maximum bound of $Pl(L_{ij})$ is given by (4) and is $T_{\text{Skew}\,ij(\text{max})} = T_{\text{CP}} - T_{\text{PD}\,\text{max}}$. Observe that in Fig. 3(a), a non-empty permissible range for each individual local data path can is obtained with a clock period $T_{\text{CP}} = 6$ time units (tu). However, no clock skew value exists that is common to the paths connecting vertices v_1 and v_3. The common value for $T_{\text{Skew}13}$ is only obtained when the clock period is increased to 8 tu.

To guarantee that a clock skew value exists for any pair of registers $v_k, v_l \in V$ within a global data path, a set of global timing constraints must be satisfied. Complete proofs of the following theorems are found in [19]. The global timing constraints (4) and (5) are used to calculate the permissible range of any global data path $P_{kl} \in V$, and are based on the permissible range of the local data paths within the respective global data path. In particular, (4) determines the minimum and maximum clock skew of a global data path with respect to v_k, while (5) constrains the clock skew of two vertices connected by multiple forward and feedback paths. These two constraints can be formally stated as:

Theorem 1. *For any global data path $P_{kl} \in V$, clock skew is conserved. Alternatively, the clock skew between any two storage elements, $v_k, v_l \in V$, is the sum of the clock skews of each local data path $L_{k1}, L_{12}, \ldots, L_{n-1l}$, where $L_{k1}, L_{12}, \ldots, L_{n-1l}$ are the local data paths within P_{kl},*

$$T_{\text{Skew}}(P_{kl}) = T_{\text{Skew}}(L_{k1}) + T_{\text{Skew}}(L_{12}) + \cdots + T_{\text{Skew}}(L_{l-1l}). \quad (4)$$

Theorem 2. *For any global data path P_{kl} containing feedback paths, the clock skew in a feedback path between any two storage elements, say v_m and $v_n \in P_{kl}$, is the negative of the clock skew between v_m and v_n in the forward path,*

$$T_{\text{Skew}}(P_{kl}) = -T_{\text{Skew}}(P_{lk}). \quad (5)$$

In the presence of multiple parallel and/or feedback paths connecting any two registers R_k and R_l, a permissible range only exists between these two registers if there is overlap among the permissible ranges of each individual parallel and feedback path connecting both registers. Furthermore, the upper and lower bounds of such a permissible range are determined from the upper and lower bounds of the permissible ranges of each individual parallel and feedback path. Formally, the concept of permissible range overlap and the upper and lower bounds of the permissible range of a global data path P_{kl} can be stated as follows:

Theorem 3. *Let $P_{kl} \in V$ be a global data path within a circuit C with m parallel and n feedback paths. Let the two vertices, v_k and $v_l \in P_{kl}$, which are not necessarily sequentially-adjacent, be the origin and destination of the m parallel and n feedback paths, respectively. Also, let $Pg(P_{kl})$ be the permissible range of the global data path composed of vertices v_k and v_l. $Pg(P_{kl})$ is a non-empty set of values iff the intersection of the permissible ranges of each individual parallel and feedback path is a non-empty set, or*

$$Pg(P_{kl}) = \left(\bigcap_{i=1}^{m} Pg(P_{kl}^{i})\right) \cap \left(\bigcap_{j=1}^{n} Pg(P_{lk}^{j})\right) \neq \emptyset. \quad (6)$$

Theorem 4. *Let the two vertices, v_k and $v_l \in P_{kl}$, be the origin and destination of a global data path with m forward and n feedback paths. If $Pg(P_{kl}) \neq \emptyset$, the upper bound of $Pg(P_{kl})$ is given by*

$$T_{\text{Skew}}(P_{kl})_{\max} = \text{MIN}\left\{ \min_{1\le i\le m} \left[T_{\text{Skew}}\left(P_{kl}^{i}\right)_{\max}\right], \min_{1\le j\le n} \left[T_{\text{Skew}}\left(P_{lk}^{j}\right)_{\min}\right]\right\}, \quad (7)$$

and the lower bound of $Pg(P_{kl})$ is given by

$$T_{\text{Skew}}(P_{kl})_{\min} = \text{MAX}\left\{ \max_{1\le i\le m} \left[T_{\text{Skew}}\left(P_{kl}^{i}\right)_{\min}\right], \max_{1\le j\le n} \left[T_{\text{Skew}}\left(P_{lk}^{j}\right)_{\max}\right]\right\}. \quad (8)$$

Two global timing constraints impose zero clock skew among the I/O storage elements and limit the permissible clock skew range that can be implemented by the fabrication technology. By constraining the clock skew among the off-chip registers to zero, race conditions are eliminated among all integrated circuits controlled by the same clock source by avoiding the propagation of a non-zero clock skew beyond the integrated circuit. This condition is represented by the following expression,

$$T_{\text{Skew}}(P_{kl})_{\min} = \text{MAX}\left\{ \max_{1\le i\le m} \left[T_{\text{Skew}}\left(P_{kl}^{i}\right)_{\min}\right], \max_{1\le j\le n} \left[T_{\text{Skew}}\left(P_{lk}^{j}\right)_{\max}\right]\right\}. \quad (9)$$

An immediate consequence of (9) is that the clock path delay from the clock source to every input and output register is equal.

Although the permissible range of a local data path is theoretically infinite, practical limitations place constraints on the minimum clock path delays that can be implemented with a given fabrication technology. These clock path delays determine the minimum clock skew that can be assigned to any two vertices in the circuit. These fabrication dependent timing constraints are

$$|T_{\text{Skew}}(L_{ij})_{\max} - T_{\text{Skew}}(L_{ij})_{\min}| \geq C_1, \qquad |T_{\text{Skew}ij}| \geq C_2, \tag{10}$$

where C_1 and C_2 are dependent on the fabrication technology and are a measure of the statistical variation of the process parameters.

2.2. *Optimal Clock Period*

The problem of determining an optimal clock period for a synchronous circuit while exploiting non-zero clock skew has been previously studied [11, 12, 16, 17]. In these approaches, clock delays rather than clock skews are calculated. Therefore, these clock delays cannot be directly used for determining the permissible range of the local clock skews. Thus, there is no process for determining the position of the scheduled clock skew within the permissible range. A technique to perform this process is described in this paper to schedule the clock skew and to prevent synchronization failures due to process parameter variations.

The determination of the minimum clock period using permissible ranges is possible by recognizing that the width of the permissible range of a local data path is dependent on the clock period [from (3)]. The overlap of permissible ranges guarantees the synchronization of the data flow between non-adjacent registers connected by multiple feedback and/or parallel paths. This technique initially guarantees the existence of a permissible range for each local data path and terminates by satisfying (6) for every data path in the circuit. The difference between the propagation delays of a local data path L_{ij} defines the minimum clock period necessary to safely latch data within L_{ij}. The largest difference among all the local data paths of the circuit defines the minimum clock period that can be used to safely latch data into any local data path. However, as shown in the example depicted in Fig. 3, in the presence of feedback and/or parallel paths, local timing constraints may not be sufficient to determine the minimum clock period (since certain global timing constraints such as (6) must also be satisfied). Nevertheless, a clock period always exists that satisfies all the local and global timing constraints of a circuit. This clock period is bounded by two terms, $T_{\text{CP}\min}$ and $T_{\text{CP}\max}$, as independently demonstrated by Deokar and Sapatnekar in [12]. The lower bound of the clock period, $T_{\text{CP}\min}$, is the greatest difference in propagation delay of any local data path $L_{ij} \in C$,

$$T_{\text{CP}\min} = \text{MAX}\left[\max_{v_{ij}\in V}(T_{\text{PD}\max ij} - T_{\text{PD}\min ij}), \max_{v_i\in V}(T_{\text{PD}\max ii})\right], \tag{11}$$

and the upper bound of the clock period, $T_{\text{CP}\max}$, is the greatest propagation delay of any local data path $L_{ij} \in G$,

$$T_{\text{CP}\max} = \text{MAX}\left[\max_{v_{ij}\in V}(T_{\text{PD}\max ij}), \max_{v_i\in V}(T_{\text{PD}\max ii})\right]. \tag{12}$$

The second term in (11) and (12) accounts for the self-loop circuit when the output of a register is connected to its input through an optional logic block. Since the initial and final registers are the same, the clock skew in a self-loop is zero and the clock period is determined by the maximum propagation delay of the path connecting the output of the register to its input. Observe that a clock period equal to the lower bound exists for circuits without parallel and/or feedback paths. Furthermore, a clock period equal to the upper bound always exists since the permissible range of any local data path in the circuit contains the zero clock skew value. Although (12) satisfies any local and global timing constraints of circuit C, it is possible to determine a lower clock period that satisfies (6).

Several algorithms for determining the optimal clock period while exploiting non-zero clock skew exist. Fishburn [11] introduced this approach with a linear programming-based algorithm that minimizes the clock period while determining a set of clock path delays to drive the individual registers within the circuit. In [12], Deokar and Sapatnekar present a graph-based approach to achieve a similar goal, followed by an optimization step to reduce the skew between registers while preserving the minimum clock period. Other works, such as Sakallah et al. [16] and Szymanski [17], also calculate the optimal clock period and clock path delay schedule using linear programming techniques.

A graph-based algorithm is implemented in C to determine the minimum clock period and a permissible range for each local data path while ensuring that all the permissible ranges in the circuit satisfy (6) [18, 19]. The initial clock period is given by (11) and, the local and global permissible ranges for each local data path are calculated assuming this clock period. If at least one data path does not satisfy (6), the clock period is increased and the permissible ranges are re-calculated. This iterative process continues until (6) is satisfied for all global data paths. The primary distinction of this algorithm is that the permissible range of each local data path $Pl(L_{ij})$ is determined rather than the individual clock path delays to registers R_i and R_j. From each permissible range a clock skew value is chosen as explained in Section 2.3. This information is crucial for maximizing the performance of a synchronous circuit while considering the effects of process parameter variations in the design of high speed clock distribution networks.

2.3. *Selecting Clock Skew Values*

Given any two vertices $v_k, v_l \in V$, the set of valid clock skew values between v_k and v_l is given by (6) and bounded by (7) and (8), as described in Section 2.2. In the presence of feedback and/or parallel paths, the resulting permissible range $Pg(P_{kl})$ is a sub-set of the permissible range of each independent global data path between v_k and v_l, as exemplified in Fig. 3. However, due to (4), $Pg(P_{kl})$ is the sum of the permissible range of each local data path for every global data path P_{kl} connecting v_k and v_l. Therefore, it is necessary to constrain the permissible range of each local data path to a sub-set of values within its original permissible range. Alternatively, if $Pl(L_{ij})$ is the permissible range of a local data path within one of the global data paths connecting v_k and v_l, $\rho(L_{ij})$ is a sub-set of values within $Pl(L_{ij})$ such that $\rho(L_{ij}) \subset Pl(L_{ij})$. This new region $\rho(L_{ij})$ is described as the *effective permissible range* of a local data path.

An example of an effective permissible range is the parallel path shown in Fig. 3(a). For $T_{CP} = 8$ tu, the permissible range $Pg(P_{13}) = [-2, -2]$. Since $Pg(P_{13}) = Pl(L_{12}) + Pl(L_{23})$, the local data paths L_{12} and L_{23} can only assume clock skew values for which the sum is within $[-2, -2]$. In this case, the permissible range of each local data path is reduced to a single value, or $Pl(L_{12}) = [1, 1]$ and $Pl(L_{23}) = [-3, -3]$, respectively.

Assume that the clock period of the circuit in Fig. 3(a) is now increased from 8 tu to 9 tu. The new permissible range $Pg(P_{13}) = [-2, 0]$ and the effective permissible range of each local data path is $\rho(L_{12}) = [1, 2]$, $\rho(L_{23}) = [-3, -2]$, and $\rho(L_{13}) = [-2, 0]$, respectively. Note that selecting a clock skew value outside the effective permissible range of a local data path may lead to a race condition since (7) is violated. Also, there is no unique solution to the selection of an effective permissible range unless $\rho(L_{ij}) = Pl(L_{ij})$. For example, $Pl(L_{12})$ could be set to [0, 2] and $Pl(L_{23})$ set to $[-2, -2]$, giving the same permissible range $Pg(P_{13}) = [-2, 0]$. Therefore, given any two vertices $v_k, v_l \in V$ with feedback and/or parallel paths connecting v_k and v_l, the selection of a clock skew schedule requires determining the effective permissible range $\rho(L_{ij})$ for each local data path between v_k and v_l, and the relative position of $\rho(L_{ij})$ within $Pl(L_{ij})$.

The effective permissible range of a local data path $\rho(L_{ij})$ may not be unique, leading to multiple solutions to the clock skew scheduling problem. It is, however, possible to obtain one solution that is most suitable for minimizing the clock period while reducing the possibility of race conditions due to the effects of process parameter variations. This solution for $\rho(L_{ij})$ is derived from the observation that the bounds of the permissible range of any two vertices $v_k, v_l \in V$ (with possible feedback and/or parallel paths connecting v_k and v_l) are maximum when determined by (7) and (8), and that the permissible $Pg(P_{kl})$ bounded by (7) and (8) is unique. Therefore, the clock skew scheduling problem can be divided into two phases. In the first phase, the permissible range of each global data path is derived from (6), with bounds given by (7) and (8). In the second phase, the clock skew schedule is solved by the following process: 1) the permissible range of a global data path $P_g(P_{kl})$ is divided equally among each local data path belonging to each global data path connecting the vertices v_k and v_l; 2) within each global data path each effective permissible range $\rho(L_{ij})$ is placed as close as possible to the upper bound of the original permissible range $Pl(L_{ij})$, thereby minimizing the likelihood of creating any race conditions; and 3) the specific value of the clock skew is chosen in the middle of the effective permissible range, since no prior information describing the variation of a particular clock skew value may exist. An algorithm for selecting the clock skew of each local data path was implemented as described in [18, 19]. From this clock skew schedule the minimum clock path delay to each register in the circuit is calculated [19].

Providing independent clock path delays for each register is impractical due to the large capacitive load placed on the clock source and the inefficient use of die area. A tree structured clock distribution network is more appropriate, where the branching points are selected according to the delay of each clock path, the relative physical position of the clocked registers, and the sensitivity of each local data path to delay variations. Such an approach for determining the structural topology of a clock distribution network is described in the following section.

3. Clock Tree Topological Design

The topology of a clock tree derived from a clock skew schedule must ensure that the clock path delays are accurately implemented while considering the effects of process parameter variations. A tree-structured topology can be based on the hierarchical description of the circuit netlist, on implementing a balanced tree with a fixed number of branching levels from the clock source to each register with a pre-defined number of branching points per node (an example of this approach is a binary tree with n levels for 2^n registers with two branching points per node), on reducing the effects of process parameter variations by driving common local data paths by the same sub-tree, or by implementing each clock path delay with pre-defined delay segments such that the layout area of the clock tree is reduced.

The topology of the clock distribution tree is built by driving common local data paths by the same sub-tree and by assigning precise delay values to each branch of the clock tree such that the skew assignment is satisfied [20]. For this purpose, each clock path delay is partitioned into a series of branches, each branch emulating a precise quantified delay value. Between any two segments, there is a branching point to other registers or sub-trees of the clock tree, where several branches with pre-defined delays are cascaded to provide the appropriate delay between the clock source (or root) and each leaf node. The selection of the branch delay is dependent upon the minimum propagation delay that can be implemented for a particular fabrication process and the inverter transconductance (or gain). An example of the topology of a clock tree is shown in Fig. 4, where the numbers in brackets are the delays assigned to each branch and the numbers in parenthesis are the clock skew assignment.

4. Circuit Design of the Clock Tree

The circuit structures are designed to emulate the delay values associated with each branch of the clock tree. Special attention is placed on guaranteeing that the clock skew *between* any two clock paths is satisfied rather than satisfying each individual clock path delay. The successful design of each clock path is primarily dependent on two factors: 1) isolating each branch delay using active elements, specifically CMOS inverters, and 2) using repeaters to integrate the inverter and interconnect delay equations so as to more accurately calculate the delay of each clock path.

The interconnect lines are modeled as purely capacitive lines by inserting inverting buffer repeaters into the clock path such that the output impedance of each inverter is significantly greater than the resistance of the driven interconnect line [21]. As a consequence, the slope of the input signal of a buffer connected to a branching point is identical to the slope of the output signal of the buffer driving that same branching point [22].

In the existing design methodology [14, 22], the delay of a branch is implemented with one or more CMOS inverters, as illustrated in Fig. 5. The delay equations of each inverter are based on the MOSFET α-power law short-channel I-V model developed by Sakurai and Newton [23].

Each inverter is assumed to be driven by a ramp signal with symmetric rising and falling slopes, selected

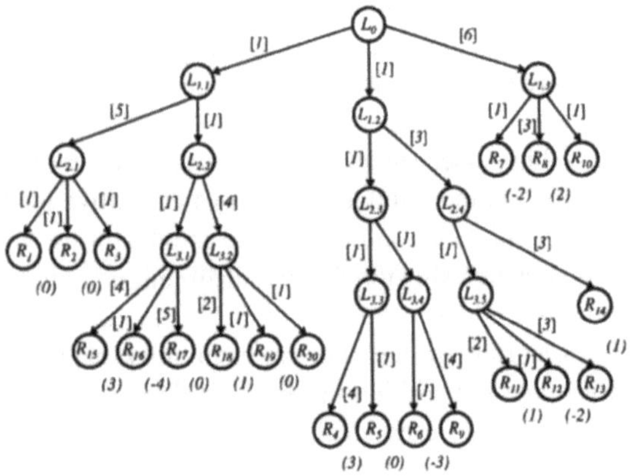

Figure 4. Topology of the clock distribution network.

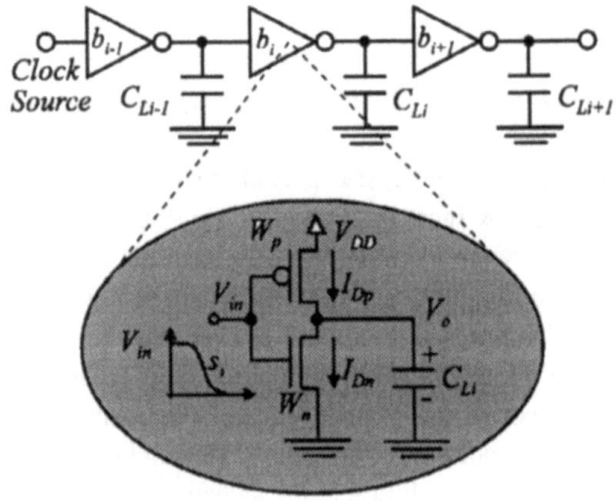

Figure 5. Design of a branch delay element.

such that during discharge (charge), the effects of the PMOS (NMOS) transistor can be neglected. The capacitive load of an inverter so as to satisfy a specific branch delay t_{di} is

$$C_{Li} = \frac{2I_{DO}}{V_{DD}}\left[t_{di} - \left(\frac{1}{2} - \frac{1-v_T}{1+\alpha}\right)t_{Ti-1}\right], \quad (13)$$

where I_{DO} is the drain current at $V_{GS} = V_{DS} = V_{DD}$, V_{DO} is the drain saturation voltage at $V_{GS} = V_{DD}$, V_{th} is the threshold voltage, α is the velocity saturation index, V_{DD} is the power supply, t_{di} is the delay of an inverter defined at the 50% V_{DD} point of the input waveform to the 50% V_{DD} point of the output waveform, $v_T = V_{th}/V_{DD}$, and t_{Ti} is the transition time of the input signal. Note that C_{Li} is composed of the capacitance of the driven interconnect line and the total gate capacitance of all b_{i+1} inverters. Since t_{di} is known, the only unknown in (13) is the transition time of the input signal t_{Ti} (provided by [23]). t_{Ti} can be approximated by a ramp shaped waveform, or by linearly connecting the points $0.1V_{DD}$ and $0.9V_{DD}$ of the output waveform. This assumption is accurate as long as the interconnect resistance is negligible as compared with the inverter output impedance.

$$\begin{aligned} t_{Ti} &= \frac{t_{0.9} - t_{0.8}}{0.8} \\ &= \frac{C_{Li}V_{DD}}{I_{DO}}\left(\frac{0.9}{0.8} + \frac{V_{DO}}{0.8V_{DD}}\ln\frac{10V_{DO}}{eV_{DD}}\right). \end{aligned} \quad (14)$$

For each clock path within the clock tree, the procedure to design the CMOS inverters is as follows: 1) the load of the initial trunk of the clock tree is determined from (13), assuming a step input clock signal; 2) the slope of the output signal is calculated from (14) and applied in (13) to determine the capacitive load of the following branch, permitting the slope of the output signal to be calculated; and 3) step 2 is repeated for each subsequent branch of the clock path. Steps 1–3 are applied to the remaining clock paths within the clock tree. Observe that if the transition time of the output signal of branch b_i does not satisfy

$$t_{Ti} \leq \frac{1}{\left(\frac{1}{2} - \frac{1-v_T}{1+\alpha}\right)}\left(t_{di+1} - \frac{V_{DD}C_{Li+1}}{2I_{DO}}\right), \quad (15)$$

(13) is no longer valid. The transition time t_{Ti} can be reduced in order to satisfy (15) by increasing the output current drive of the inverter in branch b_i. However, increasing I_{DOi} would increase the capacitive load C_{Li} in order to maintain the propagation delay t_{di} for branch b_i. Therefore, the transition time associated with branch b_i must be maintained constant as long as the propagation delay t_{di} of the branch b_i remains the same. Furthermore, the number of inverters required to implement the propagation delay t_{di} is chosen such that (15) is satisfied and the proper polarity of the clock signal driving branch b_{i+1} is maintained.

5. Increasing Tolerance to Process Parameter Variations

Every semiconductor fabrication process can be characterized by variations in process parameters. These process parameter variations along with environmental variations, such as temperature, supply voltage, and radiation, may compromise both the performance and the reliability of the clock distribution network. A bottom-up approach is presented in this section for verifying the selected clock skew values and correcting for any variations of the clock skew due to process parameter variations that violate the bounds of the permissible range.

5.1. Circuit Design Considerations

Each clock path delay can be modeled as being composed of both a deterministic delay component and a probabilistic delay component. While the deterministic component can be characterized with well developed delay models [e.g., 23], the probabilistic component of the clock path delay is dependent upon variations of the fabrication process and the environmental conditions. The variations of the fabrication process affect both the active device parameters (e.g., I_{DO}, V_{th}, μ_n) and the passive geometric parameters (e.g., the interconnect width and spacing).

The probabilistic delay component is determined for each clock path by assuming that the cumulative effects of the device parameter variations, such as threshold voltage and channel mobility, can be collected into a single parameter characterizing the gain of the inverter, specifically the output current of a CMOS inverter I_{DO} [23]. The minimum and maximum clock path delays are calculated considering the minimum and maximum I_{DO} of each inverter within a branch of the clock distribution network. The worst case variation of the clock skews is determined from the minimum and maximum clock path delays of each local data path. If at least one worst case clock skew value is outside the effective permissible range of the corresponding local data path (i.e., $T_{Skew\,ij} \not\subset \rho(L_{ij})$), a timing constraint

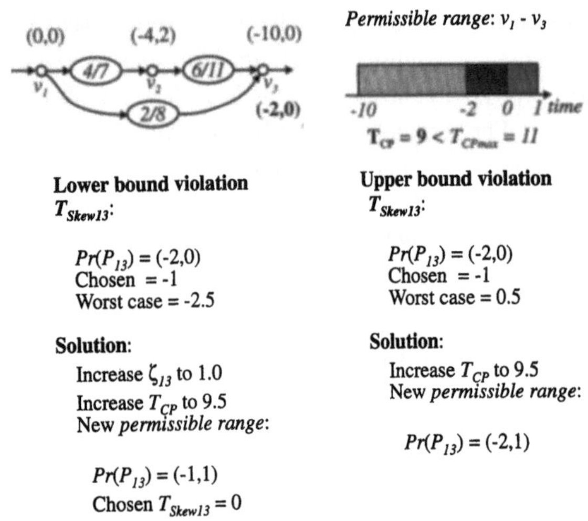

Figure 6. Example of upper and lower bound clock skew violations.

is violated and the circuit will not work properly, as illustrated in the example shown in Fig. 6.

This violation is passed to the top-down synthesis system, indicating which bound of the effective permissible range is violated. The clock skew of at least one local data path L_{ij} within the system may violate the upper bound of $\rho(L_{ij})$, i.e., $T_{\text{Skew}\,ij} > T_{\text{Skew}\,ij(\text{max})}$. Observe that if $\rho(L_{ij}) = Pl(L_{ij})$, $T_{\text{Skew}\,ij}$ does not satisfy (3), shown as region C in Fig. 2, causing zero clocking [11]. By increasing the clock period T_{CP}, the effective permissible clock skew range for each local data path is also increased ($T_{\text{Skew}\,ij(\text{max})}$ is increased due to monotonicity), permitting those local data paths previously in region C to satisfy (3). The new clock skew value may also violate the lower bound of a local data path, i.e., $T_{\text{Skew}\,ij} < T_{\text{Skew}\,ij(\text{min})}$, where $T_{\text{Skew}\,ij(\text{min})} \subset \rho(L_{ij})$. Observe that if $\rho(L_{ij}) = Pl(L_{ij})$, $T_{\text{Skew}\,ij}$ does not satisfy (2), shown as region A in Fig. 2, causing double clocking [11]. This situation can be potentially dangerous since the lower bound of $Pl(L_{ij})$ is independent of the clock frequency, causing the circuit to function improperly.

Two compensation techniques are used to prevent lower bound violations, depending upon where the effective permissible range of a local data path $\rho(L_{ij})$ is located within the absolute permissible range of the local data path, $Pl(L_{ij})$. If the worst case clock skew is in between the lower bounds of $\rho(L_{ij})$ and $Pl(L_{ij})$, $\text{MIN}[Pl(L_{ij})] < T_{\text{Skew}\,ij} < \text{MIN}[\rho(L_{ij})]$, the clock period T_{CP} is increased until the race condition is eliminated, since the effective permissible range will increase, due to monotonicity. If the worst case clock skew is less than the lower bound of the permissible range of the local data path, $T_{\text{Skew}\,ij} < \text{MIN}[Pl(L_{ij})]$, any increase in the clock period will not eliminate the synchronization failure since (2) is not dependent on the clock period. To compensate for this violation a safety term $\zeta_{ij} > 0$ is added to the local timing constraint that defines the lower bound of $Pl(L_{ij})$ [see (2)]. The clock period is increased and a new clock skew schedule is calculated for this value of the clock period. The increased clock period is required to obtain a set of effective permissible ranges with widths equal to or greater than the set of effective permissible ranges that existed before the clock skew violation. Observe that by including the safety term ζ_{ij}, the lower bound of the clock skew of the local data path containing the race condition is shifted to the right (see Fig. 2), moving the new clock skew schedule of the entire circuit away from the bound violation and removing any race conditions. This iterative process continues until the worst case variations of the selected clock skews no longer violate the corresponding effective permissible range of each local data path.

6. Simulation Results

The simulation results presented in this section illustrate the performance improvements obtained by exploiting non-zero clock skew. In order to demonstrate these performance improvements, a set of ISCAS-89 sequential circuits is chosen as benchmark circuits. The performance results are illustrated in Table 1. The number of registers and gates within the circuit including the I/O registers are shown in Column 2. The upper bound of the clock period assuming zero clock skew T_{CP0} is shown in Column 3. The clock period obtained with intentional clock skew $T_{\text{CP}i}$ is shown in Column 4. The resulting performance gain is shown in Column 5. The clock period obtained with the constraint of zero clock skew imposed among the I/O registers is shown in Column 6 while the performance gain with respect to zero I/O skew is shown in Column 7.

The results shown in Table 1 clearly demonstrate reductions of the minimum clock period when intentional clock skew is exploited. The amount of reduction is dependent on the characteristics of each circuit, particularly the differences in propagation delay between each local data path. Note also that by constraining the clock skew of the I/O registers to zero, circuit speed can be improved, although less than if this I/O constraint is not used.

Table 1. Performance improvement with non-zero clock skew.

Circuit	Size # register/# gates	T_{CP0} $T_{Skew\,ij} = 0$	T_{CPi} $T_{Skew\,ij} \neq 0$	Gain (%)	T_{CP} $T_{SkewI/O} = 0$	Gain (%)
ex1	20/-	11.0	6.3	43.0	7.2	35.0
s27	7/10	9.2	6.6	28.0	9.2	0.0
s298	23/119	16.2	11.6	28.0	11.6	28.0
s344	35/160	28.4	25.6	9.9	25.6	9.9
s386	20/159	19.8	19.8	0.0	19.8	0.0
s444	30/181	18.6	12.2	34.4	12.2	34.4
s510	32/211	19.8	17.3	13.0	17.3	13.0
s938	67/446	27.0	21.4	20.7	25.0	7.4
s1196	45/529	37.0	30.8	16.8	37.0	0.0
s1512	89/780	53.2	43.2	18.8	53.2	0.0

Table 2. Worst case variations in clock skew due to process parameter variations, $I_{DO} = 15\%$.

Circuit	T_{CP0}/T_{CPi}	Gain(%)	Permissible range	Selected clock skew	Simulated skew (ns)		Error (%)	
					Nom	Worst case	Nom	Worst case
cdn 1	11/9	18.0	$[-8, -2]$	−3.0	−3.0	−2.10	0.0	30.0
cdn 2	18/15	17.0	$[-6.8, -1.4]$	−4.2	−4.1	−3.3	2.4	21.4
cdn 3	27/18	33.0	$[-14, 2.3]$	1.1	1.14	1.3	3.6	18.2

Clock distribution networks which exploit intentional clock skew and are less sensitive to the effects of process parameter variations are depicted in Table 2. The ratio of the minimum clock period assuming zero clock skew T_{CPo} to the intentional clock skew T_{CPi} and the per cent improvement is shown in Columns 2 and 3, respectively. The permissible range most susceptible to process parameter variations is illustrated in Column 4. The selected clock skew is shown in Column 5. In Columns 6 and 7, respectively, the nominal and maximum clock skew are depicted, assuming a $\pm 15\%$ variation of the drain current I_{DO} of each inverter. Note that both the nominal and the worst case value of the clock skew are within the permissible range. The per cent variation of clock skew due to the effects of process parameter variations is shown in Columns 8 and 9. This result confirms the claim stated previously that variations in clock skew due to process parameter variations can be both tolerated and compensated.

7. Conclusions

An integrated top-down, bottom-up approach is presented for synthesizing clock distribution networks tolerant to process parameter variations. In the top-down phase, the clock skew schedule and permissible ranges of each local data path are calculated while minimizing the clock period. The process of determining the bounds of the permissible ranges and selecting the clock skew value for each local data path so as to minimize the effects of process parameter variations is described. Rather than placing limits or bounds on the clock skew variations, this approach guarantees that each selected clock skew value is within the permissible range despite worst case variations of the clock skew. Techniques for designing the topology and the CMOS-based circuit structure of the clock trees are presented. In the bottom-up phase, worst case variations of clock skew due to process parameter variations are determined from the specific clock distribution network. Variations are compensated by the proper choice of clock skew for each local data path. Results of optimizing the clock skew schedule of several MCNC/ISCAS-89 benchmark circuits are presented. A schedule of the clock skews to make a clock distribution network less sensitive to process parameter variations is presented for several example networks. An 18% improvement in clock frequency with up to a 30% variation in the nominal clock skew, and a 33% improvement in clock frequency with up to an 18% variation in the nominal clock skew are demonstrated for several example circuits.

References

1. S. Pullela, N. Menezes, J. Omar, and L.T. Pillage, "Skew and delay optimization for reliable buffered clock trees," *Proceedings of the IEEE International Conference on Computer-Aided Design*, pp. 556–562, Nov. 1993.
2. Q. Zhu, W.W.-M. Dai, and J.G. Xi, "Optimal sizing of high-speed clock networks based on distributed RC and lossy transmission line models," *Proceedings of the IEEE International Conference on Computer-Aided Design*, pp. 628–633, Nov. 1993.
3. J. Cong and K.-S. Leung, "Optimal wiresizing under the distributed elmore delay model," *Proceedings of the IEEE International Conference on Computer-Aided Design*, pp. 634–639, Nov. 1993.
4. J. Cong and C.-K. Koh, "Simultaneous driver and wire sizing for performance and power optimization," *IEEE Transactions on VLSI Systems*, Vol. VLSI-2, No. 4, pp. 408–425, Dec. 1994.
5. H.B. Bakoglu, J.T. Walker, and J.D. Meindl, "A symmetric clock-distribution tree and optimized high-speed interconnections for reduced clock skew in ULSI and WSI circuits," *Proceedings of the IEEE International Conference on Computer Design*, pp. 118–122, Oct. 1986.
6. T.-H. Chao, Y.-C. Hsu, J.-M. Ho, K.D. Boese, and A.B. Kahng, "Zero skew clock routing with minimum wirelength," *IEEE Transactions on Circuits and Systems-II: Analog and Digital Signal Processing*, Vol. CAS-39, No. 11, pp. 799–814, Nov. 1992.
7. R.-S. Tsay, "An exact zero-skew clock routing algorithm," *IEEE Transactions on Computer-Aided Design of Integrated Circuits and Systems*, Vol. CAD-12, No. 2, pp. 242–249, Feb. 1993.
8. S. Lin and C.K. Wong, "Process-variation-tolerant clock skew minimization," *Proceedings of the IEEE International Conference on Computer-Aided Design*, pp. 284–288, Nov. 1994.
9. M. Shoji, "Elimination of process-dependent clock skew in CMOS VLSI," *IEEE Journal of Solid-State Circuits*, Vol. SC-21, No. 5, pp. 875–880, Oct. 1986.
10. E.G. Friedman, *Clock Distribution Networks in VLSI Circuits and System*, IEEE Press, 1995.
11. J.P. Fishburn, "Clock skew optimization," *IEEE Transactions on Computers*, Vol. C-39, No. 7, pp. 945–951, July 1990.
12. R.B. Deokar and S. Sapatnekar, "A graph-theoretic approach to clock skew optimization," *Proceedings of the IEEE International Symposium on Circuits and Systems*, pp. 407–410, May 1994.
13. J.L. Neves and E.G. Friedman, "Design methodology for synthesizing clock distribution networks exploiting non-zero localized clock skew," *IEEE Transactions on VLSI Systems*, Vol. VLSI-4, No. 2, pp. 286–291, June 1996.
14. J.L. Neves and E.G. Friedman, "Synthesizing distributed buffer clock trees for high performance ASICs," *Proceedings of the IEEE ASIC Conference*, pp. 126–129, Sept. 1994.
15. E.G. Friedman, "Latching characteristics of a CMOS bistable register," *IEEE Transactions on Circuits and Systems—I: Fundamental Theory and Applications*, Vol. CAS I-40, No. 12, pp. 902–908, Dec. 1993.
16. K.A. Sakallah, T.N. Mudge, and O.A. Olukotun, "*CheckTc* and *minTc*: Timing verification and optimal clocking of synchronous digital circuits," *Proceedings of the IEEE/ACM Design Automation Conference*, pp. 111–117, June 1990.
17. T.G. Szymanski, "Computing optimal clock schedules," *Proceedings of the IEEE/ACM Design Automation Conference*, pp. 399–404, June 1992.
18. J.L. Neves and E.G. Friedman, "Optimal clock skew scheduling tolerant to process variations," *Proceedings of the ACM/IEEE Design Automation Conference*, pp. 623–628, June 1996.
19. J.L. Neves, "Synthesis of Clock Distribution Networks for High Performance VLSI/ULSI-Based Synchronous Digital Systems," Ph.D. Dissertation, University of Rochester, Dec. 1995.
20. J.L. Neves and E.G. Friedman, "Topological design of clock distribution networks based on non-zero clock skew specifications," *Proceedings of the IEEE Midwest Symposium on Circuits and Systems*, pp. 461–471, Aug. 1993.
21. S.Dhar and M.A. Franklin, "Optimum buffer circuits for driving long uniform lines," *IEEE Journal of Solid State Circuits*, Vol. SC-26, No. 1, pp. 32–40, Jan. 1991.
22. J.L. Neves and E.G. Friedman, "Circuit synthesis of clock distribution networks based on non-zero clock skew," *Proceedings of the IEEE International Symposium on Circuits and Systems*, pp. 4.175–4.178, May 1994.
23. T. Sakurai and A.R. Newton, "Alpha-power law MOSFET model and its applications to CMOS inverter delay and other formulas," *IEEE Journal of Solid State Circuits*, Vol. SC-25, No. 2, pp. 584–594, April 1990.

José Luis P.C. Neves received the B.S. degree in Electrical Engineering in 1986, and the M.S. degree in Computer Science in 1989 from the Federal University of Minas Gerais (UFMG), Brazil. He received the M.S. and Ph.D. degrees in electrical engineering from the University of Rochester, New York, in 1991 and 1995, respectively.

He was with the Physics Department of the UFMG as an electrical engineer from 1986 to 1987, where he managed the automation of several research laboratories, designing data acquisition equipment and writing programs for data collect and analysis. He was a Teaching and Research Assistant at the University of Rochester from 1990 to 1995. He was a computer systems administrator with the Laboratory of Respiratory Physiology in the Department of Anesthesiology, University of Rochester from 1992 to 1996, writing programs for data collect and analysis, and designing the supporting electronic equipment. He has been with IBM Microelectronics since 1996 as an advisory engineer/scientist responsible for developing and implementing clock distribution design and synthesis tools. His research interests include high performance VLSI/IC design and analysis, timing issues in VLSI design, and CAD tool and methodology development with application to the design and synthesis of clock distribution networks, low power circuits, and CMOS circuit design techniques tolerant to process parameter variations.

Dr. Neves received a Doctoral Fellowship from the National Research Council (CNPq) Brazil from 1990 to 1994. He is a member of the Technical Program Committee of ISCAS '97.
neves@ee.rochester.edu

Eby G. Friedman was born in Jersey City, New Jersey in 1957. He received the B.S. degree from Lafayette College, Easton, PA in 1979, and the M.S. and Ph.D. degrees from the University of California, Irvine, in 1981 and 1989, respectively, all in electrical engineering.

He was with Philips Gloeilampen Fabrieken, Eindhoven, The Netherlands, in 1978 where he worked on the design of bipolar differential amplifiers. From 1979 to 1991, he was with Hughes Aircraft Company, rising to the position of manager of the Signal Processing Design and Test Department, responsible for the design and test of high performance digital and analog IC's. He has been with the Department of Electrical Engineering at the University of Rochester, Rochester, NY, since 1991, where he is an Associate Professor and Director of the High Performance VLSI/IC Design and Analysis Laboratory. His current research and teaching interests are in high performance microelectronic design and analysis with application to high speed portable processors and low power wireless communications.

He has authored two book chapters and many papers in the fields of high speed and low power CMOS design techniques, pipelining and retiming, and the theory and application of synchronous clock distribution networks, and has edited one book, *Clock Distribution Networks in VLSI Circuits and Systems* (IEEE Press, 1995). Dr. Friedman is a Senior Member of the IEEE, a Member of the editorial board of *Analog Integrated Circuits and Signal Processing*, Chair of the VLSI Systems and Applications CAS Technical Committee, Chair of the VLSI track for ISCAS '96 and '97, and a Member of the technical program committee of a number of conferences. He was a Member of the editorial board of the *IEEE Transactions on Circuits and Systems II: Analog and Digital Signal Processing*, Chair of the Electron Devices Chapter of the IEEE Rochester Section, and a recipient of the Howard Hughes Masters and Doctoral Fellowships, an NSF Research Initiation Award, an Outstanding IEEE Chapter Chairman Award, and a University of Rochester College of Engineering Teaching Excellence Award.
friedman@ee.rochester.edu

Journal of VLSI Signal Processing 16, 163–179 (1997)

Useful-Skew Clock Routing with Gate Sizing for Low Power Design

JOE GUFENG XI[1] AND WAYNE WEI-MING DAI
Computer Engineering, University of California, Santa Cruz

Received September 15, 1996; Revised December 5, 1996

Abstract. This paper presents a new problem formulation and algorithm of clock routing combined with gate sizing for minimizing total logic and clock power. Instead of zero-skew or assuming a fixed skew bound, we seek to produce useful skews in clock routing. This is motivated by the fact that only *positive skew* should be minimized while *negative skew* is useful in that it allows a timing budget larger than the clock period for gate sizing. We construct an *useful-skew tree* (UST) such that the total clock and logic power (measured as a cost function) is minimized. Given a required clock period and feasible gate sizes, a set of negative and positive skew bounds are generated. The allowable skews within these bounds and feasible gate sizes together form the feasible solution space of our problem. Inspired by the Deferred-Merge Embedding (DME) approach, we devise a *merging segment perturbation* procedure to explore various tree configurations which result in correct clock operation under the required period. Because of the large number of feasible configurations, we adopt a simulated annealing approach to avoid being trapped in a local optimal configuration. This is complemented by a bi-partitioning heuristic to generate an appropriate connection topology to take advantage of useful skews. Experimental results of our method have shown 12% to 20% total power reduction over previous methods of clock routing with zero-skew or a single fixed skew bound and separately sizing logic gates. This is achieved at no sacrifice of clock frequency.

1. Introduction

Deep submicron technology is constantly pushing the performance/cost limit of VLSI systems. With increasing clock frequency and integration density, designing low power system has become a major challenge. CMOS circuits should be carefully designed to keep the rush-through current small so as to reduce the short-circuit power. Meanwhile, the dynamic power due to the capacitance switching has been considered the dominant part of system power dissipation. Carrying the heaviest load and switching at a high frequency, the clock typically dissipates 30–50% of the total power in a synchronous digital system [1, 2]. The switching logic gates contribute to the rest of the power dissipation.

On the other hand, control of clock skew and critical path timing are also critical issues in the design of high performance circuits. In this paper, we present a new approach of clock routing by combining with gate sizing to minimize total clock and logic power while achieving the required performance. We will show that this approach mitigates the unfavorable tradeoff of circuit speed and power. Savings on both logic power and clock routing cost can be achieved.

In recent years, there have been active research in the area of high-performance clock routing [3–8]. Jackson et al. first pointed out that skew should be minimized to increase clock frequency and proposed a generalized H-tree routing algorithm [3]. Tsay proposed exact *Zero-skew tree* (ZST) for clock routing under the Elmore delay model [4]. Chao et al. then found an algorithm called *Deferred-Merge Embedding* (DME) to minimize wire length of ZST for a given clock tree topology [5]. The DME algorithm was later complemented by topology generation algorithms to further minimize the total wire length of a ZST [7, 8]. Other techniques on ZST construction include wire sizing and buffer insertions [6, 9, 10]. More recently, it has been pointed out that it is almost impossible to achieve exact zero-skew in real designs [2, 10]. In fact, it is neither necessary nor desirable to achieve zero-skew [11, 12].

For low power designs, tolerable skew or a bounded-skew tree (BST) instead of ZST has been proposed to reduce clock power [2, 13, 14]. With a fixed skew bound, Cong et al. and Tsao et al. independently proposed to construct a BST which minimizes total wire length based on the DME approach [13–15].

The BST algorithms assume a fixed non-zero skew bound. However, no indication was given as to how the skew bound is derived and what appropriate value should the bound be. Moreover, if we study clock skew more closely, we would see the following properties of skew: (i) Because the logic delay varies from one block to another, the allowable skew for correct clock operation varies from one pair of clock sinks to another [11]. To use a single fixed skew bound, one has to choose the smallest skew bound of all sink pairs; (ii) Skew could be either negative or positive with respect to the logic path direction [16]. Only positive skew limits the clock period while negative skew can increase the effective clock period. The allowable negative skew is therefore considered useful skew; (iii) In addition, the allowable skew bounds can be adjusted by adjusting the logic path delays, i.e., by gate sizing.

Given the dynamic nature of clock skew, and given that clock distribution is so crucial to system power and performance, passively assuming zero-skew or a fixed skew bound in clock routing could lead to pessimistic results of the logic and clock power. Without differentiating the negative and positive skews, a bounded-skew clock tree based on a non-zero skew bound could even result in worse logic power than a zero-skew tree.

In the related area, Fishburn first proposed *clock skew optimization* to improve synchronous circuit performance or reliability by taking advantage of useful skews [11]. With allowable negative skew, the circuit can run at a clock period less than the critical logic path delay. This allows a larger timing constraint to minimize the circuit area or power. Hence, Chuang et al. incorporated *clock skew optimization* in gate sizing [17, 18]. In their work, the *optimal skews* are produced between clock sinks to give the largest timing budget possible for gate sizing. However, they assume either arbitrary or bounded skew values while overlooking the cost of clock routing and the penalty of clock power. To produce negative skews, a common approach is to insert buffers as delay elements [16, 17]. But this results in increased buffer power and process variation induced skew uncertainties [2]. Without considering the placement and routability issues, the *optimal skew* may be unrealizable or too costly to realize in a physical implementation.

For low power designs, we believe the clock routing problem should be considered together with gate sizing at the same time. Clock routing can take advantage of the allowable skew bounds given by appropriate gate sizing while the *useful skews* can be used to create a larger timing budget for gate sizing. This way, we may minimize the total power dissipation of the clock and logic gates. In this paper, we formulate and solve the **Useful-Skew Clock Routing with Gate Sizing for Power Minimization** problem. Given the feasible gate sizes and the required clock period, the negative and positive skew bounds between various sinks can be obtained. The feasible solution space of our problem is thus defined by the allowable skews and feasible gate sizes. We construct an *useful-skew tree* (UST) to produce the useful skews and minimize total clock and logic power while meeting the required clock frequency. The key difference between our *useful-skew clock routing* approach and the *clock skew optimization* approach is that we define a feasible solution space which accounts for both feasible gate sizes and clock distribution cost. To search for optimal gate sizes, we predetermine the gate sizes which result in minimum logic power under a given skew between two flip-flops. The logic power and the corresponding gate sizes for an allowable skew value are stored in a lookup table. This minimizes the time to determine the logic power and gate sizes in clock routing.

The rest of this paper is organized as follows. In Section 2, we discuss the motivation of this work and give the formulation of *Useful-skew Clock Routing with Gate Sizing for Power Minimization* (UST) problem. In Section 3, we present our solution to the UST problem. This includes the routing algorithm and a topology generation heuristic. In Section 4, we discuss the gate sizing methods used in our solution. Experimental results are given in Section 5, where five benchmark circuits are tested and compared with previous ZST and BST routing algorithms. Finally, we give some concluding remarks and briefly discuss the ongoing research about this work.

2. Problem Formulation

2.1. Clock Skew and Gate Sizing

To understand clock skew and its effects on the performance and power of digital circuits, consider a simple synchronous circuit as shown in Fig. 1. For simplicity, we assume positive edge-triggered flip-flops are used in this example and throughout this paper. Due to

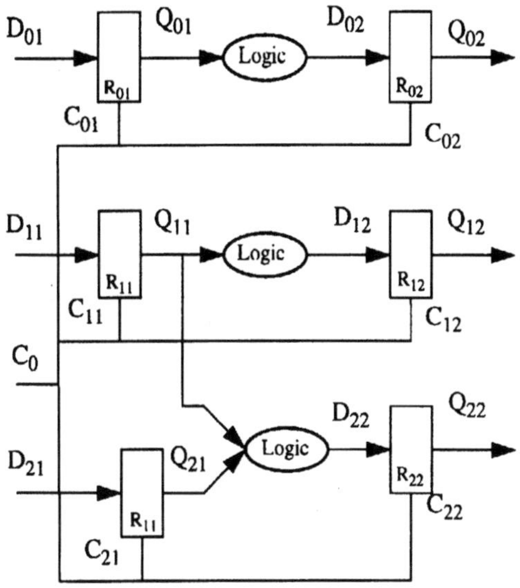

Figure 1. A synchronous circuit example.

interconnect delays, skew may result between clock terminals such as C_{01} and C_{02} of FF_{01} and FF_{02}. Figure 2 illustrates the clock operations in two cases of skews. In both cases, skews are considered allowable if correct data are produced under the given clock frequency. With excessive skew in either cases, incorrect operations may occur when data are produced either too early (known as *double-clocking*) or too late (known as *zero-clocking*) from FF_{01} to FF_{02} [11].

In general, to ensure correct clock operation under a required clock period, P, the allowable clock skews between two adjacent flip-flops, FF_i and FF_j, are:

To avoid *double-clocking* with *negative skew*, $d_i \leq d_j$:

$$d_j - d_i \leq MIN(d_{logic}) + d_{ff} - d_{hold} \quad (1)$$

To avoid *zero-clocking* with *positive skew*, $d_i \geq d_j$:

$$d_i - d_j \leq P - (d_{ff} + MAX(d_{logic}) + d_{setup}) \quad (2)$$

where d_i and d_j denote the clock arrival time, $MAX(d_{logic})$ and $MIN(d_{logic})$ denote the longest and shortest path delays of the combinational block between FF_i and FF_j.

We notice the following properties of clock skew. First, both the negative and positive skew bounds vary from one pair of clock terminals to another since combinational logic path delays vary from one to another. To use a single fixed skew bound, one has to choose the smallest skew bound of all sink pairs, both negative and positive. Secondly, *negative skew* is desirable because it does not impose direct limit on clock period. In fact, it can allow circuits to run at a clock period less than the critical path delay [11]. This can be used to either reduce the clock period or create a larger timing budget for gate sizing. Therefore, negative skew can be considered *useful skew*. Lastly, the skew bounds can

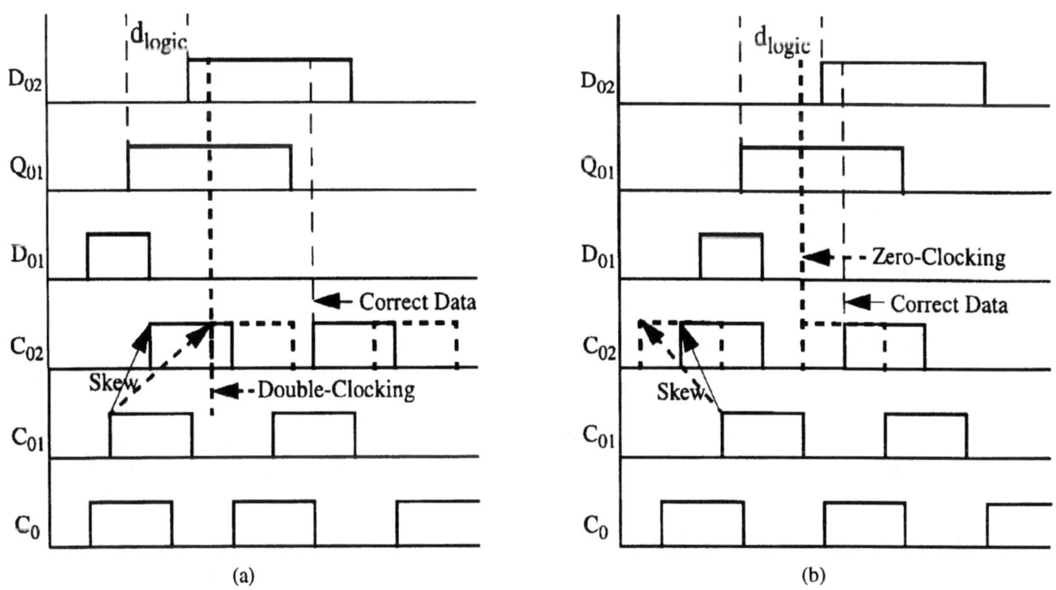

Figure 2. Clock operations with non-zero skew: (a) negative skew; (b) positive skew. With excessive skew, incorrect clock operations may occur, i.e., either double-clocking as in (a) or zero-clocking as in (b).

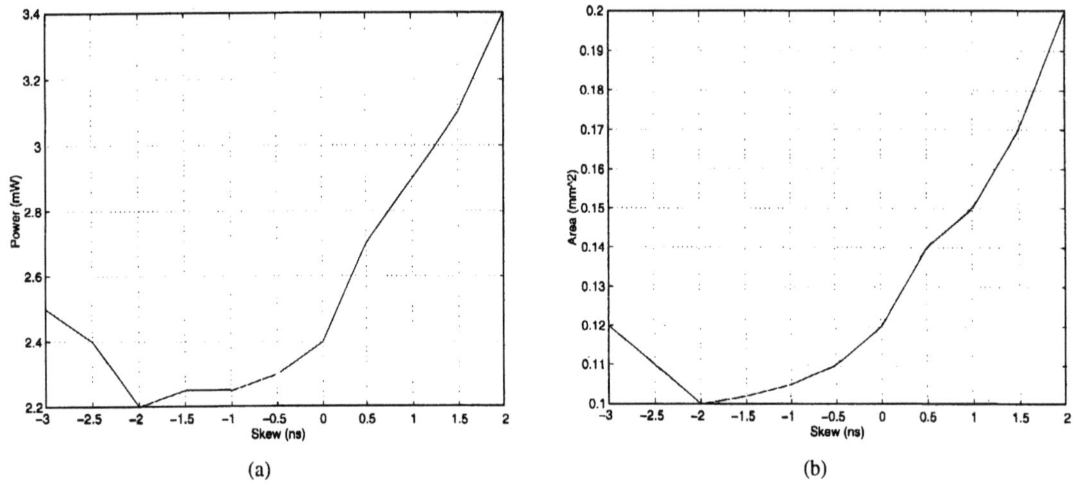

Figure 3. The minimum power and area vs. allowable skews within the negative and positive skew bounds for a combinational block between two flip-flops.

be enlarged by sizing the logic gates. The *positive skew* bound can be enlarged by sizing the gates on the long logic paths to reduce $MAX(d_{logic})$. The *negative skew* bound can be enlarged by sizing the gates on the short logic paths to increase $MIN(d_{logic})$. Increasing the logic path delays can generally be done by reducing the gate sizes which generally also reduces the dynamic power. This can be seen from Fig. 3 which shows the relationship between minimum power, area of a combinational block and the allowable skew values.

These properties motivate us to consider the clock routing problem together with gate sizing. We see that if *negative* skew and *positive* skew are treated the same, then a bounded-skew clock tree may produce skews that impose even tighter timing budget for gate sizing. Therefore, *negative* and *positive* skew bounds rather than a fixed unsigned skew bound should be considered in clock tree construction. With *negative skews* between certain clock sinks, e.g., the sinks that entail critical logic paths, a larger timing budget can allow gate sizing to further reduce logic power. If these useful skews are achieved at the expense of little increase of clock tree cost, the total logic and clock power can be minimized.

2.2. *Problem Formulation*

Assume that we are given a standard-cell based design and its required clock period, $P = 1/f$. The standard-cell library consists of a set of gates and a set of templates for each gate type. The feasible sizes for each gate are: $W_k = \{w_{k,1}, \ldots, w_{k,q}\}$. The logic netlist and placement of the cells are given. Then a set of clock sink locations $S = \{s_1, s_2, \ldots, s_n\}$, the clock source s_0 in the Manhattan plane and the initial gate sizes $X^0 = \{x_1^0, x_2^0, \ldots, x_m^0\}$ are given. Assuming we are also given a *routing topology* G which is a rooted binary tree with n leaves corresponding to the sinks in S. A clock tree T is an embedding of routing topology G, i.e., each internal node $v \in G$ is mapped to a location in the Manhattan plane. Each node v in T is connected to its parent node by an edge, e_v, with wire length, $|e_v|$ from v to its parent. The total wire length of the tree L is then the sum of all edge lengths in T. Let $D = \{d_1, d_2, \ldots, d_n\}$ be the delays from clock source s_0 to clock sinks in T. Also corresponding to each pair of clock sinks s_i and s_j are a pair of skew bounds, a *negative skew bound*, NSB_{ij} and a *positive skew bound*, PSB_{ij}. The definitions and derivations of these skew bounds are deferred to Section 3. We measure the total power dissipated by both clock and logic gates with a cost function:

$$C(T, X) = \lambda L(T) + \gamma \Phi(X) \quad (3)$$

where λ and γ are weight coefficients determined for a given technology and design. We defer the derivations of logic power $\Phi(X)$ to Section 4. We now define the *Useful-Skew Clock Routing with Gate Sizing for Power Minimization* (UST) problem:

Useful-Skew Clock Routing with Gate Sizing for Power Minimization Problem. *Given a required clock period P, a library of logic gates and their templates, $W_k = \{w_{k,1}, \ldots, w_{k,q}\}$, the set of clock sink locations*

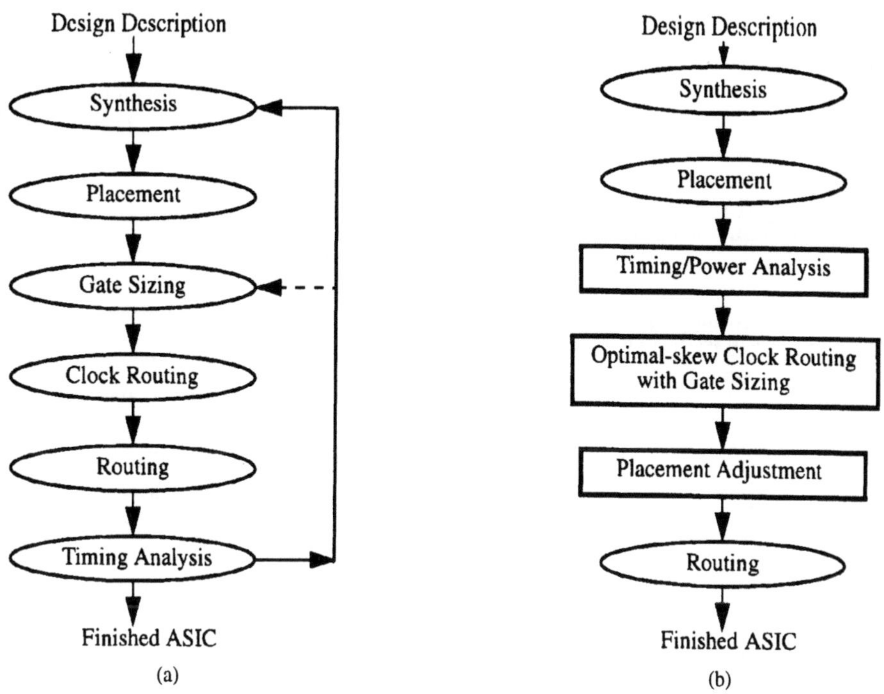

Figure 4. (a) Conventional ASIC design flow; (b) new proposed design methodology.

$S = \{s_1, s_2, \ldots, s_n\}$, *the initial logic gate sizes* $X^0 = \{x_1^0, x_2^0, \ldots, x_m^0\}$, *and a connection topology* G, *we seek a tree* T *and a set of gate sizes* $X^* = \{x_1^*, x_2^*, \ldots, x_m^*\}$. *The objective is to minimize the cost function* C *defined in* (3) *subject to the skew constraints for all sink pairs,* s_i *and* s_j: $d_j - d_i \leq NSB_{ij}$, *if* $d_i \leq d_j$, *or* , $d_i - d_j \leq PSB_{ij}$, *if* $d_i \geq d_j$.

In Section 3, we will also give a solution of generating an appropriate topology G for the UST problem, particularly under the Elmore delay model.

We envision the solution of this problem to have important impact in a practical design environment. Figure 4 contrasts a conventional ASIC design flow with our new methodology. With an initial placement, the timing and power analysis can be based on accurate estimation of interconnect delays and loads. The *Useful-Skew Clock Routing with Gate Sizing* is performed at this stage. The placement is then adjusted with new gate sizes. This would have minimal effect on clock routing results if minimal changes are made to the clock sink locations. Other placement based resynthesis techniques can also be applied here to improve delays and alleviate congestion [19]. In a conventional design flow where gate sizing and clock routing are separately optimized, design iterations are usually necessary due to the lack of common objective and constraints.

3. The UST Clock Routing Algorithm

3.1. Overview

Our useful-skew clock routing (UST) algorithm involves four tasks: (i) generating an appropriate clock tree topology; (ii) finding the locations of internal nodes of the tree; (iii) preserving the negative and positive skew bounds for correct clock operations under a given frequency and (iv) selecting the sizes of logic gates for minimum power. Figure 5 gives a high level description of our algorithm. The main idea is as follows. First, in generating a topology, we try to maximize the the feasible regions for possible internal node locations. Then, in the process of embedding the topology, we explore feasible placements of internal clock tree nodes. This is equivalent to exploring allowable skews between various clock sinks. The resulting skews are used in gate sizing to determine the minimum logic power. The nature of this problem is that there are a large number of constraints as well as a large number of feasible configurations. A simple iterative approach would easily get trapped in a local optimal configuration. Therefore, we adopt a simulated annealing approach because of its *hill-climbing* feature [20].

Our UST algorithm is inspired by the DME based algorithms which searches for internal nodes of

```
Input :
    S = set of clock sinks, n = |S|,
    P = the required clock period, P = 1/f,
    X^0 = the initial sizes of logic gates.
Output :
    an UST, T*, sizes of logic gates, X*.
PROCEDURE BuildUSTwithGateSizing (S, P, X^0) {
    X = X^0;
    G = GenerateTopology(S);
    T = BuildInitialZST(G, S); /* according to [3] */
    Prepare a dice of n - 1 facets; each represent a node in G;
    Bias the facets according to the number of sinks rooted at each node;
    t = t_0;
    while (not Frozen) {
        while (not Equilibrium) {
            Throw dice to pick a node, v;
            T' = PerformMSP(v, G, T); /* at the chosen node, perform an MSP */
            X' = GateSize(T', X, P); /* obtain the gate sizes with resulting skew */
            ΔC = C(T', X') - C(T, X);
            If (ΔC <= 0 or e^(-ΔC/(k_B t)) >= random(0, 1))
                T = T'; X = X';
        }
        t = δ(t) × t;
    }
    T* = T; X* = X;
}
```

Figure 5. High-level description of UST algorithm.

zero-skew tree (ZST) and *bounded-skew tree* (BST) in the Manhattan plane [5, 13, 14]. We use a ZST constructed by the DME algorithm [5] as the initial starting point and iteratively search for better placements of internal nodes and produce useful skews. Because it is time prohibitive to perform gate sizing at each iteration of simulated annealing, we decide to use an approximate solution of gate sizing. We predetermine the gate sizing result of each combinational logic block for a known skew value. At each iteration, the logic power in the cost function is updated with a table look-up. We defer the discussion on gate sizing to Section 4. In the following, we will analyze the negative and positive skew bounds. After reviewing some terminology, we describe the main procedure used in the UST algorithm. It is called a *Merging Segment Perturbation* (MSP) and is used to explore the optimal locations of the internal nodes. We also present a bi-partitioning heuristic to generate an appropriate connection topology and maximize useful skews.

3.2. Negative and Positive Skew Bounds

We will be using the following definitions. We say the clock sinks, s_i and s_j of two flip-flop's FF_i and FF_j, are *adjacent* if there exists a combinational logic path from FF_i to FF_j. Let d_i and d_j be the path delays from clock source to sinks s_i and s_j, the skew between s_i and s_j is *negative skew* if $d_i \leq d_j$ and is *positive skew* if $d_i \geq d_j$. We define the *negative skew bound* (NSB) between s_i and s_j as the maximum value of *negative skew* between s_i and s_j with which the clock operates correctly under a required clock frequency. Similarly, the *positive skew bound* (PSB) is the maximum value of *positive skew* with which the clock operates correctly under a given frequency. The NSB and PSB between two sinks are given by:

If s_i, s_j are *adjacent*, then,

$$NSB_{ij} = \max(MIN(d_{logic})) + d_{ff} - d_{hold} \tag{4}$$

$$PSB_{ij} = P - \min(MAX(d_{logic})) - d_{setup} - d_{ff} \tag{5}$$

where $\max(MIN(d_{logic}))$ is the maximum delay of the shortest combinational logic path that is achievable with feasible gate sizes while satisfying the long path constraint. $\min(MAX(d_{logic}))$ is the minimum delay of the longest combinational logic path that is achievable with feasible gate sizes while satisfying the short path constraint. The derivation of $\max(MIN(d_{logic}))$ and $\min(MAX(d_{logic}))$ will be given in Section 4.

If s_i and s_j are not *adjacent*, then,

$$NSB_{ij} = \infty, \quad PSB_{ij} = \infty \tag{6}$$

In addition, we also define $NSB(v)$ and $PSB(v)$ associated with each node v of a binary tree as the maximum allowable delay difference from v to its two children, a and b.

(I) If the two children nodes of v are sinks, i.e., s_i, s_j, then $NSB(v) = NSB_{ij}$ and $PSB(v) = PSB_{ij}$.

(II) If one or more of the children nodes of v are not sinks, i.e., a and b with subtrees, TS_a and TS_b, then,

$$NSB(v) = \min(d_i(a) - d_j(b) + NSB_{ij}, d_l(a) - d_k(b) + PSB_{kl}) \tag{7}$$

$$PSB(v) = \min(d_j(a) - d_i(b) + PSB_{ij}, d_k(b) - d_l(a) + NSB_{kl}) \tag{8}$$

for all sink pairs, $s_i, s_l \in TS_a$, $s_j, s_k \in TS_b$. $d_i(a)$, $d_l(a)$ and $d_j(b)$, $d_k(b)$ denote the delays from a to s_i, s_l and b to s_j, s_k, respectively.

A feasible placement of v has to satisfy $NSB(v)$ and $PSB(v)$ in order to satisfy the skew bounds between sinks rooted at v. As we will discuss later, the existence of this feasible region depends on the tree topology, the placements of v's descendent nodes.

3.3. Merging Segment Perturbation

We first review some terminology used in the DME based algorithms [5, 13–15]. A *merging segment* (MS)

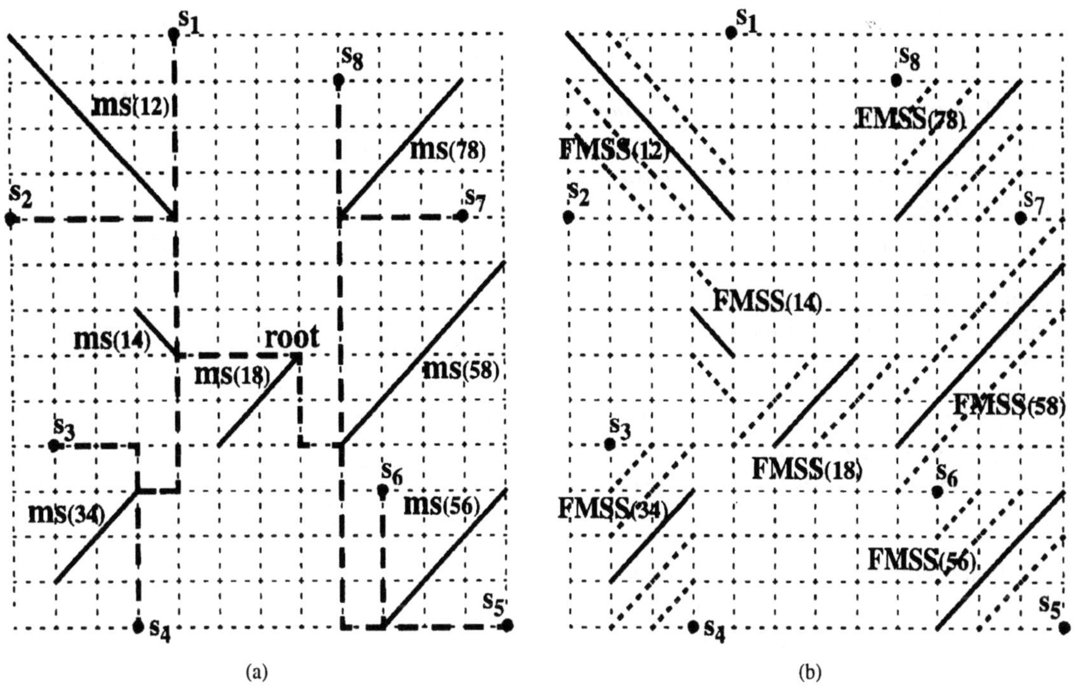

Figure 6. (a) A zero-skew tree constructed with *merging segments* using the DME algorithm, (b) the *feasible merging segment set* of each node when the lower-level merging segments are fixed.

is a line segment associated with an internal node in a clock tree and represents the loci of possible locations of this node. In the Manhattan plane, a *merging segment* is a *Manhattan Arc* which is a line segment (possibly a single point) with a slope of +1 or −1. Let $ms(v)$ be the merging segment of a node v, a and b be the children nodes of v and TS_a and TS_b be the subtrees rooted at a and b. To construct a ZST, the DME algorithm constructs $ms(v)$ from $ms(a)$ and $ms(b)$ in a bottom-up process [5]. Any point on $ms(v)$ satisfies: $\forall s_i \in TS_a, s_j \in TS_b, d_i(v) = d_j(v)$. At the same time, TS_a and TS_b are merged with minimum added wire, i.e., $|e_a| + |e_b|$ is minimized. A ZST example is shown in Fig. 6(a).

For a BST with non-zero skew bound, a *merging region* is associated with a node which contains all the feasible merging points for a given skew bound [13, 14]. A tree of merging regions is formed in a bottom-up process. To construct $mr(v)$, the *shortest-distance region* (SDR) between v's children, $mr(a)$ and $mr(b)$ is first found. The set of points within $SDR(v)$ that have minimum merging cost $|e_a| + |e_b|$ while satisfying a fixed skew bound form $mr(v)$. In the case no points in $SDR(v)$ satisfy the skew bound, $mr(v)$ is chosen as the point-set outside of $SDR(v)$ which has the minimum increase of merging cost.

In our approach, we assume the Manhattan plane is gridded and there are a discrete number of points. This reflects more of a realistic routing environment. Given two *Manhattan Arcs*, l_1 and l_2, the *shortest distance region* between l_1 and l_2, denoted $SDR(l_1, l_2)$ is the set of points that have the minimum sum of Manhattan distance to l_1 and l_2. $SDR(l_1, l_2)$ thus contains a discrete number of Manhattan Arcs. For a given topology, G, we construct a tree of *feasible merging segment sets* (FMSS). Each node $v \in G$, is associated with a $FMSS(v)$. If v is a sink, s_i, then $FMSS(v) = \{s_i\}$. If v is an internal node with children a and b and the merging segment $ms(a)$ and $ms(b)$ are chosen, then a *feasible merging segment* (FMS) of v is a Manhattan Arc which contains possible locations of v such that (i) the negative and positive skew bounds, $NSB(v)$ and $PSB(v)$ given by (7) and (8) are satisfied; (ii) the merging cost $|e_a| + |e_b|$ is minimized. Therefore, $FMSS(v)$ is defined by its children, $ms(a)$ and $ms(b)$ in a bottom-up process. For any two FMSSs, $FMSS(a)$ and $FMSS(b)$, the *shortest distance merging segments*, denoted as $SDMS(a)$ and $SDMS(b)$, are a pair of Manhattan Arcs in $FMSS(a)$ and $FMSS(b)$ which are closest to each other[2]. Figure 6(b) shows the FMSS for each node when the lower-level merging segments are fixed. Obviously, the FMSSs of all internal nodes define the feasible solutions of UST.

Lemma 1. *If every node $v \in T$ is chosen within $FMSS(v)$, then skew between any two sinks in T satisfies either their negative skew bound or their positive skew bound. In another word, the clock operates correctly under the given frequency.*

Under both the linear and Elmore delay, we have the following lemmas regarding the existence and property of $FMSS(v)$.[3]

Lemma 2. *Under both the linear and Elmore delay, the $FMSS(v)$ for any node $v \in G$ exists, i.e., there is at least one FMS, $ms(v)$, if and only if $NSB(v) + PSB(v) \geq 0$.*

Lemma 3. *Under both the linear and Elmore delay models, for any FMS within $SDR(ms(a), ms(b))$, the difference in delay from v to its two children, a and b is a linear function of the position of the FMS. If $FMSS(v)$ exists, it can be constructed in constant time.*

We construct $FMSS(v)$ from v's children, $ms(a)$ and $ms(b)$ which are either $SDMS(a)$ and $SDMS(b)$ or any FMSs within $FMSS(a)$ and $FMSS(b)$(as chosen by an MSP). First, we construct $SDR(ms(a), ms(b))$. Then we find two boundary FMSs, $ms^-(v)$ and $ms^+(v)$ which respectively equals $NSB(v)$ and $PSB(v)$. Let $\kappa = dist(ms(a), ms(b))$, x^+, x^- be the Manhattan distance from $ms(a)$ to $ms^+(v)$ and $ms^-(v)$, and y^+, y^- be the distance from $ms(b)$ to $ms^+(v)$ and $ms^-(v)$.[4] α and β are the unit length resistance and capacitance, C_a and C_b are the capacitances at a and b, and $FMSS(v)$ are computed as follows.

Case 1. If $ms^-(v)$ and $ms^+(v)$ are found within $SDR(ms(a), ms(b))$ and are parallel Manhattan Arcs, i.e., $0 \leq x^+ \leq \kappa$ and $0 \leq x^- \leq \kappa$, the parallel Manhattan Arcs between them and within $SDR(ms(a), ms(b))$ are also FMSs according to Lemma 3. Together with $ms^+(v)$ and $ms^-(v)$, they form $FMSS(v)$.

$$\begin{aligned} x^+ &= \frac{B + PSB(v)}{A}, \quad y^+ = \kappa - x^+; \\ x^- &= \frac{B - NSB(v)}{A}, \quad y^- = \kappa - x^- \end{aligned} \tag{9}$$

$$A = \alpha(C_a + C_b + \beta\kappa), \quad B = \alpha\kappa\left(\frac{1}{2}\beta\kappa + C_b\right) \tag{10}$$

Case 2. If $x^+ > \kappa$ and $0 \leq x^-\kappa$, or $x^- \leq 0$ and $0 \leq x^-\kappa$, then $FMSS(v)$ is formed by the parallel Manhattan Arcs between $ms^-(v)$ and $ms(b)$ or between $ms(a)$ and $ms^+(v)$, as well as the points on $ms(b)$ or $ms(a)$, given respectively by:

$$\kappa < x \leq \frac{\sqrt{(\alpha C_b)^2 + 2\alpha\beta PSB(v)}}{\alpha\beta}, \quad y = 0 \tag{11}$$

$$x = 0, \quad \kappa < y \leq \frac{\sqrt{(\alpha C_b)^2 + 2\alpha\beta NSB(v)}}{\alpha\beta} \tag{12}$$

Case 3. If $x^+ > \kappa$ and $x^- > \kappa$, or $x^+ < 0$ and $x^- < 0$ then $FMSS(v)$ are the set of points on $ms(b)$ or $ms(a)$, given respectively by:

$$\frac{\sqrt{(\alpha C_b)^2 + 2\alpha\beta NSB(v)}}{\alpha\beta} \leq x \leq \frac{\sqrt{(\alpha C_b)^2 + 2\alpha\beta PSB(v)}}{\alpha\beta}, \quad y = 0 \tag{13}$$

$$\frac{\sqrt{(\alpha C_b)^2 + 2\alpha\beta PSB(v)}}{\alpha\beta} \leq y \leq \frac{\sqrt{(\alpha C_b)^2 + 2\alpha\beta NSB(v)}}{\alpha\beta}, \quad x = 0 \tag{14}$$

Note in Cases 2, 3, Lemma 3 does not apply under the Elmore delay model when the merging segment is outside of $SDR(ms(a), ms(b))$. We determine $FMSS(v)$ by choosing the points on $ms(b)$ or $ms(a)$. In these cases, the merging segments have greater than minimum merging cost and wire detouring is required. When an FMS within $SDR(ms(a), ms(b))$ is used such as in Cases 1, 2, we have minimum merging cost.

A *merging segment perturbation* associated with a node v, denoted as $MSP(v)$ is a move that selects another FMS within $FMSS(v)$. Figure 7(a) shows two MSPs as examples. When selecting another merging segment of v, the FMSSs of v's parent and ancestor nodes are updated. This results in a new configuration of the clock tree, T, and hence a new set of skews. These are also allowable skews according to Lemma 1. Let p denote the parent node of v and u be the sibling of p. During an MSP, $FMSS(p)$ is redefined by the new $ms(v)$ and v's sibling. Then $SDMS(p)$ which is the closest to $ms(u)$ is found. This is chosen as the new $ms(p)$. As shown in Fig. 7(a), $SDMS(14)$ is chosen as $ms(14)$, since it is the closest to $ms(58)$. The new $ms(p)$ and $ms(u)$ form the FMSS of their parent node, i.e., the grandparent of v. This process is iterated in a bottom-up process until the root node of T is updated. According to Lemma 2, the FMSS of a node may not always be found. An $MSP(v)$ is *acceptable* if

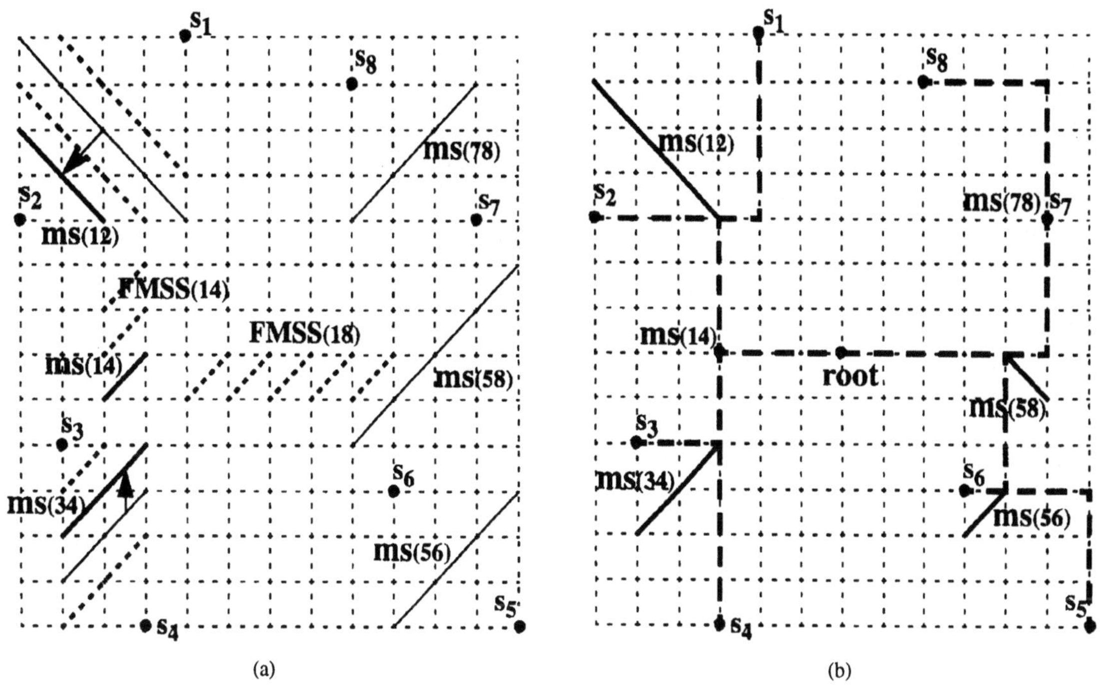

Figure 7. (a) Examples of MSP. The arrow indicate the selection of a new FMS within the *FMSS*(34) or *FMSS*(12). The new FMSS of the parent node, *FMSS*(14), is formed and *SDMS*(14) is chosen as *ms*(14) which is the closest to its sibling, *ms*(58). (b) The final UST which minimizes cost function after a sequence of MSP's.

$NSB(A) + PSB(A) \geq 0$, for all $A \in ancestors(v)$. If an $MSP(v)$ is acceptable, after updating the FMSSs of v's ancestors, a top-down process connects the merging segments by the shortest distance, analogous to DME [13, 14]. Note that the variable bounds of NSB and PSB are used at each node. Only v's ancestor nodes are updated. With a binary tree topology, one MSP takes $O(n)$ in the worst case and $O(\log n)$ in the average case. Figure 7(b) shows a final tree after a sequence of MSPs and the cost function is minimized.

The following Theorem suggests that the entire feasible solution space can be asymptotically explored.

Theorem 1. *For a given tree topology, any configuration of the clock tree that result in allowable skews (skews that allow correct clock operation under a required frequency) can be transformed to another by performing a sequence of MSPs.*

3.4. Topology Generation

From the definitions of $NSB(v)$ and $PSB(v)$, we can see that the skew constraints at higher level nodes (closer to the root) are tighter. The root node has to satisfy the smallest skew bound taken over all sink pairs rooted at its two children. If the high level nodes are given small skew budget, they will have fewer feasible merging segments. If the topology is very asymmetric, then the delay difference of two subtrees under Elmore model may become so large that feasible merging segments are limited or even can not be found according to Lemma 2.

More importantly, our objective is to produce useful skew—the negative skew. If at an internal node, v, there are two or more pairs of sinks between the two subtrees which have opposite logic path direction, then the NSB of one sink pair is constrained by the PSB of another. The negative skew of one pair of sinks results in the positive skew of another pair of sinks. The cross coupled bounds makes it difficult to achieve good results.

These observations indicate that the tree topology is very important to the success of the UST solution. Intuitively, we would like to partition the sinks into groups that have loose skew bounds with each other. Most of the adjacent sinks across two groups should have the same logic path direction (either *forward* or *backward*) such that negative skew can be maximally produced. This suggests that a top-down partitioning rather than a bottom-up clustering approach should be used since the skew bounds between sinks can be evaluated globally. We now describe a partitioning heuristic for the UST problem. It is modified from the BB bipartitioning heuristic in [5]. However, we have a distinct objective here.

We consider recursively cutting the sink set S into two subsets S_1 and S_2 in the Manhattan plane. Each cut would result in one internal node of the tree topology. At each partition, we choose a cut to (i) maximize the skew bounds for the resulting node, and (ii) maximize the number of forward (or backward) sink pairs across the cut. For a bipartition, $S = S_1 \cup S_2$, let FW_{12} and BW_{12} denote the number of sink pairs across the cut that have a logic path from S_1 to S_2 (forward) and from S_2 to S_1 (backward). The total number of adjacent sink pairs across the cut is then, $SP_{12} = FW_{12} + BW_{12}$. We define the *skew bound* between S_1 and S_2 as $SB_{12} = \min(NSB_{ij}, PSB_{kl}) + \min(PSB_{ij}, NSB_{kl}), \forall s_i, s_l \in S_1, s_j, s_k \in S_2$. We therefore use a weighted function to evaluate a cut,

$$W_{12} = w_1(SB_{12}/SP_{12}) + w_2|FW_{12} - BW_{12}| \quad (15)$$

where w_1, w_2 are determined by experiment. For lower level nodes, the partition between the two subsets should also be *balanced* to keep the delay difference small. Let $Cap(S_1)$ and $Cap(S_2)$ be the total capacitance of S_1 and S_2, respectively. $|Cap(S_1) - Cap(S_2)| \leq \epsilon$, where ϵ is gradually reduced with each level of cuts. Let $p.x$ and $p.y$ be the coordinates of a point, p. The *octagon* of S is the region occupied by S in the Manhattan plane and is defined by the eight half spaces: $y \leq \max(p.y)$, $y - x \geq \min(p.y - p.x)$, $x \geq \min(p.x)$, $y + x \geq \min(p.y + p.x)$, $y \geq \min(p.y)$, $y - x \leq \max(p.y - p.x)$, $x \leq \max(p.x)$, $y + x \leq \max(p.y + p.x), \forall p \in S$. The *octagon set* of S, $Oct(S)$, is the set of sinks in S that lie on the boundary of the octagon of S. A *reference set* is a set of $\lfloor 1/2\,|Oct(S)| \rfloor$ consecutive sinks in $Oct(S)$, denoted by REF_i, $i = 1, \ldots, |Oct(S)|$. For each sink $p \in S$, the *weight* of p relative to a reference set, REF_i, is given by $weight_i(p) = \min(dist(p, r)) + \max(dist(p, r)), \forall r \in REF_i$. Figure 8 gives a high level description of this bi-partitioning heuristic. As in [5], the time complexity is $O(n^3 \log n)$ in the worst case, and $O(n \log^2 n)$ under the more realistic circumstances.

An example of using this bipartitioning heuristic is shown in Fig. 9. Figure 9(a) shows the negative and positive skew bounds between the sinks. The clock tree using the topology generated by the clustering based

```
Input :
    S = set of clock sinks, n = |S|,
    NSB = negative skew bounds between every pair of sinks,
    PSB = negative skew bounds between every pair of sinks,
Output :
    an UST topology, G.
PROCEDURE GenerateTopology (S, NSB, PSB) {
    Compute Oct(S) and reference sets, REF = REF_i, i,= 1,...,|Oct(S)|;
    for (each REF_i) {
        S_1 = nil; S_2 = S;
        Compute weight_i(p) of each sink, p ∈ S_2;
        Sort p ∈ S_2 in ascending order of weight_i(p);
        Remove 1 sink at a time from S_2 and add to S_1;
        Each time, compute W_12, Cap(S_1), Cap(S_2); /* as given in (15) */
        Save all Cut_i^j = S_1 ∪ S_2 with |Cap(S_1) - Cap(S_2)| ≤ ε;
    }
    for (all Cut_i^j) {
        Choose Cut(S) = Cut_i^j with maximum W_12;
    }
    while (|S_1| > 2)
    GenerateTopology(S_1, NSB, PSB);
    while (|S_2| > 2)
    GenerateTopology(S_2, NSB, PSB);
}
```

Figure 8. Description of UST topology generation heuristic.

algorithm is shown in (b) [8]. It results in positive skew between s_3 and s_4 which is undesirable. In contrast, using our bi-partitioning heuristic, the final tree result in all negative skews and the routing cost is also reduced.

4. Gate Sizing

In the UST problem, we are considering power minimization of sequential circuits for standard-cell based designs. A cell library is given which consists 2–6 templates for each type of gate. The templates for a given logic gate realize the same boolean function. But they vary in size, delay, and driving capability. When discrete gate sizes are used, the delay or power minimization problem is known as NP-complete [21, 22].

Unlike previous approach of gate sizing with clock skew optimization [17, 18], our feasible solution space is defined by a clock tree with reasonable cost (measured as function of wire length) and feasible gate sizes. Our approach has two advantages: (i) With the feasible solution region controlled by clock routing, we may take into account both the logic and clock power; (ii) With known skews between each pair of flip-flops, we may decompose the sequential circuit into subcircuits which are individually combinational circuits[5].

Because gate sizing is a time consuming process [18], we predetermine the minimum power of each combinational block. The logic power for an allowable skew value and the corresponding gate sizes are stored in a look-up table. At each iteration of our UST routing

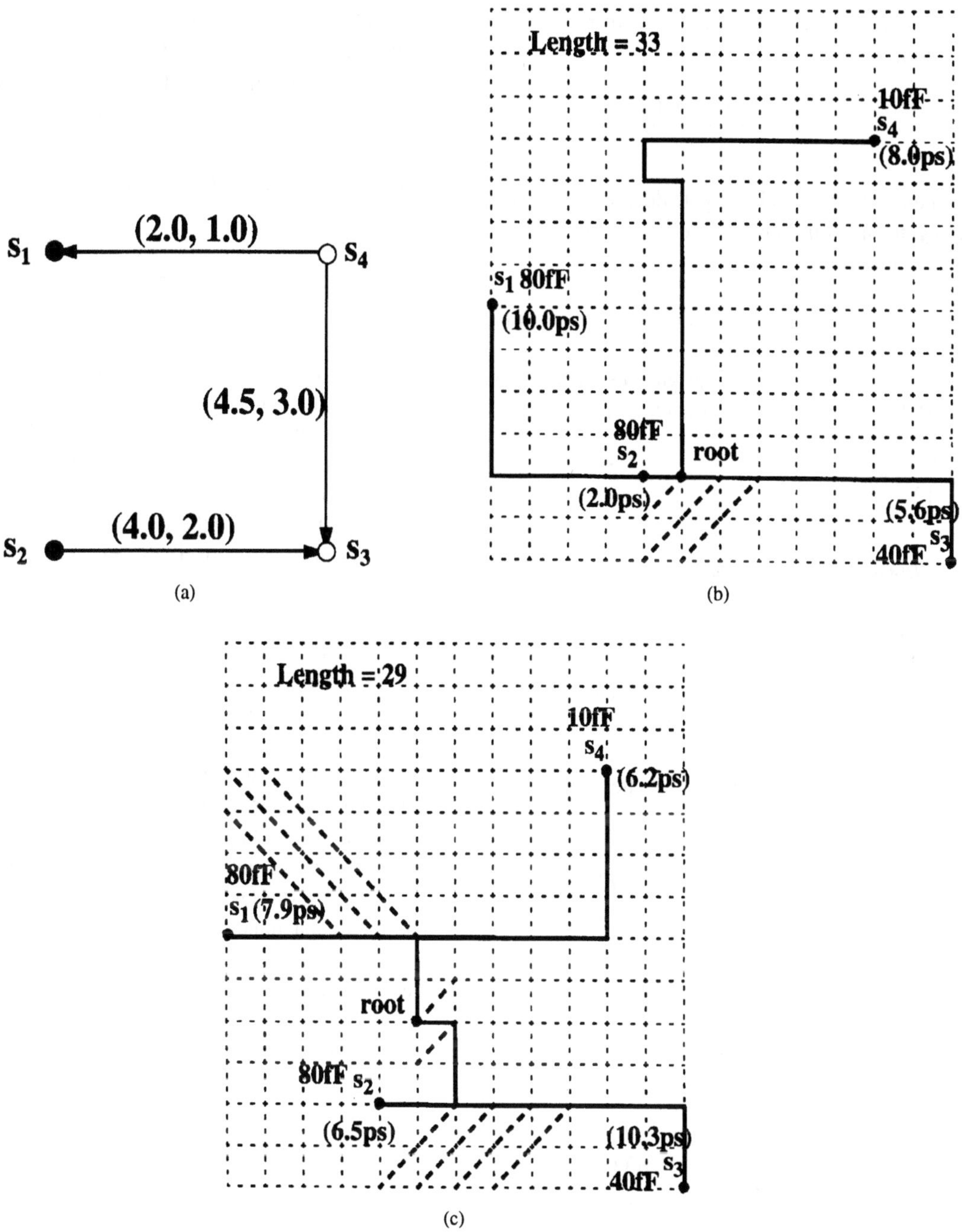

Figure 9. An example showing the effects of topology. (a) (NSB, PSB) between sinks. (b) The clock tree resulted from the clustering-based topology. s_3 and s_4 have a positive skew of 2.4. (c) The clock tree resulted from our bipartition heuristic. All sink pairs have negative skew.

algorithm, a table look-up can be done in constant time to update the cost function. Finally when the minimum cost function is achieved and the skews between each pair of flip-flops are known, the gate sizes which results in minimum power under the closest skew value are chosen. Through extensive experiments, we found this approach closely predicts the results of optimizing the entire sequential circuits [17].

We use the following delay and power models. The delay of a logic gate depends on its intrinsic delay, d_0, the total fanout load capacitance at the output, C_L, the interconnect capacitance, C_p, the gate size, x_i, and an empirical parameter, Ω characterized from SPICE simulation.

$$d_i^g = d_0 + \frac{\Omega(C_L + C_p)}{x_i} \tag{16}$$

The dynamic power of a logic gate depends on its size, the unit gate and drain capacitance, c_{gd}, and the average switching activity, a_i.

$$p_i^d = x_i c_{gd} V_{dd}^2 f a_i \tag{17}$$

The short-circuit power of a logic gate also depends on the rise/fall time of its previous gate, τ_{i-1} [23].

$$p_i^s = \frac{Ax_i}{12}(V_{dd} - V_t)^3 \tau_{i-1} f, \quad \tau_{i-1} = \frac{\Omega x_i c_g}{x_{i-1}} \tag{18}$$

4.1. Allowable Skew Bounds

As mentioned in Section 2, with a required clock period and feasible gate sizes, the allowable negative and positive skew bounds can be derived. The feasible gate sizes are referring to $w_{\min} \le w_i \le w_{\max}$, where $w_{\min}$ and $w_{\max}$ are the minimum and maximum sizes of gate templates in the library. We derive these bounds by solving the following problems.

Formulation 4.1. *Determine the feasible gate sizes, such that the maximum delay of the shortest path in a combinational logic block, denoted* $\max(MIN(d_{logic}))$ *is obtained by: maximize:* $MIN(d_{logic})$, *subject to:*

$$MAX(d_{logic}) \le P + MIN(d_{logic}) - d_{hold} - d_{setup} \tag{19}$$

where $MIN(d_{logic})$ *and* $MAX(d_{logic})$ *are the short and long path delays of the combinational block, respectively. (19) is derived from:* $d_i + MAX(d_{logic}) + d_{ff} + d_{setup} \le d_j + P$, $d_i + MIN(d_{logic}) + d_{ff} \ge d_j + d_{hold}$ *and* $d_i \le d_j$.

Formulation 4.2. *Determine the feasible gate sizes, such that the minimum delay of the longest path in a combinational logic block, denoted as* $\max(MIN(d_{logic}))$ *is obtained by: minimize:* $MAX(d_{logic})$, *subject to:*

$$MIN(d_{logic}) + d_{ff} - d_{hold} \ge 0 \tag{20}$$

where (20) *is similarly derived as* (19) *except* $d_i \ge d_j$.

Starting with minimum sizes (the smallest templates) for all gates, a static timing analysis is performed to obtain the delays for all paths. The *sensitivity* of each gate is given by $-\frac{\Delta d_i^g}{\Delta x_i}$. This is based on the decrease or increase of delay, Δd_i^g, per increment of gate size (to the next larger template), Δx_i.

$$\Delta d_i^g = -\Omega(C_L + C_p)\frac{\Delta x_i}{(x_i + \Delta x_i)x_i} + \frac{\Omega c_g \Delta x_i}{x_{i-1}} \tag{21}$$

To obtain $\max(MIN(d_{logic}))$, we first try to satisfy the constraint in (19). We iteratively increment the size of the gate on the longest path that has the largest sensitivity and is not shared by the shortest path until (19) is satisfied. The same procedure is repeated for the next longest path. Note that the short path delay, $MIN(d_{logic})$ is always increasing during this process. If in either of the following two cases, the constraint still can not be satisfied: (i) all gates except the ones on the shortest path have reached the largest templates; (ii) their sensitivities are all negative which means the increase of size will result in an increase of delay, we then size the gates of all paths. To increase the delay of $MIN(d_{logic})$, we first increase the sizes of gates on the shortest path with negative sensitivity until all of them have positive sensitivity or the largest templates have been reached. We also size the gates whose inputs are fanout of the gates on the shortest path. These gates are basically the load capacitance on the shortest path. Obtaining $\max(MIN(d_{logic}))$ is similar. We first satisfy the short path constraints by increasing the delays of paths that violate (20). Then we reduce the delays of the longest path by increasing the gate sizes on that path.

4.2. Gate Sizing with Allowable Skews

Power dissipation of a combinational circuit depends on the switching activities and therefore the input vectors. However, we may determine the average power of each combinational block by assuming an average switching activity of each gate [24]. With the required clock period and a given skew, the delay constraints of each combinational block are given. We solve the following problem for each combinational block for all $-NSB_{ij} \le d_i - d_j \le PSB_{ij}$ with a step size determined in experiment. The minimum power and the corresponding gate sizes under allowable skews within the NSB and PSB are stored in a look-up table.

Formulation 4.3. *Given d_i and d_j which are the delays from clock source to the sinks of flip-flops, FF_i and FF_j, $-NSB_{ij} \le d_i - d_j \le PSB_{ij}$, determine the minimum power of combinational logic block between FF_i and FF_j, with feasible gate sizes, subject to:*

$$d_i + MAX(d_{logic}) + d_{setup} + d_{ff} \le d_j + P \quad (22)$$

$$d_i + MIN(d_{logic}) + d_{ff} \ge d_j + d_{hold} \quad (23)$$

With minor modification, a gate sizing algorithm for combinational logic circuits with double sided constraints can be applied to this problem. In our case, we adopt the algorithm in [21]. Although this solution primarily minimizes the dynamic power dissipation, we found in experiments that the short-circuit power is also kept very small.

5. Experimental Results

The UST algorithm described in previous sections has been implemented in C in a Sun Sparcstation10 environment and has been tested on two industry circuits and three ISCAS89 benchmark circuits [25][6]. The test circuits are described in Table 1. The ISCAS89 benchmark circuits were first translated with some modifications to a 0.65 μm CMOS standard-cell library [26]. The library consists of 6 templates for inverters or buffers and 3–4 templates for each boolean logic gates. Two types of flip-flops are used with clock pin load capacitance of 70 fF and 25 fF. The cells are placed with an industry placement tool and the clock sink locations are then obtained. The clock tree is assumed to be routed on the metal2 layer. The width of all branches is chosen as 1 μm, the sheet resistance, $r = 40(\text{m}\Omega/\mu\text{m})$ and unit capacitance, $c = 0.02(\text{fF}/\mu\text{m})$. We implemented a previous standard-cell gate sizing algorithm [22] to be used with the DME based ZST and BST clock routing algorithms [5, 15][7] to compare with our UST solution. Table 2 compares the power dissipation results of UST with two other approaches: (i) ZST clock routing [5], gate sizing with zero-skew; (ii) BST clock routing [14, 15], gate sizing with a fixed skew bound. To guarantee correct clock operation, the smallest allowable skew bound (both negative and positive) of all clock sink pairs has to be chosen as the fixed skew bound in the BST/DME algorithm. We assume the clock tree is driven by a chain of large buffers at the source [2]. The power reduction varies from 11% to 22% over either ZST or BST approaches. Note that since BST does not recognize the difference between negative and positive skew, it may even produce skews that result in worse power in gate sizing. Table 3 compares the routing results of ZST, BST algorithms,

Table 1. Five circuits tested by the UST algorithm. Two industry circuits. Three ISCAS89 benchmark circuits.

Circuits	Frequency (Mhz)	# of flip flop's	# of gates	Supply (volt)
Circuit1	200	106	389	5.0
Circuit2	100	391	3653	3.3
s1423	33	74	657	3.3
s5378	100	179	2779	3.3
s15850	100	597	9772	3.3

Table 3. Comparison of wire length (μm) of clock trees on tested circuits. Also shown are the skew bounds used by BST algorithms.

Circuits	ZST	BST (Skew-bound)	UST-CL	UST-BP
Circuit1	3982	2998 (0.1 ns)	3051	2755
Circuit2	17863	16002 (0.2 ns)	16217	15924
s1423	8823	6651 (1.4 ns)	6830	6756
s5278	12967	10645 (0.3 ns)	11068	10229
s15850	30579	28348 (0.2 ns)	27369	25580

Table 2. Power reduction of UST over ZST and BST. UST-CL uses the topology generated by the clustering algorithm. UST-BP uses the bipartitioning heuristic.

	Clock power (mW)				Logic power (mW)				Reduction	
Circuits	ZST	BST	UST-CL	UST-BP	ZST	BST	UST-CL	UST-BP	UST/ZST	UST/BST
Circuit1	43.53	43.32	43.41	43.22	58.35	55.45	46.08	41.9	16%	14%
Circuit2	20.95	20.66	20.69	20.54	102.66	93.34	85.87	83.36	16%	11%
s1423	5.224	5.161	5.182	5.170	22.48	24.70	18.69	18.17	16%	22%
s5378	11.03	10.82	10.86	10.79	124.4	126.5	114.0	110.2	11%	12%
s15850	32.93	32.44	32.38	32.25	416.5	421.3	356.1	338.9	17%	18%

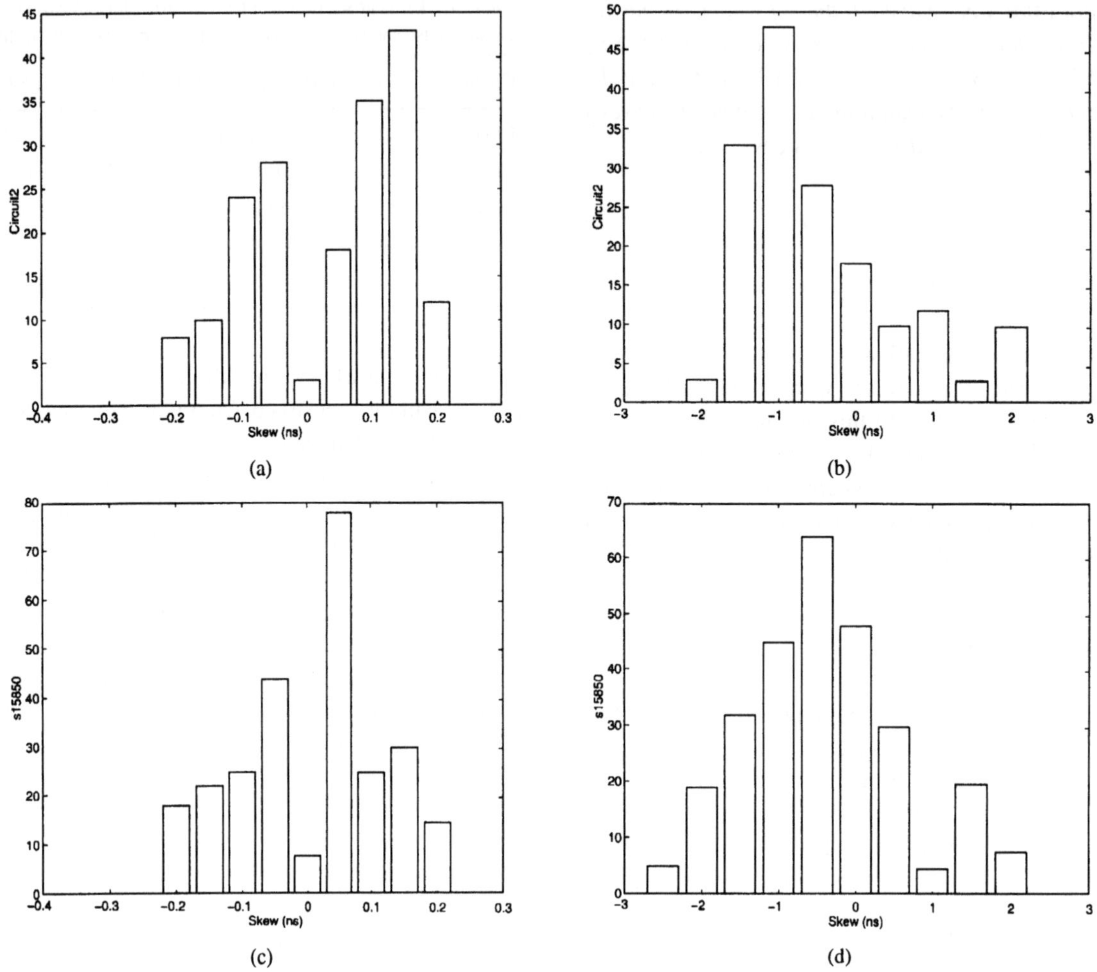

Figure 10. Comparison of negative and positive skew distributions in benchmarks, Circuit2 using BST in (a), using UST-BP in (b); and s15850 using BST in (c), using UST-BP in (d). Note that negative skew is generally useful skew.

and the UST routing results with topology generated by both the clustering-based algorithm [8] and the bi-partitioning heuristic. Because the small value of the fixed skew bound is used, BST only achieves a small savings in wire length over ZST. In contrast, the UST approach reduces wire length in all but one case.

Figure 10 shows the distribution of the negative and positive skew values in benchmarks Circuit2 and s15850 resulted from using the BST algorithm and UST algorithm. Note that negative skew is generally useful for better results in gate sizing.

In the implementation of simulated annealing, the outer loop stopping criterion (frozen state) is satisfied when the value of the cost function has no improvement for five consecutive stages. The inner loop stopping criterion (Equilibrium state) is implemented by specifying the number of iterations at each temperature. We use $n \times$ *TrialFactor* in the experiments, where $n = |S|$. For all tested cases, the *TrialFactor* ranges from 100 to 600. We choose the initial temperature as $t_0 = -\frac{\Delta C}{\ln \chi}$, where ΔC is obtained by generating several transitions at random and computing the average cost increase per generated transition and χ is the acceptance ratio. In choosing the cooling schedule, we start with $\delta(t) = 0.85$, then gradually increase to $\delta(t) = 0.95$, and stay at this value for the rest of the annealing process. For the coefficients in the cost function of (3), we set $\lambda = \beta V_{dd}^2 f 100$. This is because the wire capacitance is small and extra weight has to be used to control the wire length. γ is set to 1 in the result shown above. The results shown in the above comparisons are chosen from results obtained at CPU time ranging from 200–600 minutes. Better results are likely with more CPU time. Although the running time is large for a simulated annealing based algorithm, it is still worthwhile considering that most gate sizing approaches are time consuming especially when combined with clock skew optimization [18]. As we mentioned earlier in Section 2, the UST solution can significantly reduce design iterations. Therefore, the choice of simulated annealing is well justified.

6. Concluding Remarks and Continuing Work

Previous works in clock routing focused on constructing either zero-skew tree (ZST) or bounded-skew tree (BST) with a fixed skew bound. In contrast, we proposed an algorithm to produce useful skews in clock routing. This is motivated by the fact that negative skew is useful in minimizing logic gate power. While ZST and BST clock routing are too pessimistic for low power designs, *clock skew optimization* [11, 18] with arbitrary skew values is on the other hand too optimistic as the clock distribution cost is overlooked. We have presented a realistic approach of combining clock routing and gate sizing to reduce total logic and clock power. Included in this paper are our formulation and solutions to this complex problem. The experimental results have shown convincingly the effectiveness of our approach in power savings. In deep submicron CMOS technology, power dissipation has become a design bottleneck. We believe this work is critical in designing high-speed and low-power ICs.

We are currently investigating further improvements to the UST solution. Continuing research in this area include: More efficient and provably good clock routing algorithms; Combining clock routing with buffer insertion and buffer sizing [2] to further optimize clock skew and power as well as improve circuit reliability; More accurate approach in gate sizing to minimize both dynamic and short-circuit power dissipation.

Appendix: Proof of Lemmas

Lemma 1. *If every node $v \in T$ is chosen within $FMSS(v)$, then skew between any two sinks in T satisfies either their negative skew bound or their positive skew bound. In another word, the clock operates correctly under the given frequency.*

Proof: The proof of this lemma comes directly from the definition of $FMSS(v)$. Due to space limitation, we omit the proof here. □

Lemma 2. *Under both the linear and Elmore delay models, the $FMSS(v)$ for any node $v \in G$ exists, i.e., there is at least one feasible merging segment, $ms(v)$, if and only if $NSB(v) + PSB(v) \geq 0$.*

Proof: Let a and b be the children of v. If there exists at least one feasible merging segment, i.e., v', let the delays from v' to a and b be denoted by d_a and d_b, respectively. We have $d_a - d_b \leq PSB(v)$ and $d_b - d_a \leq NSB(v)$, which means $NSB(v) + PSB(v) \geq 0$. We use contradiction to prove the other way. If $NSB(v) + PSB(v) \geq 0$, but there exists no feasible merging segment which means either $d_a - d_b > PSB(v)$ or $d_b - d_a > NSB(v)$ or both for any merging segment. Suppose, $d_a - d_b > PSB(v)$ and $d_b - d_a \leq NSB(v)$, then since $NSB(v) + PSB(v) \geq 0$, we would have $PSB(v) \geq d_a - d_b$ which contradicts with $d_a - d_b > PSB(v)$. Similarly, contradictions would occur for other cases. Therefore, if $NSB(v) + PSB(v) \geq 0$, there must exist at least one feasible merging segment which satisfies both $d_a - d_b \leq PSB(v)$ and $d_b - d_a \leq NSB(v)$. □

Lemma 3. *Under both the linear and Elmore delay models, for any feasible merging segment within $SDR(ms(a), ms(b))$, the difference in delay from v to its two children, a and b is a linear function of the position of the feasible merging segment. If $FMSS(v)$ exists, it can be constructed in constant time.*

Proof: The case of linear delay is easily seen. We prove under the Elmore delay. Let d_a and d_b be the Elmore delay from v to its two children a and b. If a feasible merging segment can be found from within $SDR(ms(a), ms(b))$, then we have minimum merging cost: $|e_a| + |e_b| = dist(ms(a), ms(b))$ [14]. Let $x = |e_a|$, $\kappa = dist(ms(a), ms(b))$, so, $y = \kappa - x$. Then,

$$d_a = \alpha x\left(\frac{1}{2}\beta x + C_a\right),$$
$$d_a = \alpha(\kappa - x)\left(\frac{1}{2}\beta(\kappa - x) + C_a\right) \tag{24}$$

where α, β are the unit length resistance and capacitance, C_a and C_b are the load capacitances at a and b. Thus,

$$d_a - d_b = \alpha(C_a + C_b + \beta\kappa)x - \alpha\kappa\left(\frac{1}{2}\beta\kappa + C_b\right) \tag{25}$$

Because the feasible merging segment is a Manhattan Arc and every point on it has the same distance

distance to $ms(a)$ and $ms(b)$. Therefore, the difference of d_a and d_b is a linear function of the position, represented by x and $\kappa - x$ of the feasible merging segment. According to [5, 14], a merging segment or a Manhattan Arc can be computed in constant time. If $FMSS(v)$ exists within $SDR(v)$, then the boundary merging segments, $ms^+(v)$ and $ms^-(v)$ which satisfies the equality to $PSB(v)$ and $NSB(v)$ can be computed in constant time. Any parallel merging segments between them and within $SDR(ms(a), ms(b))$ also belong to $FMSS(v)$. □

Acknowledgment

We are grateful to C.W. Albert Tsao and Prof. Andrew Kahng of UCLA for providing us with the program of Ex-DME algorithms for comparisons. We also thank Prof. Jason Cong and Cheng-Kok Koh of UCLA for providing the technical reports on BST/DME algorithms.

Notes

1. Currently with Ultima Interconnect Technology, Inc. California.
2. If $FMSS(a)$ and $FMSS(b)$ overlap with each other, we arbitrarily take one pair of Manhattan Arcs as $SDMS(a)$ and $SDMS(b)$.
3. Proof of lemmas are relegated to [12].
4. In the Manhattan plane, a merging segment can be computed in constant time from the intersection of *tilted rectilinear regions* which have $ms(a)$ and $ms(b)$ as cores, x^+ and y^+ or x^- and y^- as radii, respectively [5].
5. Here, we are ignoring the primary inputs, outputs and the interactions with external circuits. We assume this approximation is acceptable in our problem formulation.
6. We were unable to use benchmarks used by [14,15] which do not have logic netlist.
7. Under Elmore delay, the BST results shown here is obtained from the BME approach described in [15].

References

1. D. Dobberpuhl and R. Witek, "A 200 mhz 64b dual-issue cmos microprocessor," in *Proc. IEEE Intl. Solid-State Circuits Conf.*, pp. 106–107, 1992.
2. Joe G. Xi and Wayne W.M. Dai, "Buffer insertion and sizing under process variations for low power clock distribution," in *Proc. of 32nd Design Automation Conf.*, June 1995.
3. M.A.B. Jackson, A. Srinivasan, and E.S. Kuh, "Clock routing for high-performance ics," in *Proc. of 27th Design Automation Conf.*, pp. 573–579, 1990.
4. R.-S. Tsay, "An exact zero-skew clock routing algorithm," *IEEE Trans. on Computer-Aided Design*, Vol. 12, No. 3, pp. 242–249, 1993.
5. T.H. Chao, Y.C. Hsu, J.M. Ho, K.D. Boese, and A.B. Kahng, "Zero skew clock net routing," *IEEE Transactions on Circuits and Systems*, Vol. 39, No. 11, pp. 799–814, Nov. 1992.
6. Qing Zhu, Wayne W.M. Dai, and Joe G. Xi, "Optimal sizing of high speed clock networks based on distributed rc and transmission line models," in *IEEE Intl. Conf. on Computer Aided Design*, pp. 628–633, Nov. 1993.
7. N.-C. Chou and C.-K. Cheng, "Wire length and delay minimization in general clock net routing," in *Digest of Tech. Papers of IEEE Intl. Conf. on Computer Aided Design*, pp. 552–555, 1993.
8. M. Edahiro, "A clustering-based optimization algorithm in zero-skew routings," in *Proc. of 30th ACM/IEEE Design Automation Conference*, pp. 612–616, 1993.
9. Jun-Dong Cho and Majid Sarrafzadeh, "A buffer distribution algorithm for high-performance clock net optimization," *IEEE Transactions on VLSI Systems*, Vol. 3, No. 1, pp. 84–97, March 1995.
10. S. Pullela, N. Menezes, J. Omar, and L.T. Pillage, "Skew and delay optimization for reliable buffered clock trees," in *IEEE Intl. Conf. on Computer Aided Design*, pp. 556–562, 1993.
11. J.P. Fishburn, "Clock skew optimization," *IEEE Transactions on Computers*, Vol. 39, No. 7, pp. 945–951, 1990.
12. Joe G. Xi and Wayne W.M. Dai, "Low power design based on useful clock skews," in Technical Report, UCSC-CRL-95-15, University of California, Santa Cruz., 1995.
13. J. Cong and C.K. Koh, "Minimum-cost bounded-skew clock routing," in *Proc. of Intl. Symp. Circuits and Systems*, pp. 322–327, 1995.
14. D.J.-H. Huang, A.B. Kahng, and C.-W.A. Tsao, "On the bounded-skew clock and steiner routing problems," in *Proc. of 32nd Design Automation Conf.*, pp. 508–513, 1995.
15. J. Cong, A.B. Kahng, C.K. Koh, and C.-W.A. Tsao, "Bounded-skew clock and steiner routing under elmore delay," in *IEEE Intl. Conf. on Computer Aided Design*, 1995 (to appear).
16. J.L. Neves and E.G. Friedman, "Design methodology for synthesizing clock distribution networks exploiting non-zero localized clock skew," *IEEE Transactions on VLSI Systems*, June 1996.
17. W. Chuang, S.S. Sapatnekar, and I.N. Hajj, "A unified algorithm for gate sizing and clock skew optimization," in *IEEE Intl. Conference on Computer-Aided Design*, pp. 220–223, Nov. 1993.
18. H. Sathyamurthy, S.S. Sapatnekar, and J.P. Fishburn, "Speeding up pipelined circuits through a combination of gate sizing and clock skew optimization," in *IEEE Intl. Conference on Computer-Aided Design*, Nov. 1995.
19. L. Kannan, Peter R. Suaris, and H.-G. Fang, "A methodology and algorithms for post-placement delay optimization," in *Proc. of 31th ACM/IEEE Design Automation Conference*, pp. 327–332, 1994.
20. S. Kirkpatrick, Jr., C.D. Gelatt, and M.P. Vecchi, "Optimization by simulated annealing," *Science*, Vol. 220, No. 4598, pp. 458–463, May 1983.
21. Pak K. Chan, "Delay and area optimization in standard-cell design," in *Proc. of 27th Design Automation Conf.*, pp. 349–352, 1990.
22. Shen Lin and Malgorzata Marek-Sadowska, "Delay and area optimization in standard-cell design," in *Proc. of 27th Design Automation Conf.*, pp. 349–352, 1990.
23. Harry, V.M. Veendrick, "Short-circuit power dissipation of static cmos circuitry and its impact on the design of buffer circuits,"

IEEE Journal of Solid-State Circuits, Vol. SC-19, pp. 468–473, Aug. 1984.
24. J. Rabae, D. Singh, M. Pedram, F. Catthoor, S. Rajgopal, N. Sehgal, and T.J. Mozdzen, "Power conscious cad tools and methodologies: A perspective," *Proceedings of IEEE*, Vol. 83, No. 4, pp. 570–593, April 1995.
25. F. Brglez, D. Bryan, and K. Kozminski, "Combinational profiles of sequential benchmark circuits," in *Proc. of IEEE Intl. Symp. on Circuits and Systems*, pp. 1929–1934, 1989.
26. National Semiconductor Corp. *cs65 CMOS Standard Cell Library Data Book*, National Semiconductor Corp., 1993.

Joe Gufeng Xi received the B.S. degree in Electrical Engineering from Shanghai Jiao Tong University, China, the M.S. degree in Computer Engineering from Syracuse University, and the Ph.D. degree in Computer Engineering from University of California, Santa Cruz, in 1986, 1988 and 1996, respectively. He is now with Ultima Interconnect Technology, Inc., Cupertino, CA. He was Senior Engineer at National Semiconductor Corp., Santa Clara, CA, where he was involved in mixed-signal IC design, behavior modeling, logic synthesis and circuit simulation. Prior to joining National, he was a design engineer at Chips and Technology, Inc., where he worked on the physical design of a microprocessor chip, including placement and routing, RC extraction and timing analysis. His research interests include VLSI circuit performance optimization, low-power design techniques of digital and mixed-signal ICs, clock distribution and system timing, and high-speed interconnect optimization. He received a nomination for the Best Paper award at the Design Automation Conference in 1995.

Wayne W.-M. Dai received the B.A. degree in Computer Science and the Ph.D. degree in Electrical Engineering from the University of California at Berkeley, in 1983 and 1988, respectively. He is currently an Associate Professor in Computer Engineering at the University of California at Santa Cruz. He was the founding Chairman of the IEEE Multi-Chip Module Conference, held annually in Santa Cruz, California since 1991. He was an Associate Editor for IEEE Transactions on Circuits and Systems and an Associate Editor for IEEE Transactions on VLSI Systems. He received the Presidential Young Investigator Award in 1990.

Journal of VLSI Signal Processing 16, 181–189 (1997)

Clock Distribution Methodology for PowerPC™ Microprocessors

SHANTANU GANGULY AND DAKSH LEHTHER
Somerset Design Center, Motorola, Austin

SATYAMURTHY PULLELA
Unified Design System Laboratory, Motorola, Austin

Received October 3, 1996; Revised November 24, 1996

Abstract. Clock distribution design for high performance microprocessors has become increasingly challenging in recent years. Design goals of state-of-the-art integrated circuits, dictate the need for clock networks with smaller skew tolerances, large sizes, and lower capacitances. In this paper we discuss some of the issues in clock network design that arise in this context. We describe the clock design methodology and techniques used in the design of clock distribution networks for PowerPC™ microprocessors that aim at alleviating some of these problems.

1. Introduction

Clock distribution design for high performance circuits is becoming increasingly challenging due to faster and more complex circuits, smaller feature sizes, and a dominant impact of interconnect parasitics on network delays. Circuit speed has increased exponentially over the years, necessitating clock distributions with much smaller skew tolerances. On the other hand, increased switching frequencies, and higher net capacitance—due to larger nets, and a stronger coupling at smaller feature sizes, have resulted in a substantial increase in the power dissipation of clock nets and often accounts for up to 40% [1] of the processor power. Consequently, in addition to the performance related goals, power optimization has become very crucial, especially for portable applications. This trade-off between power and performance adds another dimension to the complexity of designing clock distribution schemes. IC design methodologies must employ efficient techniques to focus on clock design objectives at every step of the design process.

In this paper, we discuss some of these issues and specifically address the problems due to interconnect effects on clock network design for the PowerPC™ series of microprocessors. Section 2 highlights some of the interconnect effects that adversely affect clock nets. Section 3 presents an overview of the typical clock architectures used by the PowerPC™. Section 4 presents a summary of our design flow and describes specific methods that are a part of this flow. Our methodology provides us with the flexibility to design a wide range of clock nets—ranging from nets intended for high-end desk-tops and servers to low power designs for portable applications. Section 5 summarizes the results on some of our recent designs.

2. Interconnect Effects in Clock Networks

One of the most prominent effects of interconnect on clock signal is clock skew. The impact of interconnect has become much more pronounced due to disproportionate scaling of the interconnect delay *vis-a-vis* device delay [2]. The effect of clock skew on system performance is well studied [2] and accounts for about ($\sim$10–15%) of the total cycle time.

Another important factor that contributes to the clock period is the propagation delay (or the phase delay) through the interconnect. As shown in Fig. 1(a), large phase delay compared to the cycle time results in insufficient charging/discharging of the devices thereby

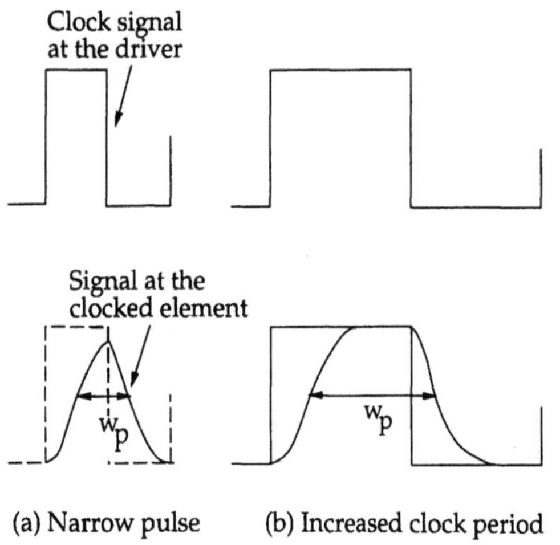

Figure 1. Effect of phase delay on clock signal. (a) The clock pulse width is not large enough to sustain the slow charging/discharging of the clock signal. (b) For effective clocking, the period must be increased.

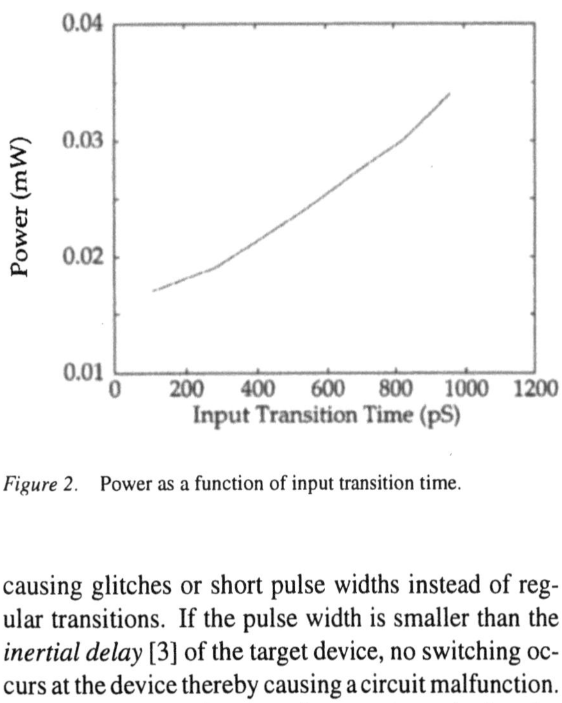

Figure 2. Power as a function of input transition time.

causing glitches or short pulse widths instead of regular transitions. If the pulse width is smaller than the *inertial delay* [3] of the target device, no switching occurs at the device thereby causing a circuit malfunction. This phenomenon forces an increase in cycle time for error free operation of the circuit as shown in Fig. 1(b).

Signal integrity becomes very critical as net sizes increase. Signal slopes must be preserved across the network for two reasons. 1) The signal slope effects the delay of the latches, and 2) poor signal slopes (large transition times) result in extra power dissipation in the latches as shown in Fig. 2.

While it is essential that these issues are addressed during clock net design, it is important to consider using accurate modeling techniques for representing the signal waveforms. The interconnect exhibits resistive shielding [4], and consequently, the signals are not crisp with a well defined delay and slope. To model these "non-digital" waveforms with sufficient accuracy, a higher order representation of the waveform is desirable, since balancing the first order network delays at the latches does not necessarily eliminate the "real skew". The techniques used as a part of our design methodology employs moments [5] of the waveform to represent the signal which allows us to optimize delays/slopes specified to any desired level of model accuracy.

3. Clock Distribution Architecture

A typical clocking network for the PowerPC™ consists of two levels of hierarchy—a primary clock distribution network, and secondary distribution networks (3). The primary clock network is a global net that distributes the clock signal to various functional blocks across the chip. One or more clock regenerators may be placed inside each of these circuit blocks, and act as regenerators of the clock signal. These regenerators in turn feed groups of latches placed in these blocks. This distribution of clock signal within a given circuit block constitutes the secondary level of the clock hierarchy.

For purposes of physical design, the primary network is further classified into two parts, a central network and a number of auxiliary networks. Each auxiliary network is a subnetwork that is fed by the central network. This hierarchical demarcation enables several designers to work on the network simultaneously. Furthermore, when automation of this task is desired, the auxiliary networks can be processed in parallel. Even

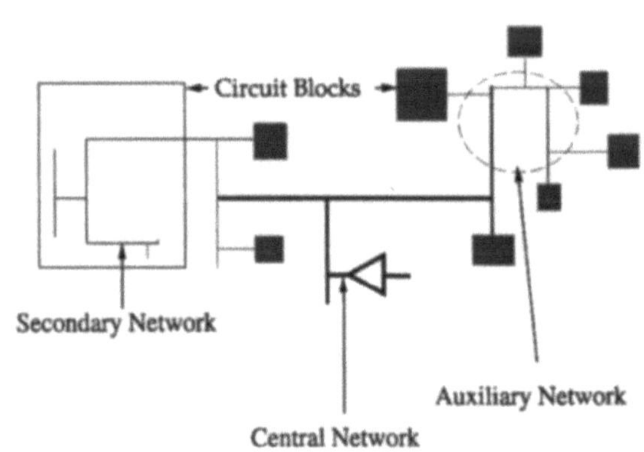

Figure 3. Clock distribution hierarchy.

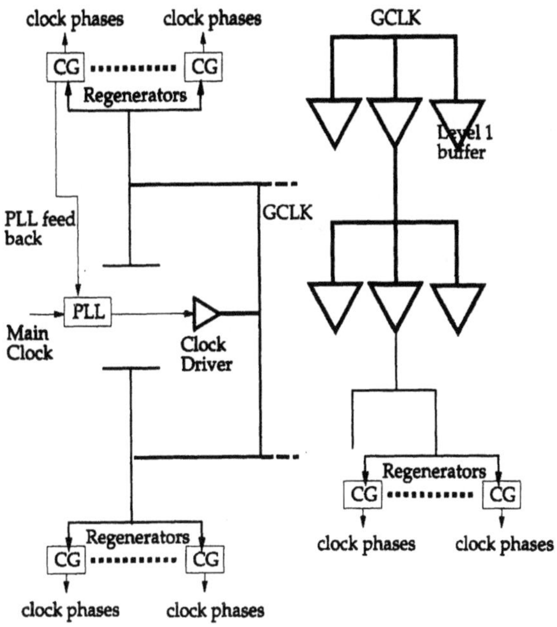

Figure 4. Typical PowerPC™ clocking scheme.

if the entire job is run on a single processor, we will later show in Section 4.2 that this hierarchical demarcation can improve the efficiency of the post processing techniques used in our methodology. Moreover, this allows parts of circuitry to run at different frequencies and phases. The external clock of the processor is fed to the processor clock net through a Phase Locked Loop. The output of this PLL is connected to a clock driver that feeds the primary network (Fig. 4). The local clock phases at the functional units are generated by the clock regenerators. Observe that this approach necessitates replication of circuitry that generates different phases from the global clock at every regenerator. Nevertheless, it has its advantages: The skew between different phases of the clock to a latch is small due to small propagation delays—therefore a complete new network is not necessary to distribute a different phase. In addition, since the net capacitance is switched at a lower frequency, it reduces the overall power. In order to guarantee a tight overall synchronization with the external clock, the differential feedback signal to the PLL is derived from one of the regenerators.

In addition to the regular network, the PowerPC™ has an additional low power network. The high performance network is in use during the normal operation of the processor, where as the low power network performs clock distribution during the power saving mode of the processor. The high performance network feeds all the functional units, hence has a higher overall load. The low power network however feeds only the essential units on the chip and has to be designed for a low value of capacitance, and performance is not a primary concern during the power saving mode. The two networks are however, required to have the same phase delay to operate at the design frequency.

4. Clock Design Flow

Clock net design starts at the logic synthesis phase. During this phase, the "logical skew", i.e., skew due to imbalances in number of loads, buffers and differential loading across buffers is minimized. Subsequently, the physical design phase eliminates the skew due to physical routing.

4.1. Synthesis

Several clock design steps are performed during synthesis. The focus here is on the control logic blocks which could potentially be synchronized to different phases of the clock. So the part of the network from clock regenerator to the latches is created during synthesis. Typically, designers instantiate a single regenerator in the hardware description of the block and associate a clock phase to this stage. At this point the following clock balancing steps are performed:

- Duplication of regenerators to eliminate nets with large fanout, since skews for these nets may be difficult to optimize during the placement or routing phases of the clock design.
- Clock buffer insertion, replication, and selection of appropriate drive levels of clock buffers to ensure that load capacitance limits on regenerator outputs and slew rates on latch inputs are met.

Clustering directives for placement tools are issued subsequently to ensure that the clusters of regenerators formed during logic synthesis phase are honored at the time of placement. This balancing results in designs where clock buffers are free from drive and slope violations under the assumed net capacitance models. The synthesis output is a logical clock distribution network that is well balanced, and meets slew rate requirements based on estimated net capacitances and the number of clocked devices, the capacitive loads, and the sizes of the buffers.

4.2. Design of Primary Clock Distribution

The design of the primary clock distribution follows the floorplanning and placement phase at which point physical information about blockage maps and routing constraints is available. This phase generates both the topology and the sizes for the primary clock network. Recall that in Section 3, we mentioned that the primary clock network consists of a central network and various auxiliary networks that it feeds. The initial topology of this network is generated by using one of two design flows, i.e., either by a semi-automatic flow or by using an automatic clock routing. The semi-automatic flow supports the design of generalized network topologies, whereas the automatic flow supports mainly a tree topology. The criterion for the choice of a specific topology depends on the size of the floor-plan, delay, signal slope, and the power goals of the specific design and the designers discretion.

Semi-Automatic Topology Design. The semi-automatic flow is tailored for designs where the topologies of the central network are generalized trees (non-binary), or meshes. Here the central network is first defined by designers and laid out manually, and a sequence of automated steps is then executed which result in a completed network topology. Most designs of the primary network (both the central and auxiliary networks) begin with an H-tree [6]. The geometrical symmetry of this structure assures a fairly well balanced clock tree in terms of the delay to the tips as well as certain amount of skew insensitivity to variations in process parameters like dielectric thickness, sheet resistance, and line widths. Modifications are made to this structure to honor placement and routing constraints, macro-blockages, as well as to ensure clock distribution to every target. Meshes are some times instantiated to ensure complete connectivity to all targets. As the layout of this primary network is simple enough, often designers choose to generate this structure manually and requires little effort.

Auxiliary Networks. The auxiliary networks are formed through a sequence of automated steps described below:

Step 1. Clustering and Load Balancing. First, clock regenerators are grouped into "clusters" that have a common source or a tapping point. The assignment of a regenerator to a cluster depends on the the physical location of the regenerator, and the estimated delay from the cluster source. Each regenerator is then assigned to a branch of the central network that has the shortest delay to the regenerator. Detailed routing of these clusters is performed later, based on this assignment. These regenerators are now deemed to belong to their respective auxiliary networks. The clustering algorithms also ensure a fairly balanced set of clusters in terms of the load.

Step 2. Routing of Auxiliary Networks. The cluster source (or the tip of the central network) and corresponding targets are maze routed at maximum width, considering the blockages in the proximity of the route. Although a maze-router performs poorly from a skew perspective, since the intra-cluster delays are very small, the skew among the clusters is acceptable. Since the wires are routed at maximum width, they can be trimmed down to the values required to meet skew, slope, and delay objectives. After sizing, the unused routing area is recovered for routing of other signal nets.

Topology Generation by Automatic Routing. For physically large nets, automatic routing techniques are available as a part of the methodology. These techniques are essentially variants of the "zero-skew" routing algorithm [7], and generate an Elmore delay [8] balanced routing. Due to physical limitations in terms of routing area and blockages, however it is difficult to achieve perfectly balanced trees. Consequently, the best possible layout in terms of skew which honors place and route constraints is generated at maximum width and post processing techniques are used to reduce the overall skew.

The automatic routing is performed bottom-up in three steps. The first step is partitioning of the chip area into clusters, using heuristics to balance delay and capacitance in each cluster. Each cluster is then routed as mentioned above to form the auxiliary networks. A zero-skew algorithm [7] based routing scheme then recursively merges the auxiliary networks to form a binary tree topology for the central network.

4.3. Optimization

The second phase of the network design optimizes the nets generated by methods described in the previous section for performance. The topology which corresponds to the clock net is extracted from the layout. This initial topology is then described to a proprietary wire width optimization tool as a set of wires in terms of their lengths, connectivity, and load capacitances. The tool sizes the wires to yield a solution that meets:

- Expected slew rate and transition time of the clock driver output.
- The required delay and slope requirements at each clock regenerator.
- Maximum skew limit.

Subject to:

- Maximum and minimum width constraints on wire segments.
- Upper and lower bound on the phase delay.
- The maximum capacitance allowable.

This wire width optimization tool uses the Levenberg-Marquardt algorithm [9] to minimize the mean square error between the desired and the actual delays and slopes. Given a set of circuit delays d_i's at the clocked elements, and the transition times (i.e., reciprocal of the slope) t_i's[1] as functions of wire widths, we find the vector **W**—the set of widths that minimize the mean square error between the desired delays $\hat{d}_i$'s and the desired transition times $\hat{t}_i$'s and those of the circuit waveforms. The solution involves repeatedly solving the equation:

$$(\mathbf{A} + \lambda\mathbf{I})\Delta\mathbf{W} = \mathbf{S}^{\mathrm{T}}(\mathbf{D} - \hat{\mathbf{D}}) \tag{1}$$

where

$$D_i = d_i, \quad 1 \leq i \leq n, \tag{2a}$$

$$= t_i, \quad n+1 \leq i \leq 2n, \tag{2b}$$

$\mathbf{A} = \mathbf{S}^{\mathrm{T}}\mathbf{S}$, and $\mathbf{S}$ is the $2n \times m$ Jacobian matrix with entries

$$S_{i,j} = \partial d_i / \partial w_j, \quad 1 \leq i \leq n \tag{3a}$$

$$= \partial t_i / \partial w_j, \quad n+1 \leq i \leq 2n \tag{3b}$$

i.e., matrix **S** describes both the sensitivities of the delay and transition time with respect to the wire widths. Equation (1) is repeatedly solved until a satisfactory convergence to the final solution is obtained. λ in (1) is the Lagrangian multiplier determined dynamically to achieve a rapid convergence to the final solution. This method combines the properties of steepest descent methods [10] during the initial stages and the convergence properties of methods based on Taylor series truncation as we approach.

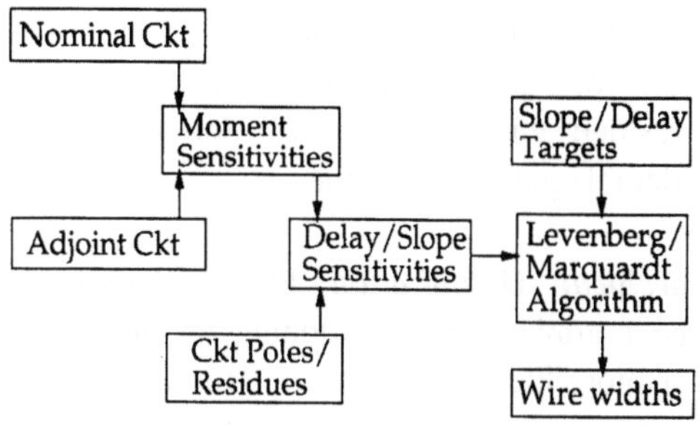

Figure 5. Skew minimization by using delay and slope sensitivities.

Critical to the success of this procedure is the efficient computation of the sensitivity matrix **S** when the size of the net is large. Sensitivities of the delay/slope with respect to the wire widths are computed by first computing the moment sensitivities at the target nodes and then transforming them to delay/transition time sensitivities as shown in Fig. 5. Computation of moment sensitivities is accomplished by using the adjoint sensitivity technique [11]. Once the moment sensitivities are computed, the poles and residues at every node must be computed to evaluate the delay/transition time sensitivity for that node [12]. Although the circuit evaluation itself is of linear complexity [13], since $2n \times m$ matrix entries are required, overall this procedure can be shown to be of $O(n^3)$ at every iteration [14] making the problem extremely complex. We use the techniques described below to reduce the problem complexity.

Problem Transformation to Target Moments. The first step towards improving the efficiency of this approach is to eliminate the need for the delay/transition time sensitivities. The problem is transformed to one of matching the circuit moments to a set of target moments. In other words instead of using delay/transition time targets along with delay/slope sensitivities with respect to the widths, we generate the target moments for a given delay and slope [15] as shown in Fig. 6. These

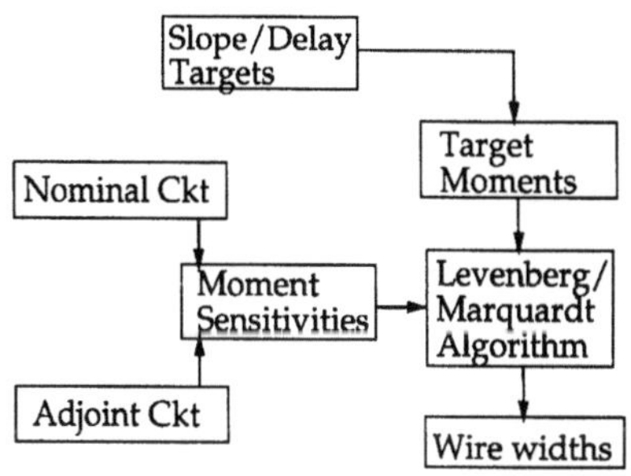

Figure 6. Skew minimization by using *moment* sensitivities.

targets need to be computed only once. This eliminates the need for translating the moment sensitivities into pole-residue sensitivities at every iteration and yields considerable gains in run time.

Hierarchical Optimization. Large clock-networks can be optimized by partitioning the problem into two or more hierarchical levels. This optimization is performed bottom-up and yields a significant improvement in the the total run-time of the tool, without observable loss in accuracy. For example, if a clock-network divided into a central network with m_C wires and m_A wires at the auxiliary network level is partitioned into k-clusters, the time complexity is proportional to $(m_C^3 + m_A^3/k^2)$ in comparison to $(m_A + m_C)^3$ for the entire network. Figure 7 illustrates the concept of hierarchical optimization, which is outlined below:

1. As described earlier, the auxiliary networks corresponding to each cluster, conveniently form the first level of the hierarchy. The regenerators at the leaves of the clusters are modeled as capacitances or even higher order load models. All auxiliary-networks are optimized individually for skew, and a specific value of delay and slope target. The widths of wires at this level are constrained to be between the initial routed width and a minimum width.
2. Each cluster is replaced by it's equivalent driving point load model [16], and the average delay for each cluster is estimated. The central network is optimized by considering the loading of clusters and their internal delays as follows. Assume a central network feeding k cluster-networks, each with average delay[2] d_{c_j}, $1 \leq j \leq k$. If the required delay for every node in the network is d_n, then the central network is optimized by setting the vector of delays $\mathbf{D} = (d_n - d_{c_1}, d_n - d_{c_2} \cdots d_n - d_{c_k})$ in (1). An equivalent π-model is used to represent the loads of

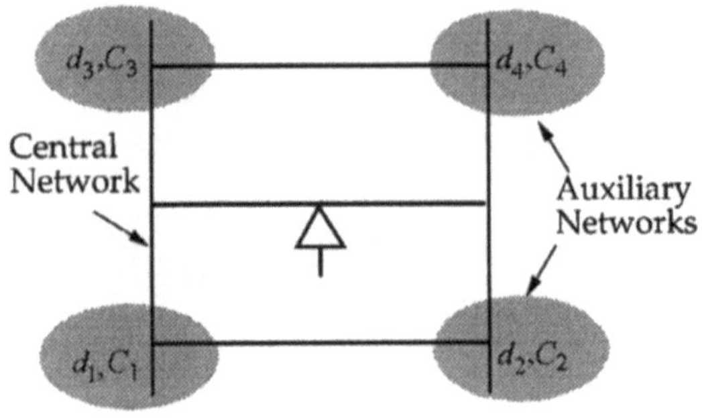

Figure 7. Hierarchical partitioning for optimization.

each of these clusters, while optimizing the central network.

Heuristics. The overall run times for large networks can be reduced substantially by using efficient heuristics:

1. Discard the insensitive wires for optimization. This results in a dramatic reduction in the size of the matrix and therefore a quick convergence is achieved [14].
2. The sensitivities at each iteration, change by a very small amount when wire sizes are changed. Therefore we recompute the moment sensitivities only once in several iterations [17].
3. During the first few iterations of the optimization we use only the first moment sensitivity since this directly influences the delay. Once the circuit delays are within a certain percent of the target delay, sensitivities corresponding to higher moments of the circuits are used so that both slope and delay targets are met.

Secondary Clock Distribution Design. The skew at the secondary clock distribution is smaller primarily due to the smaller interconnect lengths. Currently we use optimization routines to resize the buffers and regenerators to slow down or speed up the clock phases as required. Buffer and regenerator resizing is possible without invalidating the placement and wiring because all buffers and regenerators in the cell library are designed to present the same input capacitance and physical footprint.

This approach does not guarantee zero skew, due to the granularity in the power levels of the buffers and regenerators. In the future we plan to use wire-width optimization in conjunction with buffer resizing to further minimize skew.

4.4. Verification of Clock Distribution

Extraction is an essential step before clock timing can be verified. Clock nets are extracted after signal nets are routed. This allows an accurate extraction of area, fringe and coupling capacitance between the nets. Depending on the status of the design and the criticality of nets, this may involve using a variety of techniques, ranging from statistical modeling to the use of a finite element field solver for selected geometries.

The extracted parameters are stored in the data base to facilitate a chip level static timing analysis.

Verification. The clock verification tool uses STEP (a proprietary static timing tool) to generate a chip-level timing model. This timing model comprises of non-linear pre-characterized models of gates and RC models of the interconnect. The clock verifier allows the user to describe the clock network in very simple terms: a start point (a pin or a net) usually corresponding to the PLL block or the clock driver, the blocks that clocks pass through (buffers, regenerators, etc.), and blocks where the clocks stop (latches). For pass-thru and stop blocks, pins are specified that pass and stop the clock. Various timing assertions are also specified by the user, and these assertions are verified against the timing data model.

The clock verifier reads the control information, traces the desired clock network, and using STEP, obtains arrival time, rise time and fall time information at the pins of the blocks it encounters. It also verifies, for pass-thru blocks, that the paths specified through the block actually exists. By proper specification of pass-thru and stop blocks, the user can control the depth and breadth of the network to be analyzed. Figure 8 shows an example of controlling the clock hierarchy for verification.

Along with pass-thru and stop blocks, the user can also specify specific instances and nets to be ignored during network traversal. This allows the user to further prune the network, to omit non-critical elements such as scan and test clocks, and to ignore known problem blocks that will be fixed later.

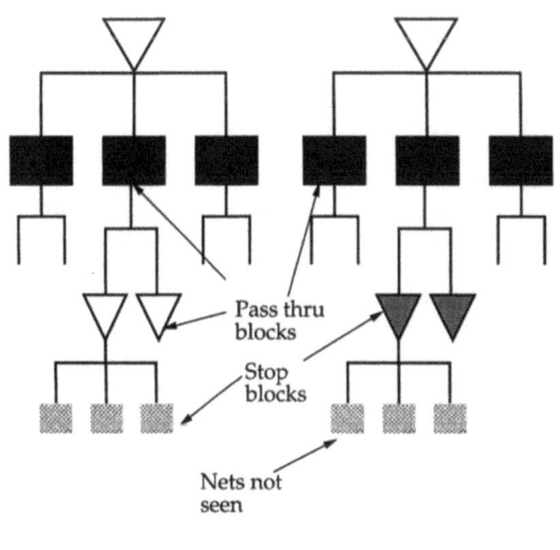

Figure 8. Defining the clock verification hierarchy.

Timing Checks. The following checks are performed during verification:

1. Early and late arrivals of the low-to-high and high-to-low transitions of the different clock phases.
2. Low-to-high and high-to-low transition time violations.
3. Set up and hold time violations.
4. Overlap between different clock phases.

5. Results

Our first example is a clock net designed using this methodology for our previous generation of microprocessors shown in Fig. 9. A small set of representative clusters for this network are shown in column 1 of Table 1. Column 2 shows the number of regenerators in these clusters and columns 4 and 5 show the internal delay and skew of the clusters respectively. Table 2 shows the statistics for the net.

A global skew of less than 50 pS was achieved with given wire-width constraints. The total run time for

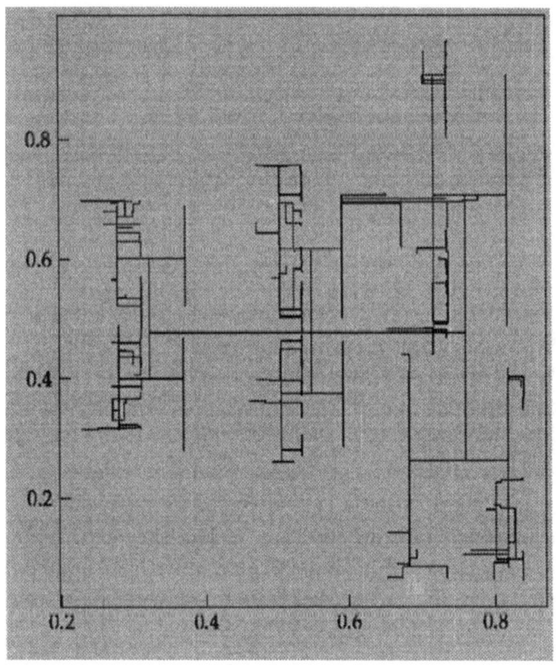

Figure 9. A primary network with a tree topology (the dimensions are normalized).

Table 1. Statistics for auxiliary networks.

Cluster name	# Regenerators	Capacitance(pF)	Delay (nS)	Skew (pS)
fxu_sw	35	6.062	0.108	1.509
fpu_se	18	3.588	0.106	17.59
fpu_sw	6	1.273	0.101	2.321
biu_lmw	20	3.381	0.102	7.101
biu_umw	16	1.501	0.102	5.796
biu_sw	16	2.090	0.102	4.430
fpu_nw	1	0.230	0.100	0.000
fpu_ne	9	1.117	0.101	4.407
fxu_nw	12	1.835	0.107	9.794

Table 2. Wire-width optimization results of laid clock nets.

Network statistics	Initial	Post-optimization
Skew(pS)	107	45
Phase delay (pS)	230	190
Transition time (pS)	235	250
Capacitance (pF)	33	35

Table 3. Results on current generation processor.

	Initial	Final	Target
Delay (pS) variation	117–230	177–189	190
10%–90% variation (pS)	166–307	249–258	250
C-total (pF)	33.27	33.60	37.00 (limit)

the entire design process described above, when performed in a hierarchical fashion was a little more than 3 hours on a IBM RISC System 6000™/Model 560. The run-time for the width optimization was less than 5 minutes on an average for the cluster networks and approximately 15 minutes for optimization of the central network with the estimated capacitance and delay. The quick turn around time of the tool has enabled the designers to experiment with different topologies and converge on to a design in a relatively short time. Table 3 shows corresponding results for a more recent processor designed to operate at 200 MHz. The methodology has been successfully used for processors of both these generations.

6. Conclusions

An overview of issues and considerations in contemporary clock design for high performance microprocessors was presented. A clock design methodology encompassing various stages of chip design and the techniques that address these problems was described here.

Notes

1. 10–90% transition time.
2. Of course we do consider slopes of the clusters as well, however we omit it here for simplicity.

References

1. D.W. Dobberpuhl, "A 200 mhz dual issue cmos microprocessor," *IEEE Journal of Solid State Circuits*, Vol. 27, pp. 1555–1567, 1992.
2. H.B. Bakoglu, *Circuits, Interconnects, and Packaging for VLSI*, Addison-Wesley Pub Co., Reading, MA, 1990.
3. Edward J. McCluskey, *Logic Design Principles*, Prentice Hall Series in Computer Engineering, New Jersey 07632, 1986.
4. J.J. Qian, Satyamurthy Pullela, and Lawrence T. Pillage, "Modeling the 'effective capacitance' of RC-interconnect," *IEEE Transactions on Computer Aided Design*, pp. 1526–1535, Dec. 1994.
5. Lawrence T. Pillage and R.A. Rohrer, "Asymptotic waveform evaluation for timing analysis," *IEEE Transactions on Computer Aided Design*, pp. 352–366, April 1990.
6. H.B. Baloglu, J.T. Walker, and J.D. Meindl, "Symmetric high-speed interconnections for reduced clock skew in ULSI and WSI circuits," in *Proceedings of the IEEE ICCD*, pp. 118–122, Oct. 1986.
7. Ren-Song Tsay, "Exact zero skew," in *IEEE International Conference on Computer Aided Design*, pp. 336–339, Nov. 1991.
8. W.C. Elmore, "The transient response of damped linear networks with particular regard to wideband amplifiers," *Journal of Applied Physics*, Vol. 19, No. 1, 1948.
9. D.W. Marquardt, "An algorithm or least squares estimation of non-linear parameters," *Journal of Society of Industrial and Applied Mathematics*, Vol. 11, No. 2, pp. 431–441, June 1963.
10. D.D. Morrision, "Methods for non-linear least squares problems and convergene proofs, tracking programs and orbit determination," in *Proceedings of the Jet Propulsion Laboratory Seminar*, pp. 1–9, 1960.
11. S.W. Director and R.A. Rohrer, "The generalized adjoint network sensitivities," *IEEE Transactions on Circuit Theory*, Vol. CT-16, No. 3, 1969.
12. Noel Menezes, Ross Baldick, and Lawrence T. Pillage, "A sequential quadratic programming approah to concurrent gate and wire sizing," in *Proceedings of the International Conference on Computer Aided Design*, pp. 144–151, Nov. 1995.
13. Curtis L. Ratzlaff, Nanda Gopal, and Lawrence T. Pillage, "RICE: Rapid interconnect circuit evaluator," in *Proceedings*

of the 28th Design Automation Conference, pp. 555–560, 1991.
14. Satyamurthy Pullela, Noel Menezes, and Lawrence T. Pillage, "Moment-sensitivity based wire sizing for skew reduction in on-chip clock nets," *IEEE Transactions on Computer Aided Design*, (to be published).
15. Noel Menezes, Satyamurthy Pullela, Floren Dartu, and Lawrence T. Pillage, "RC-interconnect synthesis—A moments approach," in *Proceedings of the IEEE International Conference on Computer-Aided Design*, pp. 418–425, 1994.
16. P.O' Brien and T.L. Savarino, "Modeling the driving-point characteristic of resistive interconnect for accurate delay estimation," in *Proceedings of the IEEE International Conference on Computer-Aided Design*, pp. 512–515, 1989.
17. Satyamurthy Pullela, Noel Menezes, and Lawrence T Pillage, "Reliable non-zero skew clock trees using wire width optimizatio," in *Proceedings of the 30th Design Automation Conference*, pp. 165–170, June 1993.

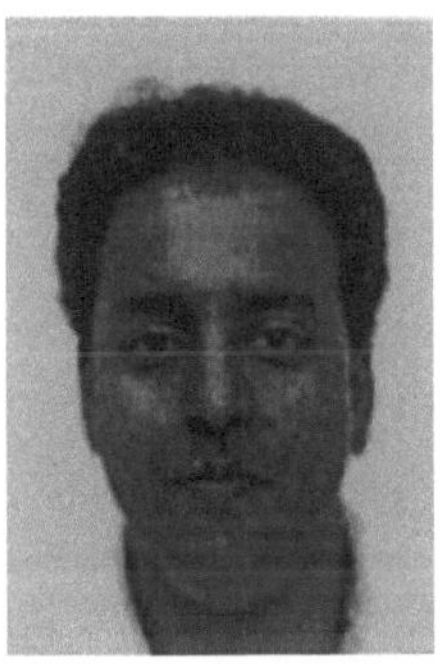

Shantanu Ganguly received the B.Tech. degree in Electrical Engineering from Indian Institute of Technology, Kharagpur, India in 1985, the M.S. and Ph.D. degrees in Computer Engineering from Syracuse University, NY in 1988 and 1991 respectively. In 1991 he joined Motorola's Sector CAD organization in Austin TX. Since 1992 he has been part of the PowerPC CAD organization in Austin TX. His interests include circuit simulation, parasitic extraction, power analysis, clock design and layout automation.
shantanu@ibmoto.com

Daksh Lehther received the B.E. degree from Anna University Guindy, Madras, India in 1991, M.S. degree from Iowa State University Ames, IA. He has been at Motorola Inc., Austin TX since August 1995. His current interests lie in developing efficient techniques for the computer-aided design of integrated circuits, with focus on areas of interconnect analysis, optimization physical design, and timing analysis.
daksh@ibmoto.com

Satyamurthy Pullela received the B. Tech. degree in Electrical Engineering from the Indian Institute of Technology, Madras in 1989, and Ph.D. in Electrical and Computer Engineering from the University of Texas at Austin, TX in 1995. He has been working in the High Performance Design Technology group in Motorola since May 1995. His interests include circuit simulation, timing analysis, interconnect analysis and optimization, and circuit optimization.
pullela@adttx.sps.mot.com

Journal of VLSI Signal Processing 16, 191–198 (1997)

Circuit Placement, Chip Optimization, and Wire Routing for IBM IC Technology

D.J. HATHAWAY
IBM Microelectronics Division, Burlington facility, Essex Junction, Vermont 05452

R.R. HABRA, E.C. SCHANZENBACH AND S.J. ROTHMAN
IBM Microelectronics Division, East Fishkill facility, Route 52, Hopewell Junction, New York 12533

Received and Revised November 22, 1996

Abstract. Recent advances in integrated circuit technology have imposed new requirements on the chip physical design process. At the same time that performance requirements are increasing, the effects of wiring on delay are becoming more significant. Larger chips are also increasing the chip wiring demand, and the ability to efficiently process these large chips in reasonable time and space requires new capabilities from the physical design tools. Circuit placement is done using algorithms which have been used within IBM for many years, with enhancements as required to support additional technologies and larger data volumes. To meet timing requirements, placement may be run iteratively using successively refined timing-derived constraints. Chip optimization tools are used to physically optimize the clock trees and scan connections, both to improve clock skew and to improve wirability. These tools interchange sinks of equivalent nets, move and create parallel copies of clock buffers, add load circuits to balance clock net loads, and generate balanced clock tree routes. Routing is done using a grid-based, technology-independent router that has been used over the years to wire chips. There are numerous user controls for specifying router behavior in particular areas and on particular interconnection levels, as well as adjacency restrictions.

Introduction

Traditionally, the goals of chip physical design have been to find placements which are legal (i.e., are in valid locations and do not overlap each other) and wirable for all circuits in a fixed netlist, and to route wires of uniform width on a small number of layers (two or three) to complete the interconnections specified in that netlist. The physical design process has been divided into two parts: placement, which is the assignment of circuits in the netlist to locations, or cells, on the chip image, and wiring, which is the generation of routes, using the available interconnection layers, to complete the connections specified in the netlist.

Recently, new technology characteristics and constraints and increased performance pressures on designs have required new capabilities from the chip physical design process. Wiring is now the dominant contributor to total net load and delay, and its contribution may vary significantly depending on the physical design solution chosen. This requires timing controls [1–4] for placement and wiring. Newer and larger chip technologies also provide more layers of wiring which must be accommodated by the wiring programs. These large chips also typically contain tens of thousands of latches, each requiring scan and clock connections. Such connections, as they appear in the input netlist to physical design, are usually somewhat arbitrary. Reordering the scan chain and rebuilding the clock distribution tree to reduce wire demand can significantly improve the physical design, since even with increased wiring layers these chips tend to be wire-limited. Clock trees must also be optimized to minimize clock skew, which has a direct impact on chip performance. Physical constraints on wire length and width to avoid electromigration failures and to limit noise must also be taken into consideration.

Hierarchical design of these large chips also imposes some new requirements on the physical design of the hierarchical components. However, in this paper we generally concentrate on the physical design of a single hierarchical component; other consequences of hierarchy are addressed in [1].

The design tools and the methodology for their use described in this paper have evolved from those used for earlier IBM technologies [2–4].

Physical Design Methodology

Many interdependencies exist among placement, clock and scan optimization, wiring, and hierarchical design planning [1]. Ordering of steps in the physical design process is required in order to give the best results and to ensure that the necessary prerequisites for each step are available. The general flow is as follows:

1. Identify connections to be optimized after placement, so that they will not influence placement. These include the scan and clock connections to latches.
2. Generate constraints for placement on the basis of a timing analysis done using idealized clock arrival times at latches and estimates of wire load and RC delay before physical design. These constraints include limits on the capacitance of selected nets and limits on the resistance or RC delay for selected connections.
3. Perform an initial placement to determine an improved basis for constraint generation, and optionally to fix the placement of large objects.
4. Generate new constraints for placement on the basis of a timing analysis done using wire load and RC delay values derived from the initial placement.
5. Perform placement.
6. Optimize the clock trees and scan connections.
7. Make logic changes, including changes to circuit power levels, to fix timing problems.
8. Legalize placement.
9. Generate new timing constraints for wiring on the basis of a timing analysis done using the actual clock tree and wire load and RC delay values derived from the final placement.
10. Perform routing.

Note that evaluation of the timing is performed at many points in this process, and the results determine whether to proceed to the next step or to go back through some of the previous steps. In particular, the user may need to iterate on constraint generation, placement, optimization, and timing until the design meets its timing goals. The user must also evaluate the wirability of the design throughout the process, and make adjustments to constraints or methodology if necessary.

Placement

Placement can be used at several points in the design process, and different algorithms are appropriate depending upon the state of the design.

Placement is often run before the logic has been finalized to obtain an early indication of the timing and wirability. At this point, the feedback may be used to influence logic changes. This may also be the time at which the locations of large objects are determined. The placement program may run more quickly by not considering such details as legality, and there may be less emphasis on achieving the best possible result. The results of this placement may be used as input to the tool which generates capacitance constraints used to drive subsequent placements.

Legality incorporates such constraints as the circuits not overlapping one another and remaining within the bounds of their placement area, being placed in valid orientations and in rows specified in the chip image, satisfying other restrictions supplied by either the user or the technology supplier, and ensuring that there are no circuit-to-power shorts (a concern in some custom circuits).

In the past, all legal location restrictions were specified to the placement programs in the form of "rules" which specify for a particular chip image and circuit type where on the chip circuits of this type can be placed. Now the program is expected, in most cases, to determine this itself, in part because of the extensive number of available chip images and the large amount of data which might be involved.

Once the logic has stabilized, more emphasis is placed on achieving a high-quality, and legal, placement. Some placement tools ignore at least some aspects of legality during the optimization phase, relying upon a separate legalization postprocessing step. Others attempt to ensure that they produce a completely legal result, while permitting such conditions as overlaps (with penalty) during the optimization.

Both clock optimization and power optimization (switching implementations of circuits in order to improve timing) can produce overlaps. These overlaps can simply be removed through a "brute force" technique, or overlap removal can be performed with some form of placement optimization. It is important to ensure that the quality of the placement is maintained: Clock skew, timing, and wirability should not worsen. It is often necessary to compromise between these conflicting factors. For example, the smallest clock skew is achieved by preventing the circuits in timing-critical clock trees from moving during overlap removal, but this can cause the other circuits to move much farther and can affect both the timing and the wirability.

The basic algorithms used in our placement programs are simulated annealing [5] and quadratic placement with iterative improvement [6, 7]. These are by no means new techniques, but the programs have been continually enhanced to give better results, in general, and to support the new specific technology-driven requirements. For example, the simulated annealing placement program now has the capability of performing low-temperature simulated annealing (LTSA). LTSA determines the temperature at which an existing placement is in equilibrium, and starts cooling from that temperature, thus effecting local improvements to a placement without disrupting the global placement characteristics.

Both simulated annealing and quadratic placement accept many controls. They include preplacement, floor-planning, specification of circuits to be placed in adjacent locations, net capacitance and source-to-sink resistance constraints, and weights for the various components of the scoring function (including net length, congestion, and population balancing).

Chip Optimization

Generally, the netlist which is the input to the physical design process contains all connections and circuits required in the design, and must be preserved exactly through the physical design process. Connections within clock trees (and other large signal-repowering trees) and latch scan chains (and other types of serial connections such as driver inhibit lines), however, may be reconfigured to improve chip wirability and performance. The best configuration of these connections depends on the results of chip placement, and thus the final construction of these types of structures must be a part of the physical design process. We call these special physical design processes *chip optimization*.

Chip optimization consists of two major parts. First, because many of the connections in the portions of the design being optimized will change after placement, they must be identified before placement is done and communicated to the placement tools so that they do not influence the placement process. We call this process *tracing*. Second, after placement is done we must actually perform the optimization of these special sections of logic. The specific optimization steps differ for clock trees and for scan chains.

Tracing and optimization of clock trees have been done for several years using separate programs. Recently these functions have been taken over by a new combined clock tracing and optimization program. The tracing function in the earlier tool is essentially the same as that in the new one. The optimization capability, however, has been significantly enhanced. The earlier clock optimization program could interchange connections of equivalent nets (as identified by the tracer) using a simulated annealing algorithm, could move dummy load circuits (terminators), and could move driving buffer circuits to the center of the sinks being driven. All of these actions were performed to reduce wiring and to balance the load and estimated RC delay on equivalent nets. In the remainder of this paper we describe the capabilities of and results from the new combined tracing and optimization program when discussing clock tree optimization.

Tracing of clock trees takes as its input a list of starting nets (the roots of the clock tree) and a description of the stopping points. Tracing proceeds forward through all points reachable in a forward trace from the starting nets and stops when latches or other explicitly specified types of circuits are reached, or when other explicitly specified stopping nets are reached. Placement is told to ignore all connections within the clock tree.

Tracing of scan chains takes as its input a list of connections to be kept and a list of points at which the chains should be broken. Tracing proceeds by finding the scan inputs of latches and tracing back from them, through buffers and inverters if present, to their source latches. These scan connections are then collected into chains. Placement is told to ignore all connections in the scan chains which will be subject to reordering, and the list of these scan chain connections and the polarity of each (the net inversion from the beginning of the scan chain) are passed as input to the scan optimization program.

A variety of styles of clock distribution network have been described in recent years. Several of these styles use a single large driver or a collection of drivers to drive a single clock net. Mesh clock distribution [8] and trunk and branch distribution [8] methods attempt to minimize clock skew by directly minimizing delay. This requires wide clock wiring (and/or many clock wires in the case of mesh distribution), thus causing a significant impact on wirability, an significant power expenditure to switch the high-capacitance clock net. H-tree [9] and balanced wire tree distribution [10–12] methods attempt to equalize the RC delay to all clock sinks using a delay-balanced binary tree distribution network. These methods tend to create long clock distribution delays owing to long electrical paths to the clock sinks. To avoid current density limitations of the clock conductors and excessive clock pulse degradation, these methods generally also require wide nets toward the root of the clock tree, again affecting wirability and power consumption. The delay problems of the single net distribution schemes are basically due to the $O(n^2)$ increase of RC delay with wire length. By limiting the length and load of any individual clock net in the clock distribution tree, this behavior is eliminated. For these reasons, our clock optimization methodology is directed toward a distributed buffer tree clock distribution network [10, 13].

The goals of the optimization vary for different levels of the clock tree. Toward the root, where the interconnection distances are large (and hence the RC delay is significant) and the number of nets is small, RC-balanced binary tree routing is used to help balance skew. Toward the leaves, where interconnection distances are very small (and hence RC delays are negligible) and where the number of nets is large, normal minimum Steiner routing is used, and the optimization goal is to balance the net loadings in order to balance the driving circuit delays. Because balanced tree routing requires more wiring resource than minimum Steiner routing, this approach tends to improve chip wirability.

Optimization of any fan-out tree always has as one goal the minimization of wiring congestion. For clock trees, and additional (and often more important) goal is the minimization of clock skew. The clock optimization performed includes the interchange of equivalent connections, the placement of circuits in the clock tree, the adjustment of the number of buffers needed in the clock tree, and the generation of balanced wiring routes for skew control. The new clock tracing and optimization program is designed as a collection of optimization algorithms which are called out by a Scheme language [14] script which is modifiable by the user. New features include the following:

- It can directly optimize a cross-hierarchical clock tree.
- It can add and delete terminators to better balance the capacitive load.
- It can make parallel copies of clock buffers. This means that the netlist can start with a skeleton clock tree that has the correct number of levels, but only one buffer at each level, and the optimizer will fill out the tree with the necessary number of buffers at each stage.
- It has an option to generate balanced wire routes for long skew-critical nets. This option creates "floorplan routes" which are subsequently embedded in detail by the wiring program. By avoiding the issues of detailed wiring in the optimizer, we eliminate the data volume required for detailed blockage information, which in turn makes it easier to perform cross-hierarchy optimization.
- It operates in several passes from the leaves to the root of the clock tree, allowing it to consider the locations of both inputs (established during the previous pass) and outputs of a block when determining its location.
- A combination of greedy initialization and iterative improvement functions offers performance improvements over the simulated annealing algorithm used in the previous clock optimization tool.

An example of the results of load balancing is shown in Fig. 1. The three parts of the figure illustrate the three levels of a clock tree on an IBM Penta technology [15] chip containing 72000 circuits and 13000 latches, and occupying 713000 image cells on a 14.5-mm image. The characteristics of the resultant trees, before addition of dummy loads for final load balancing, are shown in Table 1.

Table 1. Clock tree load-balancing results.

Tree level	Number of nets	Estimated net load (fF) Maximum	Minimum	Mean	Standard deviation
1	24	1142	731	947	112
2	123	1446	773	1078	108
3	1120	646	285	529	20

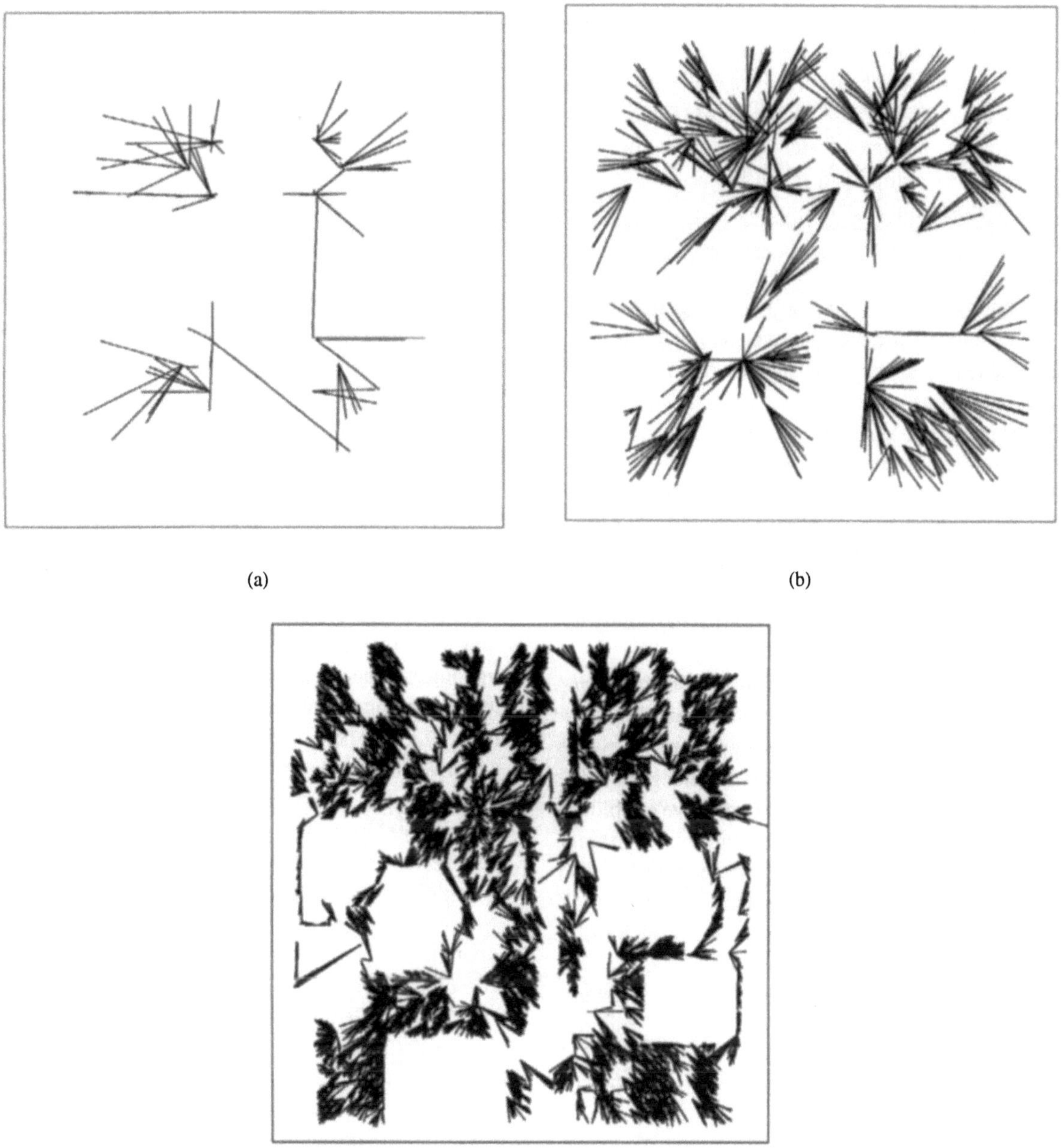

(a) (b) (c)

Figure 1. Load-balanced clock nets for level (a) 1, (b) 2, (c) 3.

Scan chain optimization is performed using a simulated annealing algorithm to reconfigure the connections in each chain in order to minimize wire length. If the user has specified breaks in the chain, the program optimizes each section of the chain separately. The program also preserves the polarity of each latch in a scan chain. Each latch is connected such that the parity (evenness or oddness) of the number of inversion between it and the start of the chain is preserved. Future work in this area will replace the simulated annealing optimization algorithm with a greedy initialization function followed by an iterative refinement step, in a manner similar to that employed in the new clock optimization program.

Routing

The routing program [16] has evolved over the years in response to a variety of pressures. With improvements in devices, routing plays an increasingly larger part in the design performance. Users need tighter control over the routing to improve the design and achieve greater productivity. The routing program has also had to handle the rapid increases in chip sizes and density.

As circuits become faster and wires become narrower, wires comprise a much larger part of path delays. Before routing, timing analysis is run using estimated paths. On the basis of this analysis, capacitance limits are generated for the critical nets and used by the routing program. In resolving congested areas, the capacitance of these critical nets is not allowed to exceed the limits. Less critical nets are rerouted around the area of congestion.

The routing program receives guidance from the clock optimization program for nets in clock and other timing-critical trees, in the form of floorplan routes. The routing program breaks each of these multi-pin nets into a group of point-to-point subnets. Each of these subnets is then routed to match the delay selected by the clock optimization program as closely as possible.

To achieve the desired electrical and noise characteristics, users can specify the wire width and spacing to be used for each net. Noise becomes a problem when the switching of one net causes a significant change in voltage on an adjacent net because of capacitive coupling. Clock nets are often given a wider width and spacing to reduce their resistance, capacitance, and noise.

High clock speeds and long narrow wires can result in a reliability problem known as electromigration. Over time, the movement of electrons can move the metal atoms and result in a break in the wire. To avoid this problem, the nets are evaluated prior to routing to determine which are susceptible to electromigration failure. These nets are then assigned capacitance limits and may be assigned a greater wire width.

Users often want to fine-tune the wires for some nets, such as clocks, and keep these wires fixed through multiple passes of engineering changes. Users would also like to stop between iterations of routing to verify that the routing of the selected nets has met all criteria before continuing. To accommodate these requirements, the routing program allows nets and wire segments to be assigned to groups. The user can specify how to treat existing wires on the basis of the group they are in. For each iteration, all existing wires in a group can be

- Fixed (not allowed to be rerouted).
- Fixed unless erroneous (segments which are invalid after an engineering change can be rerouted).
- Allowed to be rerouted if needed to complete another connection.
- Deleted (in the case of a major logic or placement change).

At the end of routing, all new wire segments are assigned to a user-specified group. The routing program makes sure that nets routed in one iteration do not prevent the remaining nets from being completed. This allows the user to have the program route just the clock nets in the first iteration. Once it has been verified that these routes meet the clock skew objectives, the wires for these nets can be fixed during the remaining iterations. A set of timing-critical nets can be routed in the second iteration. After analysis has verified that these nets meet their timing objectives, the remaining nets can be routed in the third iteration without changing the wires for the clock and timing-critical nets. This methodology allows tight clock skew and timing objectives to be met; it also allows timing problems requiring logic or placement changes to be identified quickly, before running a relatively long routing iteration on the majority of the nets.

Current chips can measure over twenty millimeters on a side and contain up to six layers of routing requiring 1600 megabytes to describe if kept in an uncompressed format. Designs can contain over a third of a million nets and a million pins which must be connected with over 300 meters of wire. The routing program uses compressed forms of the image, pin, and wire data in order to reduce system requirements and be able to handle these large designs on a workstation, even in flat mode. The 1600-megabyte chip description can be compressed to three megabytes. The data representation of 300 meters of wire, made up of over three million wire segments and two million vias, can be compressed to only 35 megabytes.

Before starting a potentially long routing run on a large design, the routing program allows the user to evaluate the design. A fast global routing step can be run to identify areas of congestion which may have to be resolved by changing the placement. The global results can also be fed to timing analysis to determine whether placement or logic changes must be made before detailed routing should be started. A single iteration of detailed routing can also be run to help identify congestion and timing problems before making a full routing run. A special iteration of routing can be made to identify pins which are inaccessible because of errors in the design rules, placement, or power routes.

Logic and placement are often changed to improve the design after the first routing run. The routing program automatically determines how these changes affect the wires and makes the required updates. This includes detecting old wires which are now shorted to new or moved circuits. The checking and update phases of the routing program run quickly when the logic and placement changes have been limited to small areas.

The user can control the cost of routing in each direction by interconnection level for up to four groups of nets. This can be used to have the short nets prefer the lower interconnection levels and the long nets use the upper interconnection levels. These weights can be set by area. This method is useful between macros where there is a high demand parallel to the edges of the macros and little demand to enter the macros.

In addition to congestion, timing, clock skew, and data volume, the routing program must handle special features of the technology. The routing program is often given multiple points at which it can connect to a pin. These points are in groups connected through high-resistance polysilicon. The routing program is prevented from routing into one group of a pin and out another, so that there is no polysilicon in the middle of a path to adversely affect timing and reliability. Unused pins must be connected to power or ground. The routing program recognizes any unused pins and ties them to the proper power bus.

If the routing program cannot resolve all of the congestion and complete all connections, a "ghost" iteration is run. This iteration completes as much of each of the remaining connections as possible and routes special wires, flagged as "ghosts", where no room can be found. The ghost wires may be replaced manually or automatically using a new set of parameters. Timing analysis can be run using these ghost wires as estimates. Display of the ghost wires can help identify congested areas.

Summary

Changes in physical design tools and methodology have been made to accommodate the higher performance requirements, larger chip sizes, and increasing importance of interconnect delay found in today's chip designs. Enhancements have been made to the placement, chip optimization, and routing tools to improve their capacity and performance and the quality of their results. Controls and options have been added to the tools to help the designer iteratively converge on a viable physical design implementation. The tools have also been enhanced to accommodate new requirement imposed by the technology.

The placement, clock optimization, and routing tool described here have been used on numerous timing-critical CMOS designs. Clocks for these designs range from 50 MHz up to 250 MHz. The clock skew due to physical design has been under 200 ps, although the skew due to process, power supply, and other variation can be ten times that. As an example, a design with 206000 objects to be placed and 205000 nets to be routed has been completed using a 15.5-mm chip image; it used more than 130 meters of wire and 1.6 million vias. Without clock and scan optimization, this design might have used more than 200 meters of wire, requiring a larger chip image.

Acknowledgments

The authors wish to acknowledge the contributions of Roger Rutter of IBM Endicott, NY, for his contributions to the chip optimization methods described here, and Chuck Meiley of IBM Almaden, CA, for his contributions to the wiring methods described here and for his assistance with the wiring portions of this paper. We also thank Bruce Winter to IBM Rochester, MN, for his assistance in providing design examples used in this paper, and both Bob Lembach of IBM Rochester, MN, and Mike Trick of IBM Burlington, VT, for their methodology descriptions.

References

1. J.Y. Sayah, R. Gupta, D. Sherlekar, P.S. Honsinger, S.W. Bollinger, H.-H. Chen, S. DasGupta, E.P. Hsieh, E.J. Hughes, A.D. Huber, Z.M. Kurzum, V.B. Rao, T. Tabtieng, V. Valijan, D.Y. Yang, and J. Apte, "Design planning for high-performance ASICs," *IBM J. Res. Develop.*, Vol. 40, No. 3, pp. 431–452.
2. R.S. Belanger, D.P. Conrady, P.S. Honsinger, T.J. Lavery, S.J. Rothman, E.C. Schanzenbach, D. Sitaram, C.R. Selinger, R.E. DuBios, G.W. Mahoney, and G.F. Miceli, "Enhanced chip/package design for the IBM ES/9000," *Proceedings of the IEEE International Conference on Computer Design*, pp. 544–549, 1991.
3. J.H. Panner, R.P. Abato, R.W. Bassett, K.M. Carrig, P.S. Gillis, D.J. Hathaway, and T.W. Sehr, "A comprehensive CAD system for high-performance 300K-circuit ASIC logic chips," *IEEE J. Solid-State Circuits*, Vol. 26, No. 3, pp. 300–309, March 1991.
4. R.F. Lembach, J.F. Borkenhagen, J.R. Elliot, and R.A. Schmidt, "VLSI design automation for the application system/400," *Proceedings of the IEEE International Conference on Computer Design*, pp. 444–447, 1991.

5. S. Kirkpatrick, C.D. Gelatt, and M.P. Vecchi, "Optimization by simulated annealing," *Science*, Vol. 220, No. 4598, pp. 671–680, May 1983.
6. K.J. Antreich, F.M. Johannes, and F.H. Kirsch, "A new approach for solving the placement problem using force models," *Proceedings of the IEEE Symposium on Circuits and Systems*, pp. 481–486, 1982.
7. R.-S. Tsay, E.S. Kuh, and C.-P. Hsu, "PROUD: A fast sea-of-gates placement algorithm," *Proceedings of the 25th ACM/IEEE Design Automation Conference*, pp. 318–323, 1988.
8. K. Narayan, "Clock system design for high speed integrated circuits," *IEEE/ERA Wescon/92 Conference Record*, pp. 21–24, 1992.
9. H.B. Bakoglu, J.T. Walker, and J.D. Meindl, "A symmetric clock distribution tree and optimized high speed interconnections for reduced clock skew in ULSI and WSI circuits," *Proceedings of the IEEE International Conference on Computer Design*, pp. 118–122, 1986.
10. K.M. Carrig, D.J. Hathaway, K.W. Lallier, J.H. Panner, and T.W. Sehr, "Method and apparatus for making a skew-controlled signal distribution network," U.S. Patent 5,339,253, 1994.
11. R.-S. Tsay, "An exact zero-skew clock routing algorithm," *IEEE Trans. Computer-Aided Design*, Vol. 14, No. 12, pp. 242–249, Feb. 1993.
12. K.D. Boese and A.B. Kahng, "Zero-skew clock routing trees with minimum wirelength," *Proceedings of the Fifth Annual IEEE International ASIC Conference and Exhibit*, pp. 17–21, 1992.
13. S. Pullela, N. Menezes, J. Omar, and L.T. Pillage, "Skew and delay optimization for reliable buffered clock trees," *Proceedings of the IEEE/ACM International Conference on Computer-Aided Design*, pp. 556–562, 1993.
14. R. Kent Dybvig, *The Scheme Programming Language*, Prentice-Hall, Inc., Englewood Cliffs, NJ, 1987.
15. C.W. Koburger III, W.F. Clark, J.W. Adkisson, E. Adler, P.E. Bakeman, A.S. Bergendahl, A.B. Botula, W. Chang, B. Davari, J.H. Givens, H.H. Hansen, S.J. Holmes, D.V. Horak, C.H. Lam, J.B. Lasky, S.E. Luce, R.W. Mann, G.L. Miles, J.S. Nakos, E.J. Nowak, G. Shahidi, Y. Taur, F.R. hite, and M.R. Wordeman, "A half-micron CMOS logic generation," *IBM J. Res. Develop.* Vol. 39, Nos. 1/2, pp. 215–227, Jan./March 1995.
16. P.C. Elmendorf, "KWIRE: A multiple-technology, user-reconfigurable wiring tool for VLSI," *IBM J. Res. Develop.*, Vol. 28, No. 5, pp. 603–612, Sept. 1984.

David J. Hathaway received the A.B. degree in physics and engineering sciences in 1978, and the B.E. degree in 1979 from Dartmouth College. In 1982 he received the M.E. degree from the University of California at Berkeley. In 1980 and 1981 he worked on digital hardware design at Ampex Corporation in Redwood City, CA. Mr. Hathaway joined IBM in 1981 at the Essex Junction development laboratory, where he is currently a senior engineer. From 1981 to 1990 he was involved in logic synthesis development, first with the IBM Logic Transformation System and later with the IBM Logic Synthesis System. From 1990 to 1993 he led the development of an incremental static timing analysis tool, and since 1993 has been working on clock optimization programs. Mr. Hathaway has three patents issued and seven pending in the U.S., and four publications. He is a member of the IEEE and the ACM.
david_hathaway@vnet.ibm.com

Rafik R. Habra received his B.S. and M.S. degrees in electrical engineering, both from Columbia University, in 1966 and 1967. He joined IBM in 1967 in the then Components Division in East Fishkill; he is currently employed there as a senior engineer. He worked first on numerical analysis applications, but soon joined the design automation effort at IBM, still in its early stages during that period. Mr. Habra led an effort to provide a chip design system comprising technology development, manual placement, and wiring, as well as shapes generation and checking. This was used for chip production during the seventies. He then became involved with providing a graphic solution to the task of embedding with checking overflow wires that proved instrumental in shortening the design cycle of chips and TCM modules. Mr. Habra holds a patent on parallel interactive wiring; a second patent on parallel automatic wiring is pending.
habra@vnet.ibm.com

Erich C. Schanzenbach received a B.S. degree in physics in 1979 from Clarkson University. He joined IBM Corporation in 1980 at the East Fishkill facility, where he is currently an advisory engineer. In 1980 and 1981, he worked on chip placement, and has spent the last fifteen years developing chip routing tools. Mr. Schanzenbach has one U.S. patent pending and one previous publication.
Schnanzen@fshvm1.vnet.ibm.com

Sara J. Rothman received the A.B. degree in mathematics in 1974 from Brown University, and the M.A. degree in mathematics from the University of Michigan in 1975. She completed her doctoral course work and taught at the University of Michigan until 1980, when she joined the IBM Corporation. Her first assignment, as part of the Engineering Design Systems organization, was to see whether the brand-new technique of simulated annealing could be used for industrial chip design; since then, she has worked on chip placement.
rothman@vnet.ibm.com

Journal of VLSI Signal Processing 16, 199–215 (1997)

Practical Bounded-Skew Clock Routing*

ANDREW B. KAHNG AND C.-W. ALBERT TSAO
UCLA Computer Science Dept., Los Angeles, CA 90095-1596

Received September 24, 1996; Revised October 11, 1996

Abstract. In Clock routing research, such practical considerations as hierarchical buffering, rise-time and overshoot constraints, obstacle- and legal location-checking, varying layer parasitics and congestion, and even the underlying design flow are often ignored. This paper explores directions in which traditional formulations can be extended so that the resulting algorithms are more useful in production design environments. Specifically, the following issues are addressed: (i) clock routing for varying layer parasitics with non-zero via parasitics; (ii) obstacle-avoidance clock routing; and (iii) hierarchical buffered tree synthesis. We develop new theoretical analyses and heuristics, and present experimental results that validate our new approaches.

1. Preliminaries

Control of signal delay skew has become a dominant objective in the routing of VLSI clock distribution networks and large timing-constrained global nets. Thus, the "zero-skew" clock tree and performance-driven routing literatures have seen rapid growth over the past several years; see [1, 2] for reviews. "Exact zero skew" is typically obtained at the expense of increased wiring area and higher power dissipation. In practice, circuits still operate correctly within some non-zero skew bound, and so the actual design requirement is for a *bounded-skew routing tree* (BST). This problem is also significant in that it unifies two well-known routing problems—the *Zero Skew Clock Routing Problem* (ZST) for skew bound $B = 0$, and the classic *Rectilinear Steiner Minimum Tree Problem* (RSMT) for $B = \infty$.

In our discussion, the *distance* between two points p and q is the Manhattan (or rectilinear) distance $d(p, q)$, and the distance between two sets of points P and Q is $d(P, Q) = \min\{d(p, q) \mid p \in P \text{ and } q \in Q\}$. The *cost* of the edge e_v is simply its wirelength, denoted $|e_v|$; this is always at least as large as the Manhattan distance between the endpoints of the edge, i.e., $|e_v| \geq d(l(p), l(v))$. *Detour wiring*, or *detouring*, occurs when $|e_v| > d(l(p), l(v))$. The cost of T, denoted $cost(T)$, is the total wirelength of the edges in T. We denote the set of sink locations in a clock routing instance as $S = \{s_1, s_2, \ldots, s_n\} \subset \Re^2$. A *connection topology* is a binary tree with n leaves corresponding to the sinks in S. A *clock tree* $T_G(S)$ is an embedding of the connection topology in the Manhattan plane, i.e., each internal node $v \in G$ is mapped to a location $l(v)$ in the Manhattan plane. (If G and/or S are understood, we may simply use $T(S)$ or T to denote the clock tree.) The root of the clock tree is the *source*, denoted by s_0. When the clock tree is rooted at the source, any edge between a parent node p and its child v may be identified with the child node, i.e., we denote this edge as e_v. If $t(u, v)$ denotes the signal delay between nodes u and v, then the *skew* of clock tree T is given by

$$skew(T) = \max_{s_i, s_j \in S} |t(s_0, s_i) - t(s_0, s_j)|$$
$$= \max_{s_i \in S}\{t(s_0, s_i)\} - \min_{s_i \in S}\{t(s_0, s_i)\}$$

The BST problem is formally stated as follows.

Minimum-Cost Bounded Skew Routing Tree (BST) Problem: Given a set $S = \{s_1, \ldots, s_n\} \subset \mathcal{R}^2$ of sink locations and a skew bound B, find a routing topology G and a minimum-cost clock tree $T_G(S)$ that satisfies $\text{skew}(T_G(S)) \leq B$.

*Support for this work was provided by Cadence Design Systems, Inc.

1.1. The Extended DME Algorithm

The BST problem has been previously addressed in [3–5]. Their basic method, called the *Extended DME* (Ex-DME) algorithm, extends the DME algorithm of [6–9] via the enabling concept of *merging region*, which is a set of embedding points with feasible skew and minimum merging cost if no detour wiring occurs[1]. For a fixed tree topology, Ex-DME follows the 2-phase approach of the DME algorithm in constructing a bounded-skew tree: (i) a bottom-up phase to construct a binary tree of merging regions which represent the loci of possible embedding points of the internal nodes, and (ii) a top-down phase to determine the exact locations of the internal nodes. The reader is referred to [4, 3, 5, 10] for more details (the latter is available by anonymous ftp). In the remainder of this subsection, we sketch several key concepts from [4, 3, 5].

Let $max\_t(p)$ and $min\_t(p)$ denote the maximum and minimum delay values (max-delay and min-delay, for short) from point p to all leaves in the subtree rooted at p. The skew of point p, denoted $skew(p)$, is $max\_t(p) - min\_t(p)$. (If all points of a pointset P have identical max-delay and min-delay, and hence identical skew, we similarly use the terms $max\_t(P)$, $min\_t(P)$ and $skew(P)$.) As p moves along any line segment the values of $max\_t(p)$ and $min\_t(p)$, along with $skew(p)$, respectively define the *delay* and *skew functions* over the segment.

For a node $v \in G$ with children a and b, its merging region, denoted $mr(v)$, is constructed from the so-called "joining segments" $L_a \in mr(a)$ and $L_b \in \mathrm{mr}(b)$, which are the closest boundary segments of $mr(a)$ and $mr(b)$. In practice, L_a and L_b are either a pair of parallel Manhattan arcs (i.e., segments with possibly zero length having slope $+1$ or -1) or a pair of parallel rectilinear segments (i.e., horizontal or vertical line segments). The set of points with minimum sum of distances to L_a and L_b form a *Shortest Distance Region* $SDR(L_a, L_b)$, where the points with skew $\leq B$ (i.e., feasible skew) in turn form the merging region $mr(v)$. [5] prove that under Elmore delay each line segment $l = \overline{p_1 p_2} \in SDR(L_a, L_b)$ is *well-behaved*, in that the max-delay and min-delay functions of point $p \in l$ are of the forms $max\_t(p) = \max_{i=1,\ldots,n_1} \{\alpha_i \cdot x + \beta_i\} + K \cdot x^2$ and $min\_t(p) = \min_{i=1,\ldots,n_2} \{\alpha'_i \cdot x + \beta'_i\} + K \cdot x^2$, where $x = d(p_1, a)$ or $d(p_2, b)$. In other words, the skew values along a well-behaved segment l can be either a constant (when $K = \alpha_i = \alpha'_i = 0$) or piecewise-linear decreasing, then constant, then piecewise-linear increasing along l. This important property enables [5] to develop a set of construction rules for computing the merging region $mr(v) \in SDR(L_a, L_b)$ efficiently in $O(n)$ time. The resulting merging region is shown to be a convex polygon bounded by at most 2 Manhattan arcs and 2 horizontal/vertical segments when L_a and L_b are Manhattan arcs, or a convex polygon bounded by at most $4n$ (with arbitrary slopes) segments where n is the number of the sinks. The empirical studies of [5] show that in practice each merging region has at most 9 boundary segments, and thus is computed in constant time.

Since each merging region is constructed from the closest boundary segments of its child regions, the method for constructing the merging region is called *Boundary Merging and Embedding* (BME). [5] also propose a more general method called *Interior Merging and Embedding* (IME), which constructs the merging region from segments which can be interior to the children regions. The routing cost is improved at the expense of longer running time. For arbitrary topology, [3] propose the Extended Greedy-DME algorithm (ExG-DME), which combines merging region computation with topology generation, following the Greedy-DME algorithm approach of [11]. The distinction is that ExG-DME allows merging at non-root nodes whereas Greedy-DME always merges two subtrees at their roots; see [3] for details.

Experimental results show that ExG-DME can produce a set of routing solutions with smooth skew and wirelength trade-off, and that it closely matches the best known heuristics for both zero-skew routing and unbounded-skew routing (i.e., the rectilinear Steiner minimal tree problem).

1.2. Contributions of the Paper

In this paper, we will show that these nice properties of merging regions and merging segments still exist when layer parasitics (i.e., the values of per-unit capacitance and resistance) vary among the routing layers and when there are large routing obstacles. Therefore, the ExG-DME algorithm can be naturally extended to handle these practical issues which are encountered in the real circuit designs. Section 2 extends the BME construction rules for the case of varying layer parasitics. We prove that if we prescribe the routing pattern between any two points, any line segment in $SDR(L_a, L_b)$ is well-behaved where L_a and L_b are two single points. Hence, the BME construction rules are still applicable. Section 3 proposes new merging region construction rules when there are obstacles in the routing plane. The solution is based on the concept of a *planar merging region*, which contains all the minimum-cost merging points when no detouring occurs. Finally, Section 4 extends our bounded-skew routing method to handle the practical case of buffering hierarchies in large circuits, assuming (as is the case in present design methodologies) that the buffer hierarchy (i.e., the number of buffers at each level and the number of levels) is given. Some conclusions are given in Section 5.

2. Clock Routing for Non-Uniform Layer Parasitics

In this section, we consider the clock routing problem for non-uniform layer parasitics, i.e., the values of per-unit resistance and capacitance on the V-layer (vertical routing layer) and H-layer (horizontal routing layer) can be different[2]. We first assume that via has no resistance and capacitance, then extend our method for non-zero via parasitics.

Let node v be a node in the topology with children a and b, and let merging region $mr(v)$ be constructed from joining segments $L_a \subseteq mr(a)$ and $L_b \subseteq mr(b)$. When both L_a and L_b are vertical segments or are two single points on a horizontal line, only the H-layer will be used for merging $mr(a)$ and $mr(b)$. Similarly, when L_a and L_b are both horizontal or are two single points on a vertical line, only the V-layer will be used for merging $mr(a)$ and $mr(b)$.[3] The original BME construction rules [5] still apply in these cases.

Corollary 1 below shows that for non-uniform layer parasitics, joining segments will never be Manhattan arcs of non-zero length. Thus we need consider only the possible modification of BME construction rules for the case where the joining segments are two single points which do not sit on a horizontal or vertical line. In this case, both routing layers have to be used for merging $mr(a)$ and $mr(b)$. One problem with routing under non-uniform layer parasitics is that different routing patterns between two points will result in different delays, even if the wirelength on both layers are the same. However, if we can prescribe the routing pattern for each edge of the clock tree, the ambiguity of delay values between two points can be avoided. Figure 1 shows the two simplest routing patterns between two points, which we call the HV and VH routing patterns. Other routing patterns can be considered, but may result in more vias and more complicated computation of merging regions.

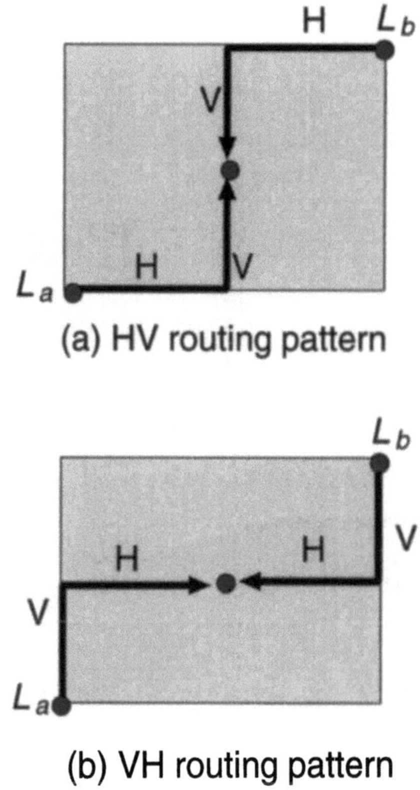

Figure 1. Two simple routing patterns between two points: HV and VH.

Theorem 1. *Let v be a node in the topology with children a and b, with the subtrees rooted at a and b having capacitive load C_a and C_b. Assume that joining segments $L_a \subseteq mr(a)$ and $L_b \subseteq mr(b)$ are two single points. Under the the HV routing pattern,* (i) *any line segment $l \in SDR(L_a, L_b)$ is well-behaved,* (ii) *merging region $mr(v)$ has at most 6 sides, and* (iii) *$mr(v)$ has no boundary segments which are Manhattan arcs of non-zero length.*

Proof: Without losing generality, we assume that L_a and L_b are located at $(0, 0)$ and (h, v) as shown in Fig. 2. Let $A(x, y)$ and $B(x, y)$ be respectively the average max-delay from a and b to p under the HV routing pattern. Let r_1, c_1 and r_2, c_2 be per-unit resistance and capacitance of the H-layer and the V-layer. We refer to the original delays and skew at point L_a as $max\_\bar{t}(L_a)$, $min\_\bar{t}(L_a)$, and $\overline{skew}(L_a)$. Similarly, we refer to the original delays/skew at point L_b as $max\_\bar{t}(L_b)$, $min\_\bar{t}(L_b)$, and $\overline{skew}(L_b)$. For point $p = (x, y) \in SDR(L_a, L_b)$,

$$\begin{aligned} A(x, y) &= max\_\bar{t}(L_a) + r_1 x(c_1 x/2 + C_a) \\ &\quad + r_2 y(c_2 y/2 + C_a + c_1 x) \\ &= K_1 \cdot x^2 + Ex + K_2 \cdot y^2 \\ &\quad + Fy + Gxy + D. \end{aligned} \tag{1}$$

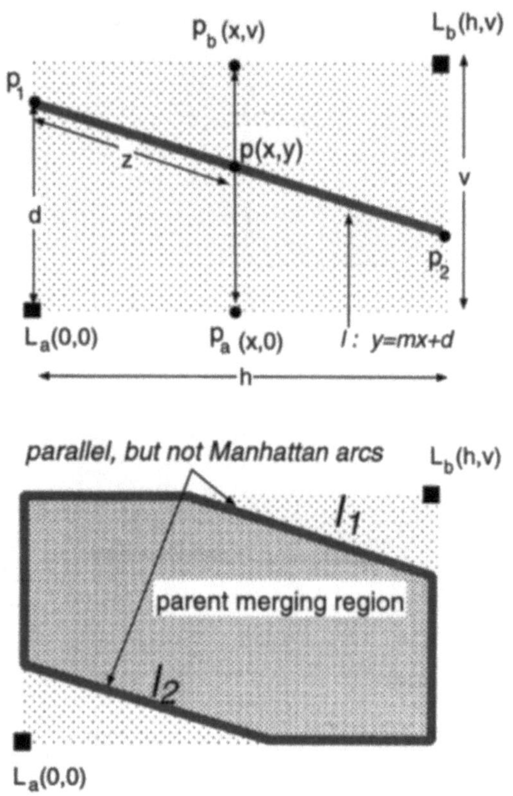

Figure 2. The merging region $mr(v)$ constructed from joining segments L_a and L_b which are single points by using the HV routing pattern for non-uniform layer parasitics.

where $K_1 = r_1c_1/2$, $E = r_1C_a$, $K_2 = r_2c_2/2$, $F = r_2C_a$, $G = r_2c_1$, and $D = max\_\bar{t}(L_a)$. Similarly,

$$\begin{aligned} B(x, y) &= max\_\bar{t}(L_b) + r_1(h - x) \\ &\quad \times (c_1(h - x)/2 + C_b) + r_2(v - y) \\ &\quad \times (c_2(v - y)/2 + C_b + c_1(h - x)) \\ &= K_1 \cdot x^2 + Jx + K_2 \cdot y^2 \\ &\quad + Ly + Gxy + M. \end{aligned} \tag{2}$$

where J, L, and M are also constants. Therefore,

$$\begin{aligned} max\_t(p) &= \max(A(x, y), B(x, y)) \\ &= \max(Ex + Fy + D, Jx + Ly + M) \\ &\quad + K_1 \cdot x^2 + K_2 \cdot y^2 + Gxy \end{aligned} \tag{3}$$

Similarly, we can prove that

$$\begin{aligned} min\_t(p) &= \min(A(x, y), B(x, y)) \\ &= \min(Ex + Fy + D', Jx + Ly + M') \\ &\quad + K_1 \cdot x^2 + K_2 \cdot y^2 + Gxy \end{aligned} \tag{4}$$

where $D' = min\_\bar{t}(L_a)$ and $M' = M - \overline{skew}(L_b)$.

If line segment $l \in SDR(L_a, L_b)$ is vertical, then for point $p(x, y) \in l$ we have

$$max\_t(p) = K_2 \cdot y^2 + \max\{F_v y + O, L_v y + P\} \tag{5}$$

$$min\_t(p) = K_2 \cdot y^2 + \min\{F_v y + O', L_v y + P'\} \tag{6}$$

where $F_v = F + Gx$, $L_v = L + Gx$, $O = D + K_1 \cdot x^2 + Ex$, $O' = D' + K_1 \cdot x^2 + Ex$, $P = M + K_1 \cdot x^2 + Jx$, and $P' = M' + K_1 \cdot x^2 + Jx$ are all constants. So, l is well-behaved.

If l is not vertical and described by the equation $y = mx + d$ where $m \neq \infty$ (see Fig. 2), then from Eqs. (1) and (2)

$$\begin{aligned} A(x, y) &= K_1 \cdot x^2 + Ex + K_2 \cdot (mx + b)^2 \\ &\quad + F(mx + b) + Gx(mx + b) + D \\ &= K \cdot x^2 + Hx + I \\ B(x, y) &= K_1x^2 + Jx + K_2(mx + b)^2 \\ &\quad + L(mx + b) + Gx(mx + b) + M \\ &= K \cdot x^2 + H'x + I', \end{aligned}$$

where K, H, I, H', and I' are all constants. Hence,

$$max\_t(p) = K \cdot x^2 + \max(Hx + I, H'x + I') \tag{7}$$

$$min\_t(p) = K \cdot x^2 + \min(Hx + Q, H'x + Q') \tag{8}$$

When $max\_t(p)$ and $min\_t(p)$ are written as functions of $z = d(p, p_1) = (1 + m)x$, they will still have the same coefficient in the quadratic term; this implies that any line segment $l \in SDR(L_a, L_b)$ is well-behaved.

Let l_1 and l_2 be the non-rectilinear boundary segments of $SDR(L_a, L_b)$ which have non-zero length. By the fact that $skew(l_1) = skew(l_2) = B$ and Eqs. (3) and (4), l_1 and l_2 will be two parallel line segments described by equations $(E - J)x + (F - L)y + D - M' = \pm B$. In practice, $|E - J| \neq |F - L|$ unless both layers have the same parasitics, i.e., $r_1 = r_2$ and $c_1 = c_2$. Thus, l_1 and l_2 will not be Manhattan arcs. □

We similarly can prove that Theorem 1 holds when the routing pattern is VH, or even when the routing pattern is a linear combination of both routing patterns such that each tree edge is routed by HV with probability $0 \leq \alpha \leq 1$ and VH with probability $1 - \alpha$. Notice that at the beginning of the construction, each node v is a sink with $mr(v)$ being a single point. Thus, no merging region can have boundary segments which are Manhattan arcs with constant delays, and we have

Corollary 1. *For non-uniform layer parasitics, each pair of joining segments will be either* (i) *parallel rectilinear line segments or* (ii) *two single points.*

Since any line segment in $SDR(L_a, L_b)$ is well-behaved for non-uniform layer parasitics, the BME construction rules are still applicable, except that (i) we have to prescribe the routing pattern for each tree edge, and (ii) the delays are calculated based on Eqs. (5), (6) for points on a vertical line $l \in SDR(L_a, L_b)$, and (7), (8) for points on a non-vertical line $l \in SDR(L_a, L_b)$, whenever L_a and L_b are two single points.

Theorem 2. *With non-zero via parasitics (per-unit resistance $r_v \geq 0$, per-unit capacitance $c_v \geq 0$), Theorem* 1 *still holds except that there will be different delay/skew equations for points on boundary segments and interior segments of $SDR(L_a, L_b)$.*

Proof: Again, without losing generality we assume the HV routing pattern. In Fig. 3(a), we assume that points L_a and L_b are both located in the H-layer. Under the HV routing pattern, most merging points p are on the V-layer except the top and bottom boundaries of $SDR(L_a, L_b)$ (e.g., point q in the figure). For point p on the V-layer, there is exactly one via in the path from p to L_a and L_b according to the HV routing pattern. Then, delay equations for merging points $p = (x, y) \in SDR(L_a, L_b)$ on the V-layer become

$$\begin{aligned} A(x, y) &= max\_\bar{t}(L_a) + r_1 x(c_1 x/2 + C_a) \\ &\quad + r_v(C_a + c_1 x + c_v/2) \\ &\quad + r_2 y(c_2 y/2 + C_a + c_1 x + c_v) \\ &= K_1 \cdot x^2 + J_1 x \\ &\quad + K_2 \cdot y^2 + L_1 y + r_2 c_1 xy + M_1, \\ B(x, y) &= max\_\bar{t}(L_b) + r_1(h - x)(c_1(h - x)/2 + C_b) \\ &\quad + r_v(C_b + c_1(h - x) + c_v/2) + r_2(v - y) \\ &\quad \times (c_2(v - y)/2 + C_b + c_1(h - x) + c_v) \\ &= K_1 \cdot x^2 + J_2 x \\ &\quad + K_2 \cdot y^2 + L_2 y + r_2 c_1 xy + M_2 \end{aligned}$$

where J_1, L_1, M_1, J_2, L_2, and M_2 are all constants. Since the quadratic terms $K_1 \cdot x^2$ and $K_2 \cdot y^2$ are the same as before, Theorem 1 holds for the merging points in $SDR(L_a, L_b)$ on the V-layer.

For merging points $q \in SDR(L_a, L_b)$ on the H-layer, the number of vias from q to L_a and L_b can be either 0 or 2. The delay calculations for merging points p and q will not be the same because of the unequal number of vias from the merging points to L_a and L_b.

Figure 3(b) shows one of the three cases where without loss of generality either point L_a or L_b is located on the V-layer. As shown in the Figure, we use point q to represent the merging point on the left or right boundary of $SDR(L_a, L_b)$ on the V-layer, point q' to represent the merging point on the top or bottom boundary of $SDR(L_a, L_b)$ on the H-layer, and point $p \in SDR(L_a, L_b)$ to represent the other merging points which are on the V-layer (but not on the right or left boundaries). In this case, the number of vias from point q, q' and p to L_a or L_b are not equal; their delay equations will not be identical, but will still have the same quadratic terms $K_1 \cdot x^2$ and $K_2 \cdot y^2$. Therefore, Theorem 1 still holds except that there will be different delay/skew equations for points on boundary segments and interior segments of $SDR(L_a, L_b)$. □

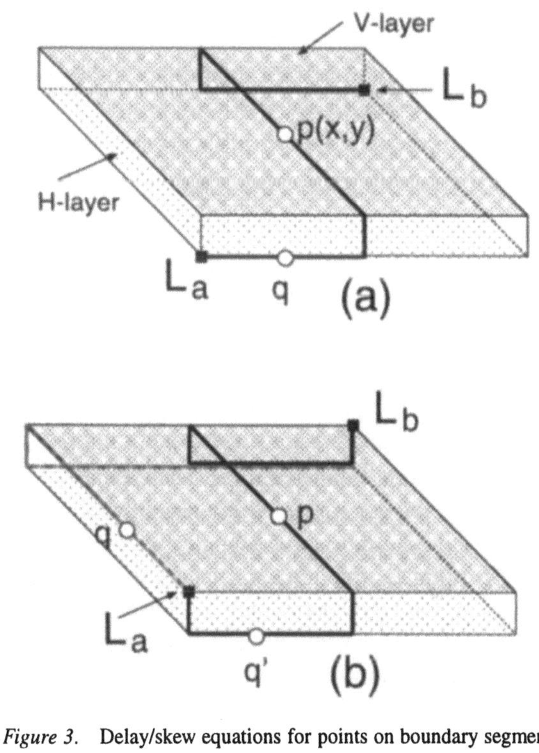

Figure 3. Delay/skew equations for points on boundary segments and interior segments of $SDR(L_a, L_b)$ are different when via resistance and/or capacitance are non-zero.

Table 1. Comparison of total wirelength of routing solutions under non-uniform and uniform layer parasitics, with ratios shown in parentheses. We mark by * the cases where the routing solution under non-uniform layer parasitics has smaller total wirelength than the solution under uniform layer parasitics.

Skew bound	r1		r2		r3		r4		r5	
	Wirelengths under non-uniform layer parasitics (normalized) wirelengths under uniform layer parasitics									
0 [11]	1253.2		2483.8		3193.8		6499.7		9723.7	
0	1332.5	(1.01)	2623.8	(1.01)	*3359.1	(0.99)	*6810.7	(0.99)	*10108.7	(1.00)
	1320.7		2603.6		3382.4		6877.5		10138.5	
1 ps	1283.5	(1.04)	2531.8	(1.05)	3207.0	(1.03)	6461.5	(1.04)	9610.8	(1.05)
	1232.2		2401.7		3118.1		6241.1		9190.7	
5 ps	1182.1	(1.05)	2333.3	(1.03)	2988.6	(1.04)	5979.8	(1.05)	8753.9	(1.05)
	1130.6		2256.2		2875.1		5715.1		8371.2	
10 ps	1158.6	(1.08)	2248.3	(1.03)	2810.7	(1.02)	5719.0	(1.05)	8482.4	(1.05)
	1069.2		2183.5		2747.6		5453.8		8063.7	
20 ps	1071.5	(1.03)	2183.4	(1.06)	2709.8	(1.05)	5474.6	(1.03)	8018.2	(1.04)
	1039.6		2069.1		2569.0		5290.1		7695.9	
50 ps	1058.6	(1.05)	2028.9	(1.06)	2557.0	(1.04)	5195.8	(1.04)	7562.9	(1.04)
	1009.3		1917.8		2459.7		5008.0		7248.2	
100 ps	989.0	(1.03)	1929.0	(1.03)	2463.9	(1.05)	4940.1	(1.03)	7193.1	(1.05)
	964.3		1880.7		2350.1		4786.1		6869.6	
200 ps	936.7	(1.05)	1886.7	(1.08)	*2356.0	(0.99)	4734.4	(1.04)	6905.9	(1.04)
	895.8		1741.6		2359.5		4540.1		6650.0	
500 ps	919.4	(1.12)	1770.9	(1.01)	2205.2	(1.01)	4635.1	(1.02)	6564.1	(1.02)
	820.4		1754.6		2187.4		4564.2		6449.3	
1 ns	830.0	(1.01)	*1664.2	(0.93)	*2156.4	(0.99)	*4500.5	(0.99)	*6395.4	(0.99)
	819.1		1709.4		2175.8		4531.4		6453.4	
10 ns	775.9	(1.00)	*1569.4	(0.97)	*2160.6	(0.98)	*4072.1	(0.97)	6168.5	(1.03)
	775.9		1613.5		2212.4		4184.2		5979.3	
∞	775.9	(1.00)	1522.0	(1.00)	1925.2	(1.00)	3838.2	(1.00)	5625.2	(1.00)
	775.9		1522.0		1925.2		3838.2		5625.2	
∞ [12]	769.3		1498.8		1902.6		3781.4		5571.1	

Experiments and Discussion

Table 1 compares the total wirelength of routing solutions under non-uniform and uniform layer parasitics for standard test cases in the literature. The per-unit capacitance and per-unit resistance for the H-layer are $c_1 = 0.027$ fF and $r_1 = 16.6\,\mathrm{m}\Omega$, respectively. For the uniform layer parasitics, the per-unit capacitance and per-unit resistance of the V-layer are equal to those of the H-layer, i.e., $c_2 = c_1$ and $r_2 = r_1$. For the non-uniform layer parasitics, we set $c_2 = 2.0 \cdot c_1$ and $r_2 = 3.0 \cdot r_1$, respectively. For simplicity, we use only the HV routing pattern and ignore the via resistance and capacitance. As shown in the Table, the solutions under non-uniform layer parasitics have larger total wirelength than those under uniform layer parasitics in most cases, especially when the skew bound is small. This may be due to the fact that merging regions under non-uniform layer parasitics tend to be smaller (and hence have higher merging cost at the next higher level) because the joining segments cannot be Manhattan arcs of non-zero length. When the skew bound is small, most of the merging regions are constructed from Manhattan arcs, and hence the solutions under non-uniform layer parasitics are more likely to have larger total wirelength. When the skew bound is infinite, no joining segments can be Manhattan arcs of non-zero length, and thus the routing solutions under non-uniform and uniform layer parasitics have identical total wirelength. In all the test cases, the wirelengths are evenly distributed among both routing layers—differences between the wirelengths on both layers are all less than 10% of the total wirelength, and less than 5% in most cases.

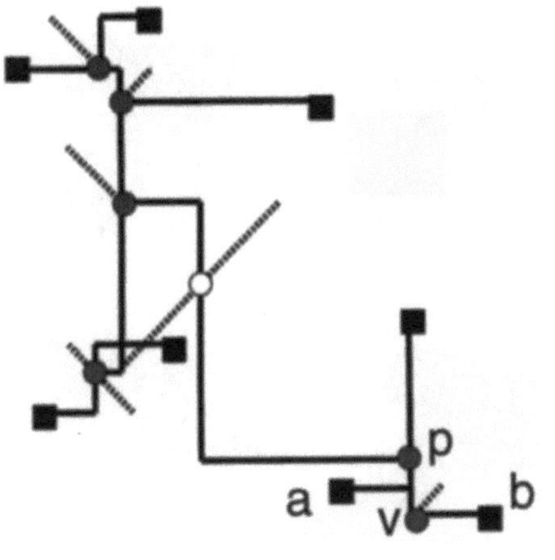

(a) Uniform layer parasitics (WL=2978 um)

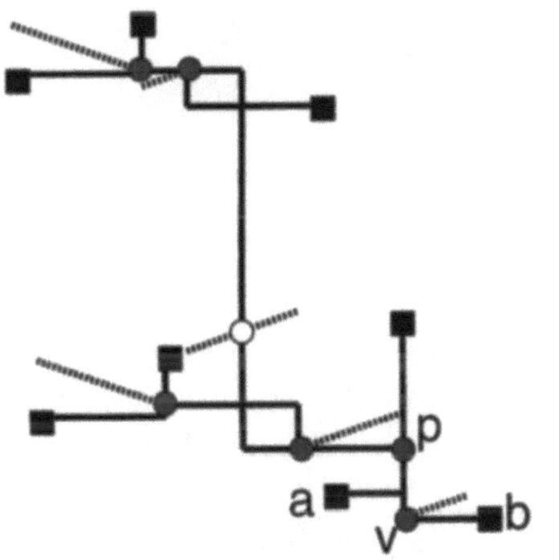

(b) Non-uniform layer parasitics (WL=2808 um)

Figure 4. Examples of 8-sink zero-skew trees for the same uniform and non-uniform layer parasitics used in Table 1. Note that the merging segments (the dashed lines) in (a) are Manhattan arcs while those in (b) are not.

We also perform more detailed experiments on benchmark r1 to compare the total wirelength of zero-skew routing for different ratios of r_2/r_1 and c_2/c_1. When $(r_2c_2)/(r_1c_1)$ changes from 1 to 10, the total wirelength of solutions only varies between $+4\%$ and -1% from that obtained for uniform layer parasitics (i.e., $(r_2c_2)/(r_1c_1) = 1$). Hence, the routing solution obtained by our new BME method is insensitive to changes in the ratio of H-layer/V-layer RC values.

Figure 4 shows examples of 8-sink zero-skew clock routing trees using the same HV routing pattern and layer parasitics that are used in the Table 1 experiments. We observe that no merging segments under non-uniform layer parasitics are Manhattan arcs and joining segments are all single points. Notice that under any given routing pattern like HV or VH, some adjacent edges are inevitably overlapped. For example, edges $\overline{av}$ and $\overline{vp}$ in Fig. 4 are overlapped because both edges are routed using the same HV patterns. If edges $\overline{av}$ and $\overline{bv}$ are routed according to the VH routing pattern, the overlapping wire can be eliminated.

Finally, we note that under uniform layer parasitics the IME method [5] is identical to the BME method for zero-skew routing since all merging segments are Manhattan arcs. However, the IME method might be better than the BME method for non-uniform layer parasitics, since merging segments are no longer equal to Manhattan arcs.

3. Clock Routing in the Presence of Obstacles

This section proposes new merging region construction rules when there are obstacles in the routing plane. Without loss of generality, we assume that all obstacles are rectangular. We also assume that an obstacle occupies both the V-layer and H-layer (this is of course a strong assumption, and current work is directed to the case of per-layer obstacles). We first present the analysis for uniform layer parasitics, then extend our method to non-uniform layer parasitics; we also give experimental results and describe an application to planar clock routing.

3.1. Analysis for Uniform Layer Parasitics

Given two merging regions $mr(a)$ and $mr(b)$, the merging region $mr(v)$ of parent node v is constructed from joining segments $L_a \subseteq mr(a)$ and $L_b \subseteq mr(b)$. Observe that a point $p \in mr(v)$ inside an obstacle cannot be the feasible merging point. Furthermore, points $p, p' \in SDR(L_a, L_b)$ may have different minimum sums of pathlengths to L_a and L_b because obstacles that intersect $SDR(L_a, L_b)$ may cause different amounts of detour wiring from p and p' to L_a and L_b. We define the *planar merging region* $pmr(v)$ to be the set of feasible merging points p such that the pathlength of the shortest planar path (without going through obstacles) from L_a through p to L_b is minimum (when the minimum pathlength from L_a to L_b is equal to $d(L_a, L_b)$, $pmr(v) \subseteq mr(v)$). Just as the merging region $mr(v)$ becomes a merging segment $ms(v)$ under zero-skew routing, the planar merging region $pmr(v)$ becomes the *planar merging segment* $pms(v)$ under zero-skew routing.

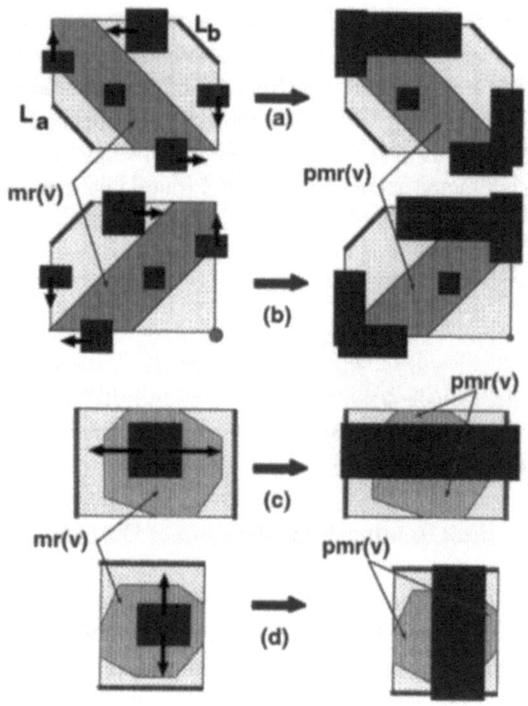

Figure 5. Illustration of obstacle expansion rules.

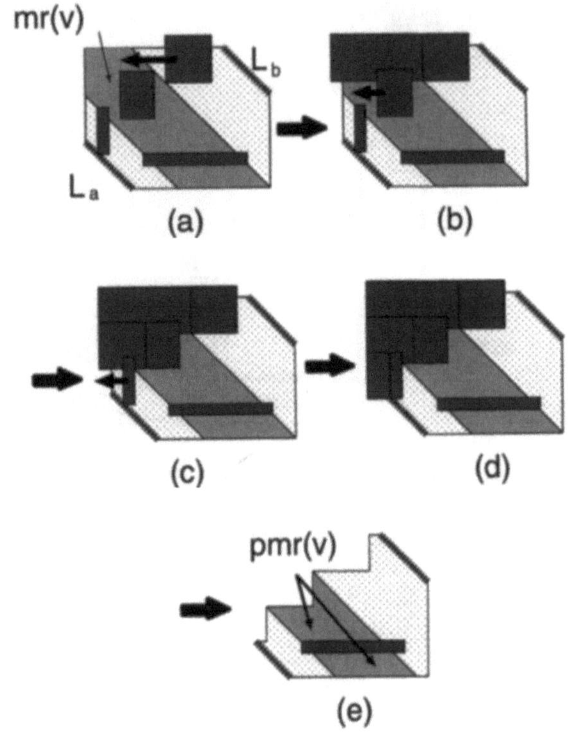

Figure 6. A "chain reaction" in the obstacle expansion.

The construction of $pmr(v)$ is as follows. If joining segments L_a and L_b overlap, $pmr(v) = mr(v) = L_a \cap L_b$. Otherwise, we expand any obstacles that intersect with rectilinear boundaries of $SDR(L_a, L_b)$ as illustrated in Fig. 5 for four possible cases; these define the *Obstacle Expansion Rules*.

Case I. (expand as in Fig. 5(a)).

1. $L_a = \{p_1\}$, $L_b = \{p_2\}$, and $\overline{p_1 p_2}$ has finite non-zero positive slope m, i.e., $0 < m < \infty$.
2. L_a or L_b is a Manhattan arc of non-zero length with slope -1.

Case II. (expand as in Fig. 5(b)).

1. $L_a = \{p_1\}$, $L_b = \{p_2\}$, and $\overline{p_1 p_2}$ has finite non-zero negative slope m, i.e., $-\infty < m < 0$.
2. L_a or L_b is a Manhattan arc of non-zero length with slope $+1$.

Case III. (expand as in Fig. 5(c)). Both joining segments are vertical segments, possibly of zero length.

Case IV. (expand as in Fig. 5(d)). Both joining segments are horizontal segments, possibly of zero length.

In Case I, an obstacle O which intersects with the top (bottom) boundary of $SDR(L_a, L_b)$ is expanded horizontally toward the left (right) side until O reaches the left (right) boundary of $SDR(L_a, L_b)$. If O intersects with the left (right) boundary of $SDR(L_a, L_b)$, then O is expanded upward (downward) until O reaches the top (bottom) boundary of $SDR(L_a, L_b)$. Case II is symmetric. In Case III, an obstacle O intersecting with $SDR(L_a, L_b)$ is expanded along the horizontal direction until O reaches both joining segments. Case IV is symmetric, with expansion in the vertical direction[4]. Finally, note that in Cases I and II an expanded obstacle O can intersect with another obstacle, which is then expanded in the same way; this sort of "chain reaction" is illustrated in Fig. 6.

With these obstacle expansion rules, we may complete the description of the planar merging region construction. For child regions $mr(a)$ and $mr(b)$ of node v, $pmr(v)$ is constructed as follows.

1. Apply the obstacle expansion rules to expand obstacles.
2. Calculate $pmr(v) = \{p \mid p \in mr(v) - \text{expanded obstacles}\}$.
3. Restore the sizes of all the expanded obstacles.

4. If $pmr(v) \neq \emptyset$ then stop; continue with next step otherwise.
5. Compute the shortest planar path P between $mr(a)$ and $mr(b)$.
6. Divide path P into a minimum number of subpaths P_i such that the pathlength of P_i, $cost(P_i)$, is equal to the (Manhattan) distance between the endpoints of P_i, i.e., if subpath $P_i = s \rightsquigarrow t$, then $cost(P_i) = d(s, t)$.
7. Calculate delay and skew functions for each line segment in P.
8. For each subpath P_i which has a point p with feasible or minimum skew, use the endpoints of P_i as the new joining segments. Then, calculate the planar merging region $pmr_i(v)$ with respect to the new joining segments, using Steps 1, 2 and 3. (Note that $pmr_i(v) \neq \emptyset$ since $p \in pmr_i(v)$).
9. $pmr(v) = \cup pmr_i(v)$, where subpath $P_i \subseteq P$ contains a point p with feasible or minimum skew.

Notice that the purpose of Step 6 is to maximize the area of $pmr(v)$. As shown in Fig. 7, if we divide subpath $P_2 = y - z - t$ into two smaller subpaths $y - z$ and $z - t$, region $pmr_2(v)$ in the Figure will shrink to be within the shortest distance region $SDR(y, z)$. Thus, like the merging regions constructed by the BME method, the planar merging regions will contain all the minimum-cost merging points when no detouring occurs. For the same reason stated in the Elmore-Planar-DME algorithm [13] the planar merging regions along the shortest planar path will not guarantee minimum tree cost at the next higher level. Thus, it is possible to construct and maintain planar merging regions along several shortest planar paths. At the same time, if an internal node v can have multiple planar merging regions, the number of merging regions may grow exponentially during the bottom-up construction of merging regions (this is the difficulty encountered by the IME method of [5]). Our current implementation simply keeps at most k regions with lowest tree cost for each internal node.

Finally, in the top-down phase of Ex-DME each node v is embedded at a point $q \in L_v$ closest to $l(p)$ (where p is the parent node of v), and that $L_v \in mr(v)$ is one of the joining segments used to construct $mr(p)$. When L_v is a Manhattan arc of non-zero length, there can be more than one embedding point for v. However, when obstacles intersect $SDR(l(p), L_v)$, some of the embedding points $q \in L_v$ closest to $l(p)$ may become infeasible because the shortest planar path from q to $l(p)$ has pathlength $> d(l(p), L_v)$. To remove infeasible embedding points from L_v, we treat $l(p)$ and L_v as two joining segments, then apply the obstacle expansion rules as in Fig. 8(b). If L'_v denotes the portion of L_v left uncovered by the expanded obstacles, the feasible embedding locations for v consist of the points on L'_v that are closest to $l(p)$.

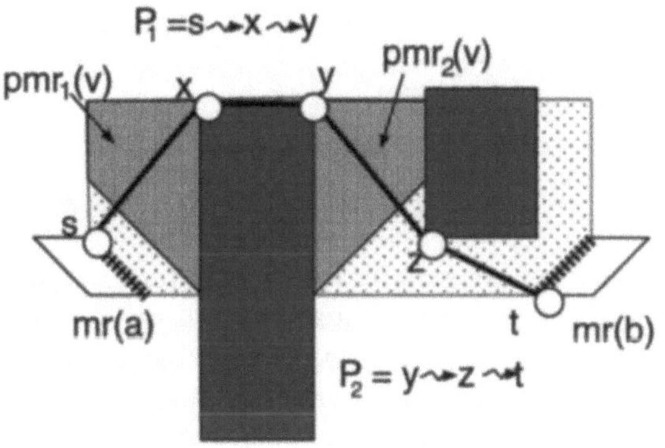

Figure 7. Construction of planar merging regions along a shortest planar path between child merging regions.

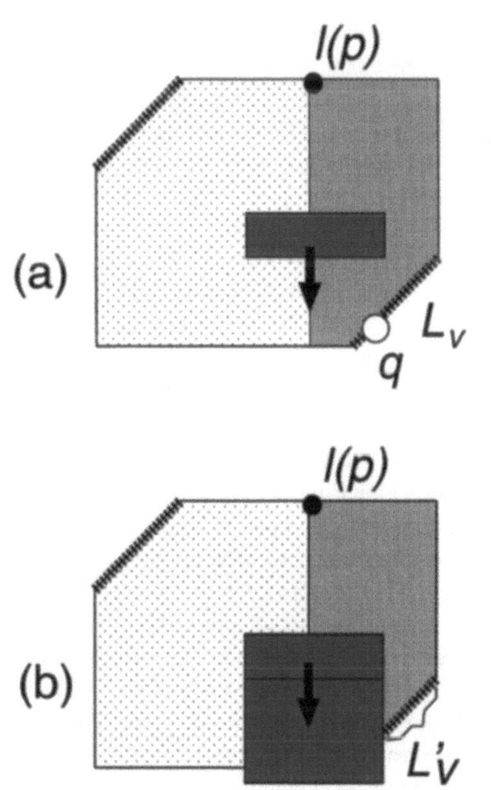

Figure 8. Modification of the embedding rule in the top-down phase of the Ex-DME algorithm when there are obstacles in the routing plane.

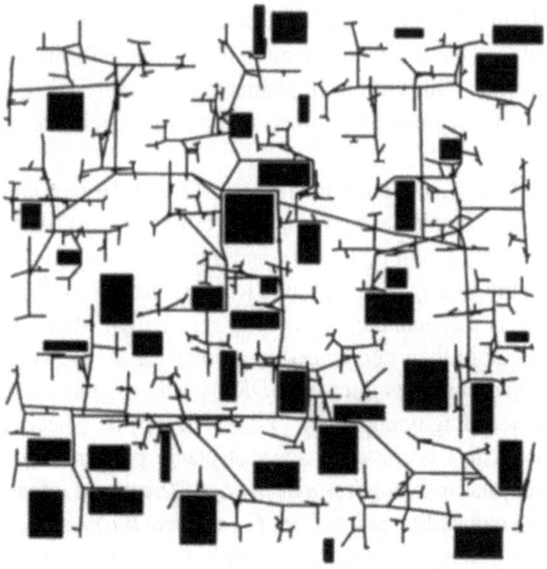

Figure 9. A zero-skew solution for the 555-sink test case with 40 obstacles.

3.2. *Experimental Results*

Our obstacle-avoiding BST routing algorithm was tested on four examples respectively having 50, 100, 150 and 555 sinks with uniformly random locations in a 100 by 100 layout region; all four examples have the same 40 randomly generated obstacles shown in Fig. 9. For comparison, we run the same algorithm on the same test cases without any obstacles. Details of the experiment are as follows. Parasitics are taken from MCNC benchmarks Primary1 and Primary2, i.e., all sinks have identical 0.5 pF loading capacitance and the per-unit wire resistance and wire capacitance are 16.6 mΩ and 0.027 fF. For each internal node, we maintain at most $k = 5$ merging regions with lowest tree cost. We use the procedure Find-Shortest-Planar-Path of the Elmore-Planar-DME algorithm [13] to find shortest planar s-t paths. The current implementation uses Dijkstra's algorithm in the visibility graph $G(V, E)$ (e.g., [14, 15]) where V consists of the source and destination points s, t along with detour points around the corners of obstacles. The weight $|e|$ of edge $e = (p, q) \in E$ is computed on the fly; if e intersects any obstacle, then $|e| = \infty$, else $|e| = d(p, q)$. The running time of obstacle-avoidance routing can be substantially improved with more sophisticated data structures for detecting the intersection of line segments and obstacles, and faster path-finding heuristic in the geometric plane. Table 2 shows that the wirelengths of

Table 2. Total wirelength and runtime for obstacle-avoiding BST algorithm, for various instances and skew bounds. Sizes and locations of obstacles are shown in Fig. 9. Numbers in parentheses are ratios to corresponding (total wirelength, runtime) values when no obstacles are present in the layout.

#Sinks	50	100	150	555
Skew bound	Wirelength: μm (normalized) CPU time: hr:min:sec (normalized)			
0	8791.1(1.06) 00:00:04(4)	11925.1(1.04) 00:00:10(2)	14747.5(1.03) 00:00:15(2)	28854.8(1.01) 00:00:34(1)
1 ps	8048.7(1.09) 00:01:09(6)	10761.4(1.04) 00:05:20(7)	13388.5(1.03) 00:11:36(3)	26240.0(1.04) 00:44:14(10)
2 ps	7831.9(1.07) 00:01:47(8)	10796.8(1.01) 00:08:17(9)	12643.0(1.02) 00:20:55(10)	25205.2(1.04) 01:30:08(13)
5 ps	7140.9(1.04) 00:04:01(13)	10493.6(1.08) 00:15:16(11)	11598.8(1.01) 00:30:34(13)	23648.0(1.04) 01:30:08(13)
10 ps	7126.2(1.06) 00:06:13(14)	9701.2(1.03) 00:19:36(12)	11426.1(1.07) 00:36:30(12)	22737.3(1.05) 01:48:06(13)
20 ps	6831.6(1.13) 00:07:40(15)	9296.4(1.03) 00:21:56(10)	11606.0(1.10) 00:40:39(3)	21641.7(1.05) 03:42:52(24)
50 ps	6468.4(1.12) 00:10:36(15)	8739.6(1.09) 00:26:47(11)	10194.4(1.10) 01:00:50(13)	22167.1(1.15) 02:18:20(14)
100 ps	6484.7(1.20) 00:13:51(18)	8588.2(1.11) 00:30:16(9)	9295.6(1.02) 01:03:00(15)	19086.6(1.01) 03:06:23(17)
1 ns	6484.7(1.24) 00:16:20(18)	8115.1(1.13) 00:36:52(11)	9265.8(1.10) 01:18:36(15)	17166.8(.99) 07:24:38(12)
10 ns	6484.7(1.24) 00:16:19(18)	8115.1(1.13) 00:36:43(11)	9265.8(1.10) 01:20:07(15)	16698.3(.99) 03:18:20(7)
∞	6484.7(1.24) 00:16:43(18)	8115.1(1.13) 00:36:52(11)	9265.8(1.10) 01:20:25(13)	16698.3(1.02) 03:21:11(7)

routing solutions with obstacles are very close to those of routing solutions without obstacles (typically within a few percent). Runtimes (reported for a Sun 85 MHz Sparc-5) are significantly higher (by factors of up to 18 for the 50-sink instance) when the 40 obstacles are present; we believe that this is due to our current naive implementation of obstacle-detecting and path-finding. Figure 9 shows the zero-skew clock routing solution for the 555-sink test case.

3.3. *Extension to Non-Uniform Layer Parasitics*

When the layer parasitics are non-uniform, no joining segment can be a Manhattan arc, so Cases I.2 and II.2 of the obstacle expansion rules are inapplicable. In Cases III and IV, only one routing layer will be used to merge the child regions, so the construction of planar merging regions will be the same as with uniform layer parasitics. Hence, the construction of planar merging

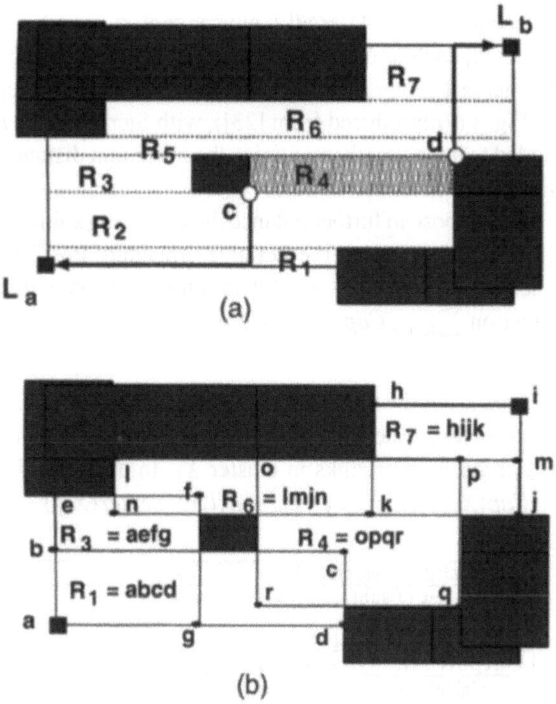

Figure 10. Obstacle-avoidance routing for non-uniform layer parasitics when joining segments L_a and L_b are single points not on the same vertical or horizontal line.

regions changes only for Cases I.1 and II.1, i.e., when the joining segments L_a and L_b are two single points which are not on the same vertical or horizontal line.

Since larger merging regions will result in smaller merging costs at the next higher level, a reasonable approach[5] is to maximize the size of the merging region constructed within each rectangle $R_i \subseteq SDR(L_a, L_b)$, by expanding R_i as shown in Fig. 10(b). After expansion, "redundant" rectangles contained in the expansions of other rectangles (e.g., rectangles R_2 and R_5 in Fig. 10 are contained in the union of expansions of R_1, R_3, R_4, R_6 and R_7) can be removed to simplify the computation. The merging region construction for Cases I.1 and II.1 with non-uniform layer parasitics is summarized as follows.

1. Divide $SDR(L_a, L_b)$ into a set of disjoint rectangles R_i by extending horizontal boundary segments of the (expanded) obstacles in $SDR(L_a, L_b)$.
2. Expand each rectangle R_i until blocked by obstacles.
3. Remove rectangles R_i that are completely contained by other rectangles.
4. For each rectangle R_i do:
   - Let $c \in R_i$ and $d \in R_i$ be the corner points which are closest to joining segment L_a and L_b. Apply prescribed routing patterns from c to L_a and from d to L_b.
   - Calculate delays at c and d.
   - Construct the merging region from points c and d as as described in Section 2.

Finally, we notice that in planar clock routing, all wires routed at a lower level become obstacles to subsequent routing at a higher level. Also, in the obstacle-avoidance routing, if some obstacle blocks only one routing layer, then the routing over the obstacle must be planar. In such cases, we may apply the concept of the planar merging region to improve the planar clock routing. In particular, we improve the Elmore-Planar-DME algorithm [13, 16] by (i) constructing the planar merging segment $pms(v)$ for each internal node v of the input topology G, and (ii) replacing the Find-Merging-Path and Improve-Path heuristics of Elmore-Planar-DME by construction of a shortest planar path P connecting v's children s and t via v's embedding point $l(v) \in pms(v)$. Total wirelength can be reduced because $l(v)$ is now selected by the DME method optimally from $pms(v)$ instead of being selected heuristically by Find-Merging-Path and Improve-Path in Elmore-Planar-DME. Our experiments [17] show that Elmore-Planar-DME is consistently improved by this technique.

4. Buffered Clock Tree Synthesis

Finally, we extend our bounded-skew routing method to handle the practical case of buffering hierarchies in large circuits. There have been many works on buffered clock tree designs. [18–20] determine the buffer tree hierarchy for the given clock tree layout or topology. [21, 22] design the buffer tree hierarchy and the routing of the clock net simultaneously. However, the prevailing design methodology for clock tree synthesis is that the buffer tree hierarchy is pre-designed before the physical layout of the clock tree (e.g., see recent vendor tools for automatic buffer hierarchy generation, such as Cadence's CT-Gen tool). In practice, a buffer hierarchy must satisfy various requirements governing, e.g., phase delay ("insertion delay"), clock edge rate, power dissipation, and estimated buffer/wire area. Also, the placement and routing estimation during chip planning must have reasonably accurate notions of buffer and decoupling capacitor areas, location of wide edges in the clock distribution network, etc. For these reasons, buffer hierarchies are typically "pre-designed" well in advance of the post-placement buffered clock tree synthesis. So our work starts with a given buffer hierarchy as an input; this defines the number of buffer levels and the number of buffers at each level. We use the notation $k_M - k_{M-1} - \cdots - k_0$ to represent a buffer hierarchy with k_i buffers at level i, $0 \leq i \leq M$. For example, a 170-16-4-1 hierarchy has 170 buffers at level 3, 16 buffers at level 2, etc. Note that we always have $k_0 = 1$ since there is only one buffer at the root of the clock tree. As in [19, 20, 22], to minimize the skew induced by the changes of buffer sizes due to the process variation, we assume that identical buffers are used at the same buffer level. (From the discussion of our method below, we can see that our method can work without this assumption by minor modification.)

We propose an approach to bounded-skew clock tree construction for a given buffer hierarchy. Our approach performs the following steps at each level of the hierarchy, in bottom-up order.

1. Cluster the nodes in the current level (i.e., roots of subtrees in the buffer hierarchy, which may be sinks or buffers) in the current level into the appropriate number of clusters (see Section 4.1).
2. Build a bounded-skew tree for each cluster by applying the ExG-DME algorithm under Elmore delay [5].
3. Reduce the total wirelength by applying a buffer sliding heuristic (see Section 4.2).

4.1. Clustering

The first step is to assign each node (e.g., sink or buffer) in the current level i of the buffer hierarchy to some buffer in level $i - 1$. The set of nodes assigned to a given level $i - 1$ buffer constitute a *cluster*. If there are k buffers in the next higher level of the buffer hierarchy, then this is a k-way clustering problem. Numerous algorithms have been developed for geometric clustering (see, e.g., the survey in [23]); our empirical studies show that the K-Center technique of Gonzalez [24] tends to produce more balanced clusters than other techniques. Furthermore, the K-Center heuristic has only $O(nk)$ time complexity (assuming n nodes at the current level). The basic idea of K-Center is to iteratively select k cluster *centers*, with each successive center as far as possible from all previously selected centers. After all k cluster centers have been selected, each node at the current level is assigned to the nearest center. Pseudo-code for K-Center is given in Fig. 11 (reproduced from [23]), with Steps 0 and 3*a* added to heuristically maximize the minimum distance among the k cluster centers.

We propose to further balance the clustering solution from K-Center using the iterative procedure PostBalance in Fig. 12, which greedily minimizes the objective function $\sum_{i=1,k} Cap(X_i)^w$. Here,

- $Cap(X_i)$ is the estimated total capacitance of the BST (to be constructed in the second major step of our approach) over sinks in cluster X_i. In other words, $Cap(X_i) = \sum_{v \in X_i} c_v + d(l(v), center(X_i)) \cdot c$,

Algorithm k-center($S, X_1 \ldots, X_k, k$)
Input: Set of subtree roots (e.g., sinks or buffers) S; number of clusters k
Output: Sets of clusters $\{X_1, X_2, \ldots, X_k\}$
0. Calculate $V = S \bigcup U$, where $U = \{ u \mid u =$ a grid point of $\sqrt{\lvert S \rvert}$ uniformly spaced horizontal and vertical lines inside $bbox(S)$ $\}$ 1. Initialize W, a set of cluster *centers*, to empty. 2. Choose some random v from V and add it to W. 3. **while** $\lvert W \rvert \leq k$, find $v \in V$ s.t. $d_W = \min_{w \in W} d(v, w)$ is maximized, and add it to W. 3a. **while** $\exists v_1 \in W, v_2 \in V - W$ s. t. d_W can be increased by swapping v_1 and v_2, then swap v_1 and v_2 (i.e., $W = W + \{v_2\} - \{v_1\}$). 4. Form clusters $X_1, X_2, \ldots, X_k$ each containing a single point of W; place each $v \in S$ into the cluster of the closest $w_i \in W$.

Figure 11. Pseudocode for a modified K-center heuristic.

Procedure PostBalance($X_1 \ldots, X_k$)
Input: Sets of clusters $\{X_1 \ldots, X_k\}$ s.t. $X_i \bigcap X_j = \emptyset \quad \forall 1 \leq i \neq j \leq k$
Output: Sets of clusters $\{X_1, \ldots, X_k\}$ s.t. $X_i \bigcap X_j = \emptyset \quad \forall 1 \leq i \neq j \leq k$
1. Calculate $S = \bigcup_{i=1,k} X_i$ 2. **do** 3. Sort clusters in increasing order of estimated load capacitance 4. **for** each cluster X_i in the sorted order 5. $n\_move = 0$ 6. Let $V = \{ v \mid v \in S - X_i \}$ 7. Sort nodes $v \in V$ in increasing order of $d(v, center(X_i))$ 8. **for** each node $v \in V$ in the sorted order Suppose $v \in X_j$, $1 \leq j \neq i \leq k$ 9. **if** $\sum_{i=1,k}(Cap(X_i))^5$ decreases by moving v to cluster X_i 10. Move v to cluster X_i (i.e., $X_i = X_i + \{v\}$, $X_j = X_j - \{v\}$) 11. $n\_move = n\_move + 1$ 12. **if** $n\_move > 3$ **Go To** 4 13. **while** there is any sink moved in current iteration.

Figure 12. Procedure PostBalance.

where c_v is the input capacitance of node v and $center(X_i)$ is the Manhattan center of the nodes in cluster X_i as defined in [25, 16].[6]

- The number w is used to trade off between balance among clusters and the total capacitive load of all clusters. A higher value of w favors balanced clustering, which usually leads to lower-cost routing at the next higher level but can cause large total capacitive load at the current level. On the other hand, $w = 1$ favors minimizing the total capacitive load at the current level without balancing the capacitive load among the clusters. Based on our experiments, we use $w = 5$ to obtain all the results reported below; this value seems to reasonably balance the goals of low routing cost at both the current and next higher levels[7].

4.2. *Buffer Sliding*

Chung and Cheng [20] shift the location of a buffer along the edge to its parent node to reduce or eliminate excessive detouring. The motivation for their technique is straightforward. In Fig. 13, subtree T_1 rooted at v_1 is driven by buffer b_1, and subtree T_2 rooted at v_2 is driven by buffer b_2. Let t_2 be the delay from parent node p to child node v_2, and let t_2' be the delay from parent node p to child node v_2 after buffer b_2 slides toward node p over a distance of x units. Let $l = d(l(p), l(v_2))$. We now have

$$\begin{aligned} t_2 &= rl(cl/2 + c_b) + t_b + r_b \cdot Cap(T_2) \\ t_2' &= r(l-x)(c(l-x)/2 + c_b) + t_b \\ &\quad + r_b(cx + Cap(T_1)) + rx(cx/2 + Cap(T_2)) \\ t_2' - t_2 &= rcx^2 + r_bcx + r(Cap(T_2) - cl - c_b)x \qquad (9) \end{aligned}$$

Notice that the coefficient of the last term in Eq. (9), $Cap(T_2) - cl - c_b$, is always positive in practice because (i) the total wirelength of T_2 is larger than that of the parent edge of T_2, and (ii) the sum of sink capacitances in T_2 is larger than the input capacitance of a buffer, so that $t_2' > t_2$. Also, as buffer b_2 is moved closer to its parent node p, delay t_2' will increasingly exceed t_2. In the case where t_1 is so much larger than t_2 that detour wiring is necessary, we can slide buffer b_2 so that delay balance is achieved at point p using less detour wiring (see Fig. 13(a)). Even when no detour wiring is necessary, the buffer sliding technique can still be used to reduce routing wirelength at the next higher level of the hierarchy. In Fig. 13(b), we reduce the wirelength by constructing a minimal Steiner tree over b_1 and b_2. Suppose the delay from p' to buffer b_1 is larger than that from p' to buffer b_2, we can slide buffer b_2 toward the left, thus increasing the delay from p' to b_2 such that p' can become the delay balance point.

There is a similar idea in [21], which reduces wirelength by inserting an extra buffer. However, adding a buffer will cause large extra delay and power dissipation. Indeed, when T_a and T_b have similar delays, excessive detour wirelength is inevitable when a buffer is added at the parent edge of just one subtree. Hence, the technique of [21] will be effective in reducing power dissipation and wirelength only when the delays of T_a and T_b are very different. ([21] also consider buffer insertion only for the zero-skew case.)

We now give a buffer sliding heuristic, called **H3** (see Fig. 14) that does not add any extra buffers and that can handle any skew bound (we find, however, that it is less effective for large skew bounds; see Section 4.3). H3 builds a low-cost tree T_{opt} over a set of of buffers $S = \{b_1, \ldots, b_k\}$ as follows. First, we construct a BST $\bar{T}$ under a new skew bound $\bar{B} \geq B$ without buffer sliding. Next, we calculate the delay d^i_{max} (d^i_{min}) which is the maximum (minimum) delay along any root-sink path in $\bar{T}$ that passes through buffer b_i (Line 7). We then calculate $d_{max} = \max_{i=1,k}\{d^i_{max}\}$ at Line 8. At Line 10, we slide each buffer b_i such that the min-delay at its input is increased by $\max\{0, d_{max} - d^i_{min} - B\}$ and $skew(\bar{T})$ is reduced toward B. Finally, we build a new tree T by re-embedding the topology of $\bar{T}$ according to the original skew bound B (Line 9); this will minimize

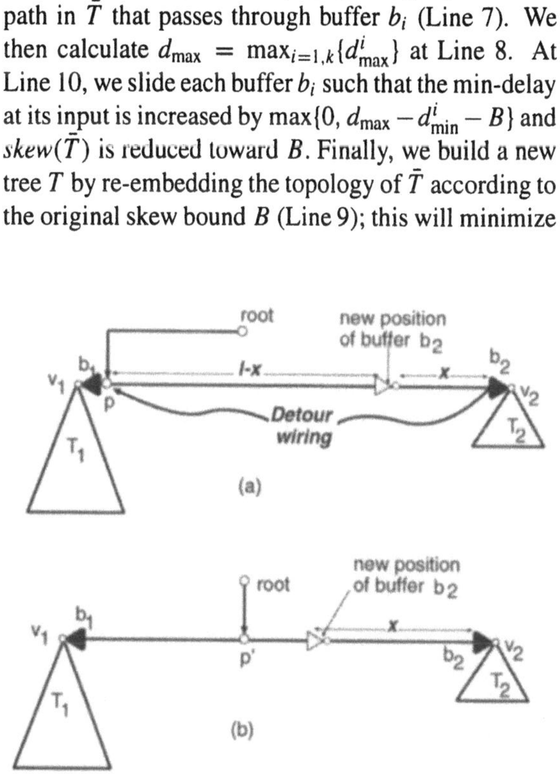

Figure 13. Two examples showing how the buffer sliding technique can eliminate (a) detour wiring or (b) routing wirelength at higher levels of the buffer hierarchy.

Procedure H3(S)

Input: Set of buffers $S = \{b_1, \cdots, b_k\}$; Skew Bound B; Set of subtrees T_i driven by buffer b_i with $skew(T_i) < B$;

Output: Tree T_{opt} with $skew(T_{opt}) \leq B$; Set of wirelength $L_i \geq 0$ inserted between buffer b_i and its subtree root $root(T_i)$.

$min\_cost = \infty$
Set new skew bound $\bar{B} = B$
do
  Build tree $\bar{T}$ over buffers in S with new skew bound $\bar{B}$ (no buffer sliding)
  for $i = 1$ to k **do**
  /* Let $max\_\bar{t}(b_i)$ ($max\_t(b_i)$) be max-delay from input of buffer b_i to sinks which are descendants of b_i before (after) buffer sliding */
    Calculate x = delay from $root(\bar{T})$ along the unique path in $\bar{T}$ to b_i
    Calculate $d^i_{max} = max\_\bar{t}(b_i) + x$, and $d^i_{min} = min\_t(b_i) + x$
  Calculate $d_{max} = \max\{d^i_{max}\}$
  for $i = 1$ to k **do**
    Calculate the length of wire L_i between b_i and $root(T_i)$ s.t. $min\_t(b_i) = min\_\bar{t}(b_i) + \max\{0, d_{max} - d^i_{min} - B\}$
  Build tree T by re-embedding topology of $\bar{T}$ under original skew bound B with wire of length L_i inserted between b_i and $root(T_i)$, $\forall i = 1, \cdots, k$
  if $cost(T) < min\_cost$
    $T_{opt} = T$
    $min\_cost = cost(T)$
  $\bar{B} = \bar{B} + 3ps$
while min_cost ever decreased in last 10 iterations

Figure 14. Procedure H3 (buffer sliding).

any potential increase in tree cost $cost(T) - cost(\bar{T})$. The above steps are iterated for different skew bounds $\bar{B} > B$, and the tree T with smallest total wirelength is chosen as T_{opt}. In general, when the new skew bound $\bar{B}$ is increasing, $cost(\bar{T})$ will be decreasing. However, the length of the wire inserted between each buffer and its subtree root will increase when the $\bar{B}$ becomes too large, and $cost(T)$ will stop decreasing after a certain number of iterations. In all of our experiments, the procedure stops within 50 iterations.

4.3. Experimental Results

For the sake of comparison, we have also implemented the following buffer sliding heuristics.

H0 No buffer sliding.

H1 Slide buffers to equalize $max\_t(b_i)$ for all $1 \leq i \leq k$, i.e., the max-delay from the input of each buffer b_i to sinks which are the descendants of b_i. This is the buffer sliding technique used in [19, 22].

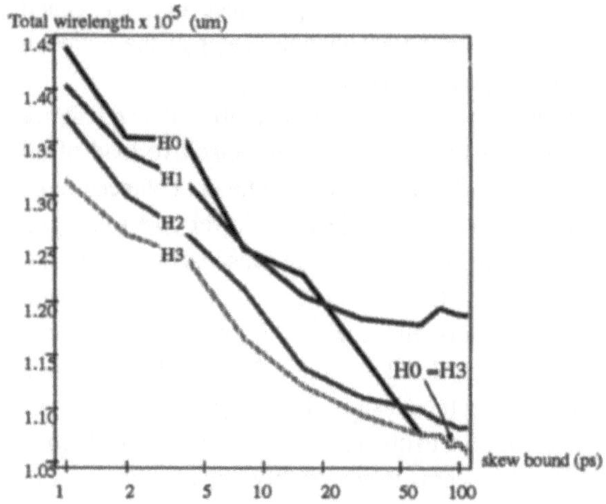

Figure 15. Total wirelength achieved by different buffer sliding heuristics on benchmark circuit r1 with a 32-1 buffer hierarchy. The wirelength unit is 100 μm. Buffer parameters are output resistance $r_b = 100\ \Omega$, input capacitance $c_b = 50$ fF, and internal delay $t_b = 100$ ps. Note that the X axis is on a logarithmic scale.

H2 Slide buffers to equalize $max\_t(b_i)$ and $max\_t(b_j)$ where b_i and b_j are the sibling buffers.

Figure 15 shows the total wirelength reduction achieved by the various buffer sliding heuristics on benchmark circuit r1 with a 32-1 buffer hierarchy. H3 is consistently better than other heuristics for the skew bound from 0 to 50 ps. When the skew bound B is larger than 50 ps, the tree cost reduction $cost(T) - cost(\bar{T})$ is very slight for any $\bar{B} > B$, and hence when we push $skew(\bar{T})$ back to B by buffer sliding, there is almost no gain in the total wirelength. Therefore, heuristic H3 will be the same as H0 when the skew bound is sufficiently large. A more detailed comparison of total wirelength reduction achieved by different buffer sliding heuristics is given in Table 3, which shows that H3 is consistently better than other heuristics for different skew bounds and buffer hierarchies. In the table, we also report ratios of tree costs, averaged over the five test cases, for each heuristic versus H3 (i.e., we normalize the tree costs against the H3 tree cost). For the zero skew regime, the heuristics H0, H1 and H2 respectively require 6.9%, 10.6% and 3.0% more wirelength on average than our heuristic H3. And for the 50 ps skew regime, the heuristics H0, H1 and H2 respectively require 3.1%, 17.0% and 1.1% more wirelength on average than our heuristic H3. Notice that heuristic H1, the method used in [19, 22], actually has the largest total wirelength in most cases.

Table 3. Detailed comparison of total wirelength achieved by different buffer sliding heuristics on benchmark circuits r1–r5 with two types of 2-level buffer hierarchy and one type of 3-level buffer hierarchy. The wirelength unit and buffer parameters are the same as those in Fig. 15.

	r1	r2	r3	r4	r5	r1	r2	r3	r4	r5
	Skew bound = 0					Skew bound = 10 ps				
	Buffer hierarchy: $2\sqrt{n}-1$									
$H0$	1,486	2,984	3,728	7,718	11,193 (1.059)	1,242	2,446	3,095	6,279	9,102 (1.061)
$H1$	1,483	3,207	3,855	8,829	11,567 (1.119)	1,232	2,850	3,175	7,710	9,785 (1.164)
$H2$	1,458	2,941	3,651	7,408	10,852 (1.032)	1,185	2,400	3,012	6,018	8,773 (1.025)
$H3$	1,404	2,802	3,553	7,261	10,589 (1.000)	1,172	2,314	2,921	5,907	8,549 (1.000)
	Buffer hierarchy: $\sqrt{n}-1$									
$H0$	1,497	2,923	3,733	7,476	11,185 (1.044)	1,231	2,516	3,053	6,128	9,128 (1.060)
$H1$	1,447	2,896	3,848	7,661	11,418 (1.051)	1,219	2,408	3,207	6,297	9,470 (1.073)
$H2$	1,450	2,825	3,646	7,319	10,878 (1.015)	1,170	2,419	2,982	5,972	8,820 (1.023)
$H3$	1,432	2,797	3,584	7,175	10,713 (1.000)	1,159	2,340	2,893	5,872	8,597 (1.000)
	Buffer hierarchy: $n^{2/3}-n^{1/3}-1$									
$H0$	1,626	3,259	4,017	7,971	11,989 (1.104)	1,306	2,693	3,375	6,713	9,816 (1.112)
$H1$	1,558	3,297	4,168	8,912	12,982 (1.149)	1,258	2,926	3,547	7,870	10,954 (1.198)
$H2$	1,556	2,989	3,808	7,594	11,368 (1.042)	1,234	2,470	3,136	6,379	9,204 (1.040)
$H3$	1,476	2,877	3,636	7,374	10,921 (1.000)	1,193	2,361	3,020	6,152	8,813 (1.000)
	Skew bound = 20 ps					Skew bound = 50 ps				
	Buffer hierarchy: $2\sqrt{n}-1$									
$H0$	1,185	2,375	2,950	6,021	8,736 (1.059)	1,074	2,168	2,780	5,630	8,245 (1.018)
$H1$	1,182	2,626	3,127	7,323	9,401 (1.155)	1,200	2,565	3,110	7,061	9,021 (1.174)
$H2$	1,147	2,245	2,893	5,816	8,397 (1.021)	1,109	2,170	2,745	5,530	7,918 (1.010)
$H3$	1,112	2,216	2,845	5,695	8,231 (1.000)	1,073	2,158	2,736	5,477	7,822 (1.000)
	Buffer hierarchy: $\sqrt{n}-1$									
$H0$	1,196	2,404	2,971	5,937	8,708 (1.058)	1,127	2,224	2,772	5,504	8,416 (1.064)
$H1$	1,153	2,370	3,116	6,116	9,248 (1.077)	1,127	2,169	2,920	5,706	8,855 (1.089)
$H2$	1,146	2,280	2,944	5,743	8,397 (1.022)	1,061	2,115	2,685	5,404	8,005 (1.020)
$H3$	1,135	2,228	2,839	5,617	8,271 (1.000)	1,053	2,080	2,607	5,261	7,854 (1.000)
	Buffer hierarchy: $n^{2/3}-n^{1/3}-1$									
$H0$	1,267	2,538	3,125	6,396	9,350 (1.089)	1,132	2,344	2,913	5,823	8,551 (1.011)
$H1$	1,256	2,780	3,494	7,532	10,806 (1.206)	1,262	2,891	3,439	7,735	11,182 (1.248)
$H2$	1,191	2,401	2,969	6,077	8,811 (1.030)	1,145	2,301	2,856	5,786	8,475 (1.003)
$H3$	1,158	2,339	2,893	5,864	8,536 (1.000)	1,112	2,312	2,937	5,684	8,327 (1.000)

5. Conclusions

In this work, we have extended the bounded-skew routing methodology to encompass several very practical clock routing issues: non-uniform layer parasitics, non-zero via resistance and/or capacitance, existing obstacles in the metal routing layers, and hierarchical buffered tree synthesis. For the case of varying layer parasitics, we prove that if we prescribe the routing pattern between any two points, merging regions are still bounded by well-behaved segments except that no boundary segments can be Manhattan arcs of non-zero length. Our experimental results show that taking into account non-uniform layer parasitics can be accomplished without significant penalty in the clock tree cost. Our solution to obstacle-avoidance routing is based on the concept of a *planar merging region* which contains all the feasible merging points p such that the shortest planar path between child merging regions via p is equal to the shortest planar path between child merging regions taking into consideration the given obstacles. Again, our experimental results are quite promising: even for the relatively dense obstacle layout studied, obstacle-avoidance clock routing seems achievable without undue penalty in clock tree cost. Finally, we extend the bounded-skew routing approach to address buffered clock trees, assuming (as is the case in present design methodologies) that the buffer hierarchy (i.e., the number of buffers at each level and the number of levels) is given. A bounded-skew buffered clock tree is constructed by performing three steps for each level of the buffer hierarchy, in bottom-up order: (i) cluster sinks or roots of subtrees for each buffer; (ii) build a bounded-skew tree using the ExG-DME algorithm under Elmore delay [5] for each cluster; and (iii) reduce the total wirelength by the H3 buffer sliding heuristic. Our experimental results show that H3 achieves very substantial wirelength improvements over the method used by [19, 22], for a range of buffer hierarchy types and skew bounds.

Notes

1. One minor caveat is that the "merging region" of [3–5] is not a complete generalization of the DME merging segment: when detour wiring occurs or when sibling merging regions overlap, the merging region may not contain all the minimum-cost merging points.
2. We assume that there are only two routing layers. Our approach can easily extends to multiple routing layers.
3. However, when detouring occurs, both the H-layer and V-layer will be used for the detour wiring. It is easy to calculate the extra wirelength on both layers if we prescribe the routing pattern for detour wiring.
4. Strictly speaking, there can be joining segments with slopes other than ± 1, 0, and ∞ although they are not encountered in practice. For the case of joining segments with slopes m with $|m| > 1$ ($|m| < 1$), we expand obstacles as in Case III (IV).
5. The simplest approach is to divide $SDR(L_a, L_b)$ into a set of disjoint rectangles R_i that contains no obstacles, as shown in Fig. 10(a). Let $c \in R_i$ and $d \in R_i$ be the corner points closest to joining segments L_a and L_b. If prescribed routing patterns are assumed for the shortest paths from c to L_a and from d to L_b, delays at c and d are well-defined. Since there are no obstacles inside R_i, the planar merging region can be constructed from points c and d for non-uniform layer parasitics using the methods of Section 2.
6. More accurate models for estimating the load capacitance of a cluster are of course possible, but have surprisingly little effect. Indeed, we implemented the best possible model (which is to actually execute the BST construction whenever a BST estimate is required) but this did not result in noticeable performance improvement.
7. We also investigated less greedy iterative methods that have the same general structure as the classic KL-FM partitioning heuristics. For example, an analog of the KL-FM pass might always expand the cluster with smallest estimated load capacitance by shifting the closest "unlocked" node in another cluster; as in KL-FM, a node that is moved becomes locked for the remainder of the pass to prevent cycling. In our experience, such more complicated heuristics do not achieve noticeably different results from the simple method we describe.

References

1. A.B. Kahng and G. Robins, *On Optimal Interconnections for VLSI*, Kluwer Academic Publishers, 1995.
2. E.G. Friedman (Ed.), *Clock Distribution Networks in VLSI Circuits and Systems: A Selected Reprint Volume*, IEEE Press, 1995.
3. J.H. Huang, A.B. Kahng, and C.-W.A. Tsao, "On the bounded-skew clock and steiner routing problems," in *Proc. ACM/IEEE Design Automation Conf.*, pp. 508–513, 1995. Also available as Technical Report CSD-940026x, Computer Science Dept., UCLA.
4. J. Cong and C.-K. Koh, "Minimum-cost bounded-skew clock routing," in *Proc. IEEE Intl. Symp. Circuits and Systems*, Vol. 1, pp. 215–218, April 1995.
5. J. Cong, A.B. Kahng, C.-K. Koh, and C.-W.A. Tsao, "Bounded-skew clock and Steiner routing under Elmore delay," in *Proc. IEEE Intl. Conf. Computer-Aided Design*, pp. 66–71, Nov. 1995.
6. K.D. Boese and A.B. Kahng, "Zero-skew clock routing trees with minimum wirelength," in *Proc. IEEE Intl. Conf. on ASIC*, pp. 1.1.1–1.1.5, 1992.
7. T.-H. Chao, Y.C. Hsu, J.M. Ho, K.D. Boese, and A.B. Kahng, "Zero skew clock routing with minimum wirelength," *IEEE Trans. Circuits and Systems*, Vol. 39, No. 11, pp. 799–814, Nov. 1992.
8. T.-H. Chao, Y.-C. Hsu, and J.-M. Ho, "Zero skew clock net routing," in *Proc. ACM/IEEE Design Automation Conf.*, pp. 518–523, 1992.
9. M. Edahiro, "Minimum skew and minimum path length routing in VLSI layout design," *NEC Research and Development*, Vol. 32, No. 4, pp. 569–575, 1991.
10. J. Cong, A.B. Kahng, C.-K. Koh, and C.-W.A. Tsao, "Bounded-skew clock and Steiner routing under Elmore delay," Technical Report CSD950030, Computer Science Dept., University of California, Los Angeles, Aug. 1995. Available by anonymous ftp to ftp.cs.ucla.edu, also available at http://vlsicad.cs.ucla.edu/~tsao.
11. M. Edahiro, "A clustering-based optimization algorithm in zero-skew routings," in *Proc. ACM/IEEE Design Automation Conf.*, pp. 612–616, June 1993.
12. M. Borah, R.M. Owens, and M.J. Irwin, "An edge-based heuristic for rectilinear Steiner trees," *IEEE Trans. Computer-Aided Design*, Vol. 13, No. 12, pp. 1563–1568, Dec. 1994.
13. A.B. Kahng and C.-W.A. Tsao, "Low-cost single-layer clock trees with exact zero Elmore delay skew," in *Proc. IEEE Intl. Conf. Computer-Aided Design*, 1994.
14. T. Asano, L. Guibas, J. Hershberger, and H. Imai, "Visibility-polygon search and euclidean shortest paths," in *Proc.*

IEEE Symp. Foundations of Computer Science, pp. 155–164, 1995.
15. E. Welzl, "Constructing the visibility graph for n line segments in $o(n^2)$ time," *Information Processing Letters*, Vol. 20, pp. 167–171, 1985.
16. A.B. Kahng and C.-W.A. Tsao, "Planar-dme: A single-layer zero-skew clock tree router," *IEEE Trans. Computer-Aided Design*, Vol. 15, No. 1, Jan. 1996.
17. C.-W.A. Tsao, "VLSI Clock Net Routing," Ph.D. thesis, University of California, Los Angeles, Oct. 1996.
18. J.G. Xi and W.W.-M. Dai, "Buffer insertion and sizing under process variations for low power clock distribution," in *Proc. ACM/IEEE Design Automation Conf.*, pp. 491–496, 1995.
19. S. Pullela, N. Menezes, J. Omar, and L.T. Pillage, "Skew and delay optimization for reliable buffered clock trees," in *Proc. IEEE Intl. Conf. Computer-Aided Design*, pp. 556–562, 1993.
20. J. Chung and C.-K. Cheng, "Skew sensitivity minimization of buffered clock tree," in *Proc. IEEE Intl. Conf. Computer-Aided Design*, pp. 280–283, 1994.
21. A. Vittal and M. Marek-Sadowska, "Power optimal buffered clock tree design," in *Proc. ACM/IEEE Design Automation Conf.*, San Francisco, June 1995.
22. Y.P. Chen and D.F. Wong, "An algorithm for zero-skew clock tree routing with buffer insertion," in *Proc. European Design and Test Conf.*, pp. 652–657, 1996.
23. C.J. Alpert and A.B. Kahng, "Geometric embeddings for faster (and better) multi-way partitioning," in *Proc. ACM/IEEE Design Automation Conf.*, pp. 743–748, 1993.
24. T.F. Gonzalez, "Clustering to minimize the maximum intercluster distance," *Theoretical Computer Science*, Vol. 38, Nos. 2–3, pp. 293–306, June 1985.
25. A.B. Kahng and C.-W. Albert Tsao, "Planar-dme: Improved planar zero-skew clock routing with minimum pathlength delay," in *Proc. European Design Automation Conf. with with EURO-VHDL*, Grenoble, France, pp. 440–445, Sept. 1994. Also available as Technical Report CSD-940006, Computer Science Dept., UCLA.

Andrew B. Kahng received the A.B. degree in applied mathematics and physics from Harvard College, and the M.S. and Ph.D. degrees in computer science from the University of California at San Diego. He joined the computer science department at UCLA in 1989, and has been an associate professor there since 1994. His honors include NSF Research Initiation and Young Investigator awards. He is General Chair of the 1997 ACM International Symposium on Physical Design, and a member of the working group that is defining the Design Tools and Test portion of the 1997 SIA National Technology Roadmap for Semiconductors. Dr. Kahng's research interests include VLSI physical layout design and performance verification, combinatorial and graph algorithms, and the theory of iterative global optimization. Currently, he is Visiting Scientist (on sabbatical leave from UCLA) at Cadence Design Systems, Inc.
abk@cs.ucla.edu

Chung-Wen Albert Tsao received the B.S. degree from National Taiwan University in 1984, and the M.S. degree from National Sun Yat-Sen University in 1988, both in electrical engineering. With assistance from a Fellowship from the Ministry of Education, Taiwan, he received the M.S. degree and Ph.D. in Computer Science from UCLA in 1993 and 1996, majoring in Theory with minors in Architecture/VLSI CAD and Network Modeling/Analysis. He is currently working at Cadence Design Systems, Inc., San Jose, California. His Ph.D. research focused on VLSI clock net routing. His current research interests include VLSI routing, partitioning and placement, computational geometry, and delay modeling.
tsao@cadence.com

in computer science from the University of California at San Diego. He joined the computer science department at UCLA in 1989 and [illegible] [illegible]

[illegible] Foundations of Computer Science, pp. 155–164, [illegible]

[illegible] Information Processing Letters, Vol. 20 [illegible]

[illegible]

Journal of VLSI Signal Processing 16, 217–224 (1997)

A Clock Methodology for High-Performance Microprocessors

KEITH M. CARRIG, ALBERT M. CHU, FRANK D. FERRAIOLO AND JOHN G. PETROVICK
IBM Microelectronics Division, Essex Junction, Vermont

P. ANDREW SCOTT
Cadence Design Systems, San Jose, California

RICHARD J. WEISS
First PASS, NW Palm Bay, Florida

Received November 15, 1996; Revised December 15, 1996

Abstract. This paper discusses an effective clock methodology for the design of high-performance microprocessors. Key attributes include the clustering and balancing of clock loads, multiple clock domains, a balanced clock router with variable width wires to minimize skew, hierarchical clock wiring, automated verification, an interface to the Cadence Design Framework II^{TM} environment, and a complete network model of the clock distribution, including loads. This clock methodology enabled creation of the entire clock network, including verification, in less than three days with approximately 180 ps of skew.

Introduction

As the performance of microprocessors increase, it is essential to proportionally improve the clock design and distribution because clock skew subtracts from the machine's cycle time. Increased chip size and latch count, and the constant need to reduce development time further aggravates the clock distribution. Hence, the overall clock performance of today's microprocessors must improve while reducing development time and allowing for increased complexity. As a result, the design and analysis of clock distribution networks has understandably received considerable attention in the literature [1].

The hierarchical clock design methodology described here is intended to improve clock design accuracy while reducing development time, particularly in the late stages of the design cycle where time is crucial. It was initially conceived for high-performance microprocessor design but is applicable to ASIC designs, especially for chips with cores and/or custom circuits. Hierarchical clock design, for example, allows the final clocks to be designed, verified and analyzed in large custom circuits or floorplan blocks well before integration of the entire chip.

This methodology was applied to a single-chip PowerPC microprocessor fabricated in a 0.35-micron ($L_{\text{eff}} = 0.25$ microns), 2.5V CMOS technology with six levels of metal, five for global wiring and one for local interconnect. The resultant die size is 10.4 mm × 14.4 mm and contains 6.5 million transistors (Fig. 1). The chip contains 48 KB of cache and 121 custom macrocells, as well as 67,000 standard cell circuits. Part of the challenge was to design a clock distribution network that serves a large number of macrocells (51) along with approximately 32,000 master/slave latch-pairs, 25,300 of which are in the custom macrocells with the remainder in the random logic at the global chip level.

Clock Generation and Distribution

Figure 2 depicts the clock generation and distribution logic for one of four clock trees used on the chip. A phase-locked-loop (PLL) is used to synthesize the

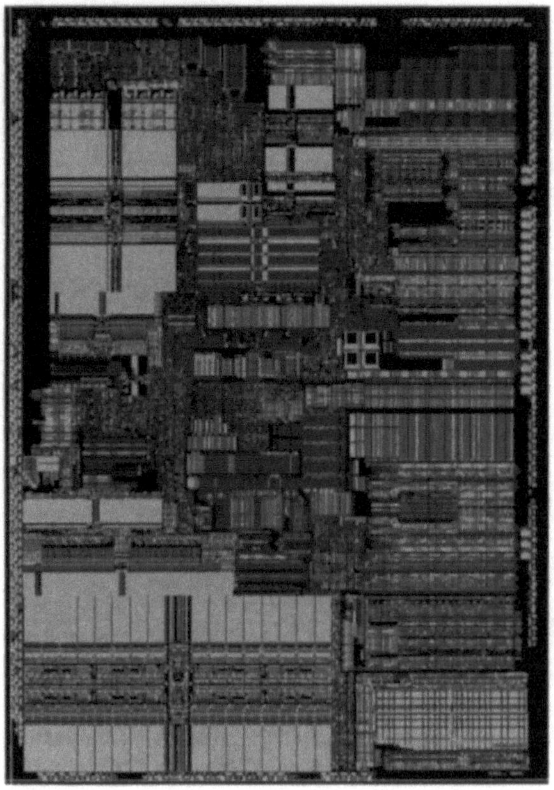

Figure 1. Chip layout (metal layers not shown).

internal processor clock from a reference clock input. The PLL multiplication factor is programmable to allow the processor to operate over a wide range of internal cycle times. The PLL also provides latency correction for the delay through the clock distribution. The PLL phase aligns the "bus" or I/O clock with the external reference clock input. The bus clock is used to latch input data and drive data off the chip. The "system" clock drives the majority of the latches on the chip.

The system clock distribution network contains three stages, a large global clock buffer (GCB), approximately 20 regional clock buffers (RCBs) and several hundred local clock buffers (LCBs). Control signals on the LCB inputs are used for clock gating and test clock generation. A single phase of the clock is distributed to the inputs of all LCBs. The LCBs generate two pairs of complementary clocks, one pair to the master latches and the other to the slave latches. Most latches have pass-gate inputs to improve performance and reduce loading on the clock. The complimentary clock inputs to the pass-gate are generated in the local clock buffer to reduce overall clock power. The LCBs typically drive 0.7 pF loads and have 100 ps/pF of delay sensitivity.

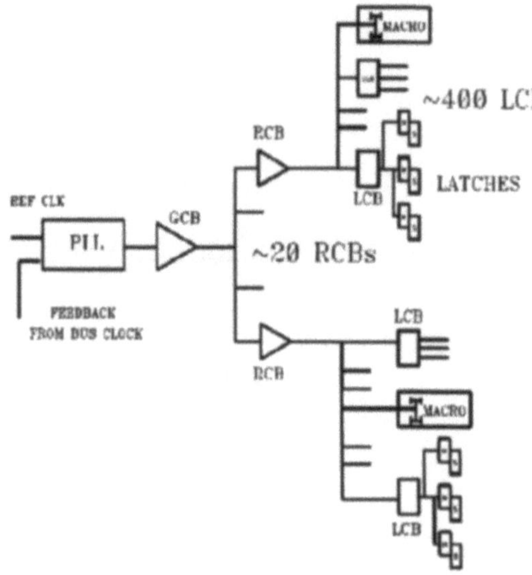

Figure 2. Clock generation and distribution logic.

In the initial netlist, the LCBs have non-overlapping mid-cycle clocks to avoid flushing data from the master to the slave latches. Selectable end-of-cycle clock overlap allows for a minimum of buffering for fast data paths while providing a variable launch time for the start of the clock cycle. The selection of LCBs allows for last minute timing correction or engineering-change capability with minimal mask changes.

Clock Methodology

Clock design is a major part of the overall chip integration methodology and often a critical path to fast design turnaround. The hierarchical clock design methodology, shown in Fig. 3, streamlines this process while minimizing clock skew.

The portion of clock trees within the custom macrocells are designed and routed as part of the macrocell layouts. As Fig. 4 illustrates, the clock interface to a macrocell layout consists of only one input pin. This simplifies global (top-level) clock routing and helps identify inaccessible clock pins. Variable width clock routing of macrocells occurs concurrently with floorplanning so that chip-level information is used to guide the optimization of the macrocell clock tree.

After floorplanning, placement and power routing is complete, clock tree synthesis and optimization is performed at the global chip level to generate and place clock buffers. These buffers are then snapped to legal placement locations and any cell overlaps resolved.

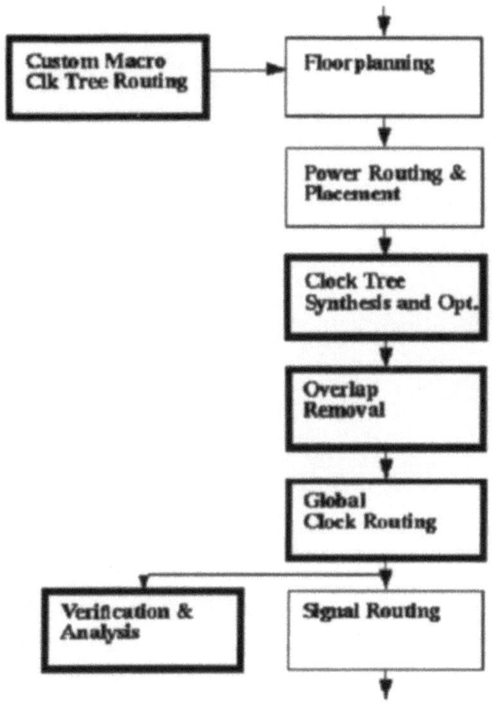

Figure 3. Clock design methodology.

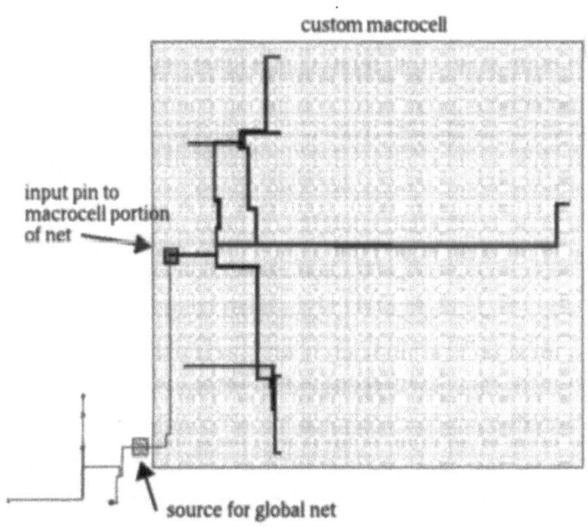

Figure 4. Example of global and macrocell (shaded region) clock wiring.

Variable width clock routing is then performed on all but the lowest (LCB to latch) levels in the clock tree; the LCB-to-latch levels are routed with minimum width wires using Cadence Cell3^{TM}.

Once the clock design is complete, signal routing takes place. At the same time, the clock nets are extracted, verified and analyzed to ensure that the skew objective is met. To ensure a functional design, early mode (fast path) analysis using the extracted SPICE netlist is performed and any problems are fixed either by replacing the LCB with one that has a late launch clock or by adding delays to the fast paths.

One of the major advantages of using this methodology over an H-tree or a grid/mesh scheme is that it minimizes routing congestion and power dissipation [2].

A. *Clock Tree Synthesis and Optimization*

An IBM clock optimization tool, CO2, was used to build the four clock trees on the chip [3]. The tool traverses the clock trees, identifying and reconfiguring equivalent nets as well as adding parallel copies of buffers to minimize clock skew. Each initial clock tree consists of the correct number of buffering levels, a global level driven by one GCB, a regional level driven by one RCB and a local level driven by one LCB. Only one buffer per level is typically required for the initial clock tree since CO2 will make copies of buffers as needed. However, multiple LCBs are initially specified when clock gating is done at the "local" level.

Clock nets are ignored during placement optimization because of their high fanouts. This ensures that the placement solution is optimized for timing and routing congestion. Once placement is complete, clock optimization is done using the initial latch placement. CO2 performs the following functions on each clock tree:

- **Clock Trace**—Traces the clock trees from root node to leaf nodes. This tracing step identifies the structure of the clock tree and defines equivalent nets whose sinks can be interchanged during optimization.
- **Initial Optimization**—Optimization begins from the leaf nodes of the clock tree. Clock sinks are clustered locally until the maximum capacitive load target is met. A clock buffer is duplicated to drive the newly formed cluster and placed at the geometric center of the cluster.
- **Refinement**—Detail optimization is performed by interchanging clock sinks of the duplicated nets to minimize worst-case RC and capacitance variation.

Buffer locations are legalized and optimized for routing; this is needed because CO2 has no knowledge of the power bus to correctly place the buffers during clock optimization. LCBs are placed at the RC centroid of the latch cluster [4]. The placement of clock buffers by

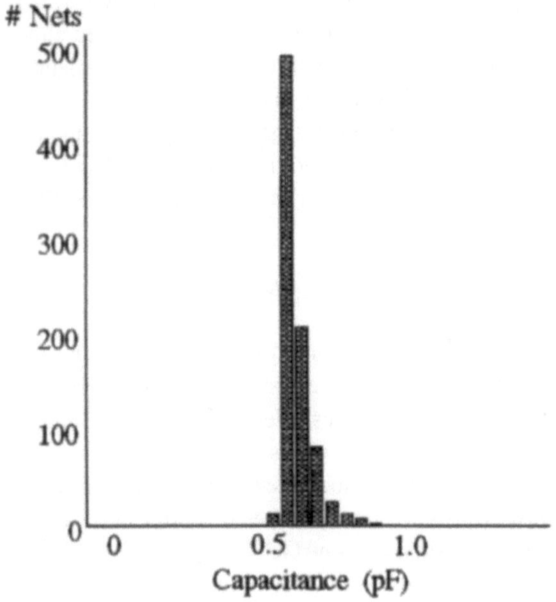

Figure 5. LCB net capacitance distribution.

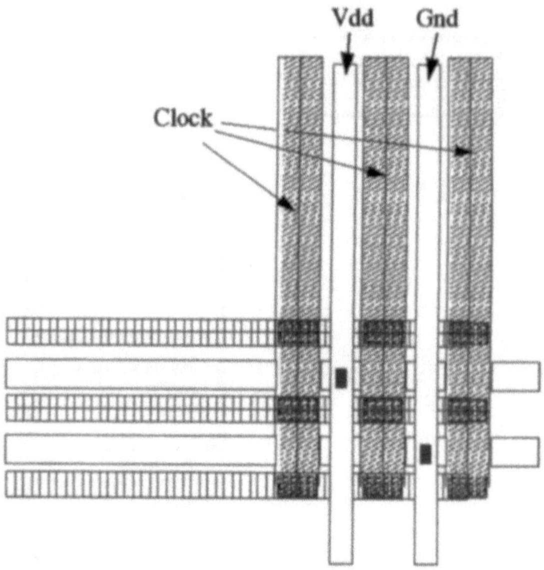

Figure 6. Example of how CLOCKTREE gains width by straddling a periodic supply buss.

CO2 causes overlaps with other standard cells, which are resolved using Cell3. The clock buffers are considered immovable while the overlapping standard cells are allowed to move to resolve the overlaps. After all overlaps are eliminated, global routing is performed on the LCB to latch nets, with accurate capacitance and RC reports generated. The capacitance report is used to add capacitive cells to the LCB nets to balance capacitive loads. Cells are placed near the LCBs to minimize RC effects. Figure 5 shows the capacitance distribution of the LCB-to-latch nets after clock optimization.

B. Global Clock Routing

A key component of the global and macrocell (see Section C) clock routing in the clock methodology is IBM CLOCKTREE, a two-layer balanced router that can vary the width of each wire segment. CLOCKTREE possesses several important features that were exploited for this chip:

- Delays at lower levels of the hierarchy are compensated for elsewhere in the clock network, a feature that enabled this hierarchical clock methodology. For example, at the global level it compensates for delays in the custom macrocell circuits, resulting in low skew throughout the entire clock network.
- Clock nets can be balanced to a target delay value which is necessary for matching net-to-net delays (e.g., across different clock domains) and for generating early or late clocks—late clocks were used for some cache circuits on the chip.
- The tool simulates the clock network as it routes, leading to the very accurate prediction of final results.
- When widening wires to meet skew and delay targets, periodic power supply lines are accounted for by running parallel connected segments between the power supply lines to gain more width (see Fig. 6).
- To reduce coupling with adjacent wires, a version of the clock wires with expanded widths is also created and used as blockage during signal routing. These wires are later replaced by those with the correct widths.

CLOCKTREE can also model a clock network with inductance; inductance effects can be significant for very wide clock wires and is becoming increasingly important as chip geometries shrink [1].

Clock routing at the global chip level is a multistep process controlled by a custom CAD tool that interfaces CLOCKTREE to Cadence Design Framework II (DFII). The first step is to create a pin and blockage map for each clock region; a clock region is a rectangular area defined by the pins that a clock net is designed to connect together. Forty-two clock regions were used for this chip, some of which overlap. To prevent the CLOCKTREE from routing over clock pins for other regions and small signal pins, blockage shapes are generated for these pins.

Because some regions overlap, the clock regions were organized into non-overlapping sets. The regions in a set are routed concurrently using CLOCKTREE; the routed wires are then imported as blockage for subsequent sets. Metal layers four (M4) and five (M5) are used for all global clock wiring. As noted earlier, loading and delay data for macrocells are passed to CLOCKTREE so that it can compensate for them at this level. Once all clock regions are routed and both skew and delay targets met, the wires are imported into Cadence DFII and a series of verification and analysis steps (described in Section D) are performed. The clock wires are then exported from DFII to Cell3 in preparation for signal routing.

C. *Clock Distribution in Custom Macrocells*

Custom macrocells are designed with one or more LCBs. There are two major techniques for reducing skew within a macrocell: tuning the LCB-to-latch circuits and balancing the wire from the clock input pin to the LCBs.

LCBs are tuned by an automated process to eliminate mid-cycle clock skew (skew from master to slave latches) caused by on-chip process variations and to match the delay of all LCB-to-latch circuits [5].

After tuning and verification, a macrocell is routed from the clock input pin to the LCBs using CLOCKTREE. The location of the clock input pin can be biased towards the top, bottom, left, right or center of the macrocell based on where the macrocell is placed in the floorplan relative to the RCB that drives it. The clock pin is placed on M4 to facilitate routing to it at the global level. To ensure that the pin is not blocked, the power router cuts windows in the M4 power routes if the clock pin is floorplanned so that it is under a wide M5 power buss.

As with the global wiring, macrocell clock routing is a multi-step process controlled by a custom CLOCKTREE to the Cadence DFII interface. The steps are similar to the global case, the primary differences being that: (a) the clock input pin location is automatically selected based on a bias directive from the designer and criteria to determine a suitable unblocked area, (b) there is only one clock region per macrocell so routing over clock pins from other regions is not an issue, and (c) the wiring is done on metal layers three and four to reduce the impact on global clock routing.

D. *Verification and Analysis*

Once the clock network has been routed at either the global or custom macrocell level, comprehensive verification and analysis steps are performed:

- **Verification**—Rigorous processes were implemented for ensuring that cells and chips pass both logical and physical verification steps. At the cell level, these include DRC, LVS and "methodology" checks (e.g., pins are on grid). At the chip level, DRC, LVS and Boolean equivalence checks (formal verification) of the netlists are done.
- **Quick SPICE and AS/X Netlist Extraction**—SPICE and AS/X (an IBM circuit simulator) netlists are extracted and used for full-chip static timing analysis and circuit-level simulation, respectively, to verify the quality of the clock network. The netlist extraction is based on an algorithm that traces the network from the source to all its sinks, assuming fully populated wiring tracks above, below, and adjacent to each net, and generates a π model for each wire segment.
- **AS/X Simulation**—Once AS/X netlists of the clock regions are generated, simulations are carried out to validate the delay and skew values reported by CLOCKTREE. A custom CAD tool automates this process of building an AS/X input deck for each region, submitting them for simulation, reading the simulation output files and compiling a detailed delay/skew report.

Results

Figure 7 shows the entire clock wire network at the global level. The capacitance for the entire network including macrocells is 558 pF. This compares favorably with the mesh scheme values of 1400 pF and 2000 pF reported in [6]. Internal experiments also demonstrate that this approach produces more than 40% improvement over H-trees. All 42 clock regions were routed with less than 50 ps skew from the latency targets. A 3-D plot of skew across the chip is shown in Fig. 8.

Fifty-one custom macrocells were routed with maximum skew of 29.6 ps (see Table 1). This was accomplished even though macrocell areas ranged from 0.01 to 31.75 mm^2 and their number of clock pins ranged from 1 to 112. Variations in the effective capacitance (0.04 to 6.62 pF) and delay values (0 to 93.7 ps) are acceptable because these values are passed

Figure 7. Global clock wires.

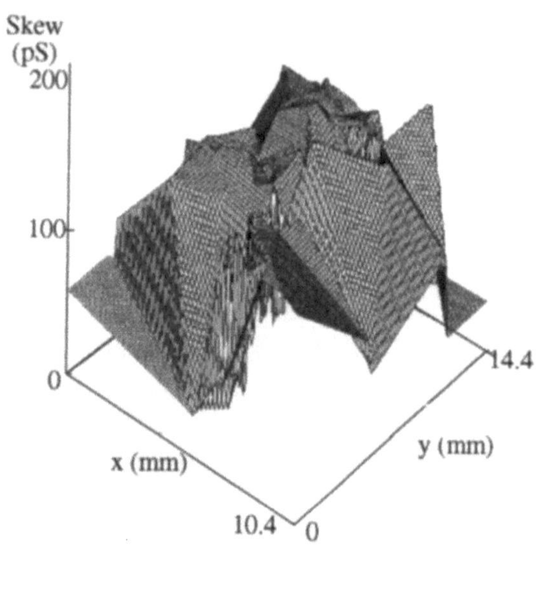

Figure 8. Skew versus chip coordinates.

Table 1. Skew results for each custom macrocell.

Macrocell name	# of pins	Area (mm)	Ceff (pf)	RC delay (ns)	Skew (ns)
b_data	21	3.35	2.71	32.10	20.00
bi_rq	9	2.31	2.79	29.20	10.60
bsbac	6	1.38	1.72	6.10	9.40
cbidp	2	0.39	1.14	6.10	1.50
cdc	1	31.75	1.13	0.00	0.00
cdtag	11	7.89	3.60	23.60	4.50
cic	29	12.02	1.50	49.70	8.90
citag	11	4.13	1.87	21.40	20.40
clk_switch_etal	1	0.09	0.08	0.00	0.00
cmpshft_unit	2	4.09	1.07	8.40	0.50
cpsel	6	1.28	1.58	12.30	11.10
csform	1	1.15	0.73	0.00	0.00
csprmux	1	0.12	0.10	0.00	0.00
ctwdp	10	1.59	1.96	16.40	4.00
dbdata_cia	6	0.98	2.92	43.60	12.80
dbdata_cntl	12	1.70	2.56	15.80	15.50
dbdata_imm	1	0.40	0.50	0.00	0.00
dcbcntl_bs1	5	0.26	0.58	1.80	0.80
dcbcntl_bs2	11	0.50	1.08	3.80	2.40
dcbdata	54	3.15	1.98	29.50	3.60
dcbdatatop	1	0.35	0.42	0.00	0.00
dcbreg	87	2.64	4.42	56.10	13.50
dcrom	6	0.98	0.32	1.30	0.60
dlsc	23	2.17	3.69	15.40	9.20
dlsc_ctl	1	0.43	0.22	0.00	0.00
dopsd	1	0.78	0.15	0.00	0.00
drgfile	112	9.53	6.62	93.70	22.60
drnc_flags	13	0.46	0.57	1.20	0.70
drnr	25	3.20	2.31	13.20	15.00
drscd	47	0.49	0.55	2.90	1.30
drsd	12	4.45	6.13	21.50	3.40
estk	12	0.65	0.35	2.00	0.10
fpa	3	8.56	5.58	31.50	13.80
fpmdiv	26	10.08	4.19	78.40	29.60
fxdlsu_all	16	7.85	4.66	32.30	6.20
fxdlsu_all	6	7.85	4.89	50.00	5.70
istk	1	0.44	0.06	0.00	0.00
lsfu	5	1.49	1.13	7.10	2.20
xa_bvbuf	3	0.32	0.48	1.60	0.50
xastk	6	1.41	1.24	9.30	6.70
xd_fclpla	4	0.52	0.61	1.90	0.20

(Continued on next page)

Table 1. (*Continued.*)

Macrocell name	# of pins	Area (mm)	Ceff (pf)	RC Delay (ns)	Skew (ns)
xd_fulipla	4	0.78	0.61	1.60	0.50
xd_imodpla	4	0.10	0.33	1.33	0.20
xd_nomodpla	4	0.20	0.49	1.30	0.40
xe_kidmux	1	0.01	0.04	0.00	0.00
xe_misr	2	0.66	0.55	1.00	0.50
xe_shaddr	2	0.08	0.28	1.00	0.00
seustk	1	0.17	0.07	0.00	0.00
xs_cam	1	0.43	0.31	0.00	0.00
xsnpd	1	0.24	0.19	0.00	0.00
sxtk	36	4.57	3.29	64.70	5.20
Total	666	150.44	86.35		
Average	13	3.13	1.80	16.47	5.50
Maximum	112	31.75	6.62	93.70	20.00
Minimum	1	0.01	0.04	0.00	0.00

to CLOCKTREE and compensated for at the global chip level. Total skew for the chip was less than 180 ps. Experience has shown that the skew can be improved by as much as an order of magnitude by doing additional routing iterations. However, this amount of skew was considered acceptable for meeting the team's turnaround time objective of one day for global clock routing.

Summary

A comprehensive clock methodology is presented that is well suited for microprocessor design or any large integrated function that contains smaller sub-blocks or cores. It offers excellent overall clock performance, minimizes design time without consuming large amounts of wiring tracks or adding needless wiring capacitance. A balanced router ensures quick turnaround time and low skew at both the macrocell and chip levels. Hierarchical clock wiring accommodates large variations in clock loading and different phase arrivals of the clock.

Acknowledgments

The authors wish to recognize Dave Hathaway for his work in the development of CO2, and Phil Restle and Peter Cook for their work in the development of CLOCKTREE. Finally, the authors wish to acknowledge their many colleagues whose diligent work made the entire program possible.

References

1. E.G. Friedman (Ed.), *Clock Distribution Networks in VLSI Circuits and Systems*, IEEE Press, Piscataway, NJ, 1995.
2. D.W. Dobberpuhl et al., "A 200-MHz 64-b dual-issue CMOS microprocessor," *IEEE J. Solid-State Circuits*, Vol. 27, No. 11, pp. 1555–1567, Nov. 1992.
3. D.J. Hathaway et al., "Circuit placement, chip optimization, and wire routing for IBM IC technology," *IBM J. Research and Development*, Vol. 40, No. 4, pp. 453–460, July 1996.
4. K.M. Carrig et al., "A methodology and apparatus for making a skew-controlled signal distribution network," U.S. patent no. 5,339,253, filed June 14, 1991, issued Aug. 16, 1994.
5. M. Shoji, "Elimination of process-dependent clock skew in CMOS VLSI," *IEEE J. Solid-State Circuits*, Vol. SC-21, No. 5, pp. 875–880, Oct. 1986.
6. M.P. Desai, R. Cvijetic, and J. Jensen, "Sizing of clock distribution networks for high performance CPU chips," in *Proc. Design Automation Conf.*, pp. 389–394, June 1996.

Keith Carrig is an Advisory engineer and scientist at IBM Microelectronics Division in Essex Junction, Vermont. He has been with IBM for 18 years. He is currently assigned to ASICS Architecture and Methodology specializing in clock distributions. He holds a BSEE degree from Rochester Institute of Technology, Rochester NY, and holds an MSEE degree from the University of Vermont in Burlington, VT. He has also completed the IBM System Research Institute program.

Albert Chu received the B.S.E.E. from North-eastern University, Boston, MA, and the M.S.E.E. from University of Vermont, Burlington, VT in 1985 and 1990 respectively. He joined IBM Microelectronics in 1985 where he worked in ASIC product development until 1995. He then worked in PowerPC microprocessor development where he was involved in the chip integration of the process design. He is currently engaged in the design of clocking system for memory products.

Frank Ferraiolo is a Senior Engineer at IBM Microelectronics Division in Essex Junction, Vermont. He joined IBM in 1982 working in fiber optic communications in Poughkeepsie, N.Y. He currently works in microprocessor development focused primarily on clocking and high speed data communication.

John Petrovick received the B.S.E.E. degree from Arizona State University, Tempe, AZ, and the M.S.E.C.E degree from the University of Massachusetts, Amherst, MA, in 1983 and 1985, respectively. He joined IBM Microelectronics in 1985 where he worked in ASIC product development until 1991. He then worked in X86 and PowerPC microprocessor development where he managed chip integration and design methodology. He is currently manager of 64Mb Synchronous DRAM development.

P. Andrew Scott received his B.Sc. (Eng.) and Ph.D. degrees in Electrical Engineering from Queen's University, Kingston, Ontario, in 1983 and 1993, respectively. In 1988 he joined the Canadian Microelectronics Corporation where he worked on development and implementation of IC and system-level design methodologies. Since 1995 he has been working for Cadence Design Systems in a technical consulting role.

Rick Weiss is president of First Pass Inc., providing personalized EDA solutions to companies nation wide. He also works part time for Florida Institute of Technology where he is a technical liaison between FIT and DOD handling EDA related contracts. Rick received his BS in computer science from SUNY at Stony Brook in 1991 and his MS in computer science from Florida Institute of Technology. His research focused primarily on parallel programming.

Journal of VLSI Signal Processing 16, 225–246 (1997)

Optical Clock Distribution in Electronic Systems

STUART K. TEWKSBURY AND LAWRENCE A. HORNAK
Department of Electrical and Computer Engineering, West Virginia University, Morgantown, WV 26506

Received July 15, 1996; Revised November 5, 1996

Abstract. Techniques for distribution of optical signals, both free space and guided, within electronic systems has been extensively investigated over more than a decade. Particularly at the lower levels of packaging (intra-chip and chip-to-chip), miniaturized optical elements including diffractive optics and micro-refractive optics have received considerable attention. In the case of optical distribution of data, there is the need for a source of optical power and a need for a means of modulating the optical beam to achieve data communications. As the number of optical data interconnections increases, the technical challenges of providing an efficient realization of the optical data interconnections also increases. Among the system signals which might be transmitted optically, clock distribution represents a substantially simplified problem from the perspective of the optical sources required. In particular, a single optical source, modulated to provide the clock signal, replaces the multitude of optical sources/modulators which would be needed for extensive optical data interconnections. Using this single optical clock source, the technical problem reduces largely to splitting of the optical clock beam into a multiplicity of optical clock beams and distribution of the individual clocks to the several portions of the system requiring synchronized clocks. The distribution problem allows exploitation of a wide variety of passive, miniaturized optical elements (with diffractive optics playing a substantial role). This article reviews many of the approaches which have been explored for optical clock distribution, ranging from optical clock distribution within lower levels of the system packaging hierarchy through optical clock distribution among separate boards of a complex system. Although optical clock distribution has not yet seen significant practical application, it is evident that the technical foundation for such clock distribution is well established. As clock rates increase to 1 GHz and higher, the practical advantages of optical clock distribution will also increase, limited primarily by the cost of the optical components used and the manufacturability of an overall electronic system in which optical clock distribution has been selectively inserted.

1. Introduction

This paper provides a broad overview of the several approaches which have been investigated for use of optical clock distribution within electronic systems. Such a review can proceed from two starting points. One starts at the higher level (optical connections among racks of electronics or among locally distributed computers) and proceed to optical connections among printed circuit boards (PCBs) through interconnections on a PCB to low level interconnections at the multichip module (MCM) and integrated circuit (IC) levels. Such a review would reflect an evolutionary extension of fiber-optics techniques such as the optical ribbon cable approaches currently being developed for commercial use (e.g., early exploratory approaches [1] through representative, recent parallel optical links [2, 3]).

A second approach, used here, starts with the more risky optical interconnect technologies needed at the lowest levels of packaging (e.g., the intra-IC interconnections) and proceeds to successively higher levels of packaging (intra-MCM, inter-MCM, inter-PCB, etc.). This approach highlights several of the exciting innovations which have been investigated by a large number of researchers over more than a decade. The importance of diffractive micro-optics (see, for example, the special issues [4, 5]) in miniaturized forms suitable for low levels of packaging clearly emerges as a primary direction for the optical "wires". Moving from the IC to the intra-MCM level, folded diffractive optics becomes a clear contender, avoiding the need to place the optical lenses, beam splitters, couplers and other elements well above the plane of the MCM. At the inter-PCB interconnection level, longer distances and less precise alignment of the electrical components (relative to the intra-IC devices and ICs mounted on MCMs) lead to significant adaptions of the folded diffractive optical techniques and the possibility of using more conventional optical elements (GRIN lenses rather than diffractive lenses) appears. In addition, as one progresses from the lowest level to the higher levels of packaging, the importance of implementing parallel data connections between components (rather than serial data connections) becomes an increasing priority.

Much of the research and exploratory development of optical interconnections at these lower levels of packaging have focussed on data interconnections, requiring an optical source (e.g., laser or LED) and a corresponding optical receiver for each data link provided. For this reason, several of the techniques discussed in this review are drawn from such optical data interconnections. However, the basic components (source, optical elements for the path taken by light, and receiver) are common also to distribution of an optical clock. The main differences between optical data and optical clock distribution are (i) the simplification of eliminating the many optical sources, using instead a single internal or external source, and (ii) the complication of having to convert that single optical clock signal into the multiplicity of clock signals to be distributed to various points in the system.

The technologies and design issues associated with optical interconnections (particularly at the lower levels) are not only rather complex but also depend critically on the specific set of elements selected for the interconnections. For this reason, the review does not attempt to provide an overview of all the detailed issues which must be treated when designing optical interconnections. Instead, the focus is on the diversity of approaches which have been extensively explored, providing a large set of references where more detail regarding technology and design issues can be found by the interested reader. The topic of optical interconnections at the lower level of packaging is a source of several innovative directions in the underlying technologies and in the architectures which might benefit from such interconnections. In addition to the considerable literature on optical interconnections, there has been considerable study of technologies and optical interconnections in the area of optical computing. In fact, several of the advances in optical interconnects were developed by those who were also exploring optical computing. In this review, the optical interconnection is taken as a passive function, and techniques using switching, such as reconfigurable optical interconnection networks, are not discussed here.

Section 2 provides some general background on the limitations of clock distribution within VLSI integrated circuits, followed by a discussion of techniques through which optical clock distribution might be provided. Several of the techniques discussed for intra-IC interconnections are also applicable to intra-MCM interconnections, which are considered in Section 3. Section 4 considers optical interconnections among MCMs. Section 5 discusses optical backplanes and direct free space optical interconnections between PCBs. A general overview of electrical and optical interconnections in electronic systems is provided in [6], with a collection of papers on such interconnections in [7]. Overviews of the considerable early work related to optical interconnections are provided in [8, 9].

2. Clock Delivery to Multiple Sites on an Integrated Circuit

2.1. Limitations of Electrical Distribution of Clock Edges

Clock signals in silicon VLSI circuits must be distributed for an external clock connection to each flip-flop within the IC. The dense population of flip-flops throughout the IC leads to a complex clock distribution network. The small feature size of interconnections used in submicron silicon VLSI imposes a significant resistance per unit length on the clock lines In addition, the minimum capacitance of the electrical interconnections is bounded by a lower limit of about 0.6 pF/cm due to fringing fields (i.e., as the line width is reduced, the electric fields tend toward the fields of a zero width line above the ground plane). The high resistance of the interconnection combined with the minimum limit on line capacitance imposes a significant RC delay factor for the long clock lines, limiting clock rates. The large current required to rapidly charge and discharge the total capacitance of the clock network, which is by far the longest signal interconnection in the IC, limits the maximum clock rate due to power dissipation (power dissipated for clock distribution can be a substantial portion of the overall IC power dissipation). The above effects limit the maximum data rate. In principle, if all clock paths to flip-flops are of equal length, the clock

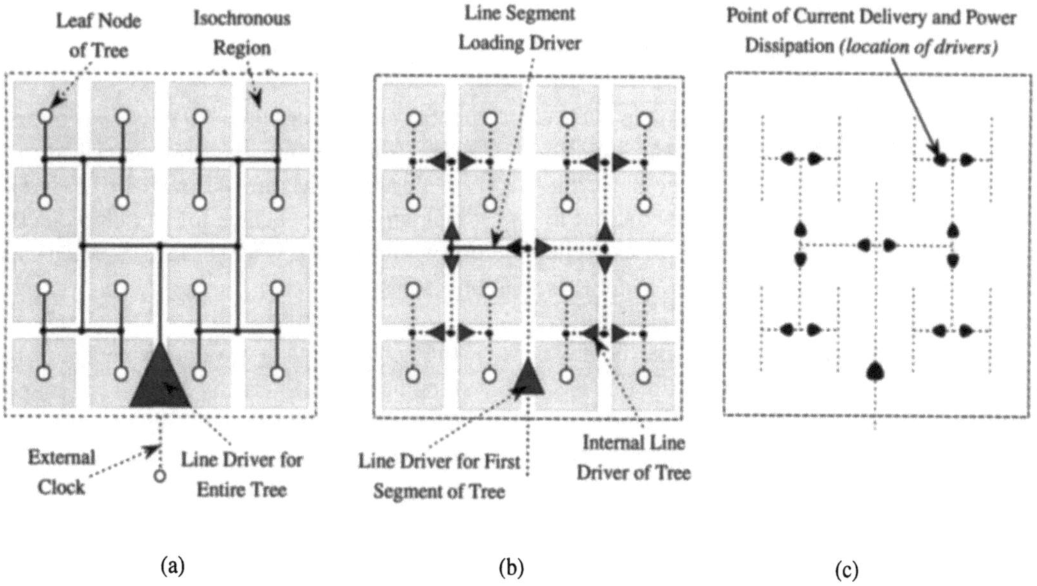

Figure 1. H-Tree distribution of optical clock. (a) H-Tree with single driver driving entire clock net. (b) Distributed drivers, separating net into shorter line segments. (c) Distribution of clock net driver power for case in (b).

skew would be zero. However, the clock signal received at different flip-flops is influenced by the routing of the clock interconnections among a dense set of data interconnections, switching increasingly rapidly as data rates evolve to Gb/s and higher data rates on ICs. These neighboring, high speed data lines couple a significant amount of crosstalk onto the clock lines, the specific crosstalk appearing at any single flip-flop depending on the specific path of the clock line to that flip-flop and the specific activity of data lines coupled to that path.

The H-Tree approach shown in Fig. 1(a) has become an increasingly useful approach for clock distribution. The equal length of each path from the external connection to each terminal point of the net provides the potential for zero clock skew under idealized conditions (i.e., no data lines on the IC and equal loading of the line segments). Addition of drivers along the H-Tree network, as illustrated in Fig. 1(b), provides several performance improvements.

- A faster rise time of the clock at the terminal end is obtained by limiting the RC delay term to only that of the line segment between the drivers (which regenerate a fast rise time signal).
- The amount of crosstalk introduced by nearby data lines is limited to the length of the line segment between drivers, with the drivers restoring the noisy signal to a clean signal. Decreasing the length of the line segment allows the crosstalk noise to be reduced to negligible levels, with virtually full elimination of the crosstalk added along the segment.
- By introducing drivers, each driver supplies only the current necessary to drive the next leg of the H-Tree network. As a result, the maximum current appearing on any clock line is greatly reduced (and the crosstalk coupled to neighboring data lines is correspondingly reduced).

The H-Tree network is not extended until each flip-flop of the VLSI circuit is included at a separate leaf node of the network. Instead, delivery of a clock signal to a region (isochronous region shown shaded in Fig. 1) of the IC is adequate, as illustrated by the modest number of terminal points of the H-Tree network in Fig. 1(a). Within each isochronous region, clock signals can be distributed without precise equalization of interconnection lengths to flip-flops due to the small propagation delays and rise-time delays within that region. At sufficiently short line lengths, given the clock rate, the interconnection behaves as a static RC interconnection, with an RC delay τ_{RC} given by $\tau_{RC} = (R_{dr} + R_l \cdot L)(C_l \cdot L + C_{ld})$, where R_{dr} is the driver resistance, R_l and C_l are the line resistance and capacitance, respectively, per unit length, L is the line length, and C_{ld} is the load capacitance imposed by the gates driven by the line. The propagation delay ($\tau_{\rm pr} = c/\sqrt{\epsilon}$, where c is the speed of light and ϵ is the relative dielectric constant of the insulator material—here SiO_2) is negligible over the short distances encountered in the isochronous region (e.g., for $\epsilon \approx 4$, $\tau_{\rm pr} \approx 60$ psec/cm). It is therefore the RC delay which dominates the area of the isochronous regions.

Similarly, it is the *RC* delay along the H-Tree network's line segments which dominates the delay of the clock to the leaf nodes of the tree.

As clock rates in VLSI ICs increase, it is necessary to provide not only higher speed clocks but also smaller clock skews, a difficult challenge since the cross sectional area of the interconnections continue to decrease in smaller feature VLSI technologies while the overall length of the clock distribution network remains proportional to the chip dimensions, rather than becoming smaller as in the case of specific digital functions on the IC. Optical clock distribution has, for several years, been considered a particularly important application of optical interconnections to provide the higher clock rates and smaller clock skews needed.

2.2. *General Approaches for Optical Clock Distribution*

Optical clock distribution to ICs has been studied for several years, with early work described in [10–12]. General approaches are illustrated in Fig. 2 (approaches which are also applicable to inter-chip clock distribution within MCMs, where ICs are mounted unpackaged on a substrate).

- In Fig. 2(a), optical waveguides replicate the H-Tree topology, with externally applied optical power distributed to the isochronous regions of the IC where, following conversion to electrical clock signals, electrical interconnections continue the distribution of the clock.

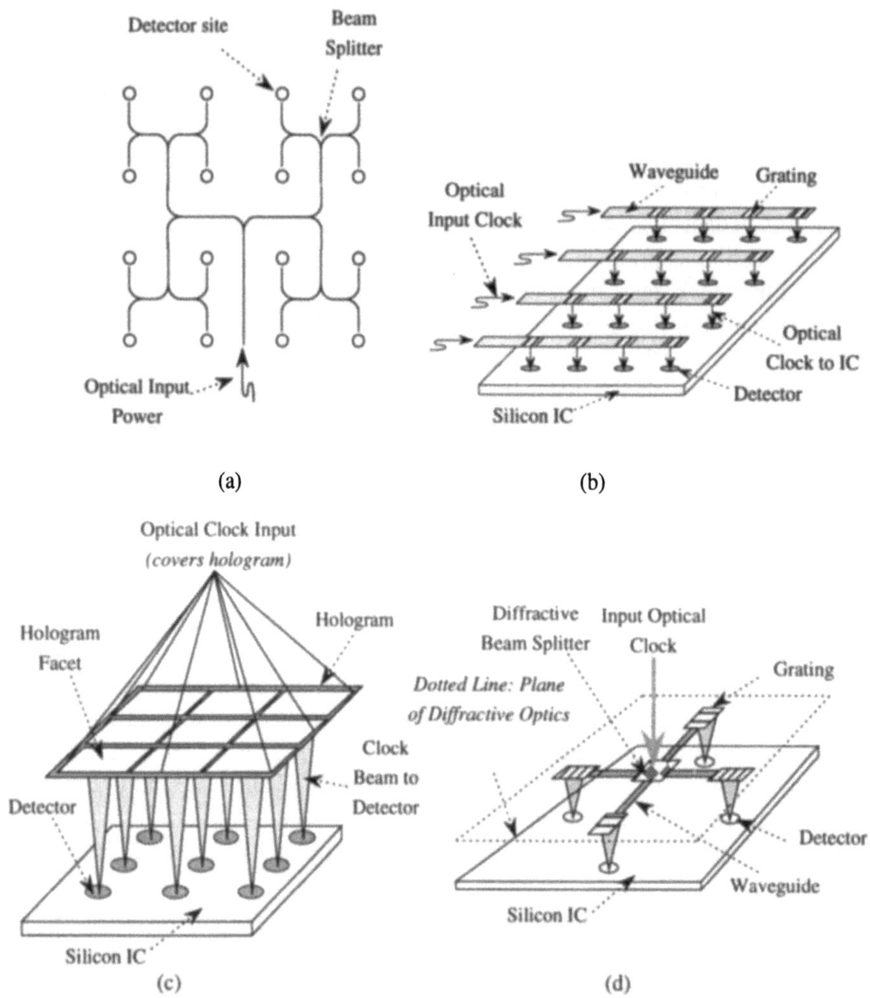

Figure 2. General approaches for optical clock distribution over the area of an IC. (a) Waveguide emulating H-Tree. (b) Tapped waveguides. (c) Hologram distribution of clock. (d) Planar diffractive optics providing vertically optical clocks.

- In Fig. 2(b), a set of parallel waveguides traverse the IC, with diffractive elements redirecting a portion of the light onto underlying detectors. The resulting array of detector sites mimics the arrangement in Fig. 2(a), but with the condition of precisely equalized paths removed. A related approach uses a single waveguide plate covering the IC and having grating couplers to direct light to selected sites on the IC. Within the waveguide, light travels at a speed $v = c/n_r$, with typical indices of refraction $n_r \approx 1.5$ leading to a clock skew between the two ends of the waveguide of about 50 psec for a 1 cm long waveguide.
- In Fig. 2(c), the optical clock signals are applied vertically (avoiding the need for fabricating waveguides on the IC surface) through use of a hologram. A related approach (more realistic for inter-IC connections within an MCM) places the optical source on the same plane as the optical detectors. As discussed later, the hologram must be separated from the IC surface by approximately 1 cm, adding significant volume to the basic IC circuit.
- In Fig. 2(d), optical clock signals are again applied vertically. However, this example uses a planar optical unit (with diffractive optics, mirrors, etc.) through which the light is first distributed in the plane of the unit and then redirected vertically to the IC (with zero skew along the optical paths).

The approaches illustrated in Fig. 2 assume a significant number of clock lines into the IC, favoring the use of silicon photodetectors compatible with the VLSI technology. Silicon is transparent to longer wavelengths (e.g., 1.3 μm and 1.5 μm used for lightwave communications), making optical power detection difficult. For this reason, shorter wavelengths (e.g., about 850 nm) are often used. Representative silicon photodetectors include the photoresistor, the photodiode (particularly the PIN diode), the avalanche photodiode, and the phototransistor. The PIN diode has been most extensively used for optical interconnections terminating on silicon ICs. At sufficiently high optical powers, the photodetector's output voltage can drive CMOS circuitry directly. However, receiver amplifiers associated with the detector allow the optical power to be greatly reduced (for a given clock rate).

2.3. *H-Tree Optical Clock Distribution using Waveguides*

Two distinct classes of waveguide are multi-mode waveguides (characterized by cross section dimensions substantially larger than the wavelength) and single mode waveguides (with cross sections comparable to the wavelength).

Multi-Mode Waveguides: Several polymers (e.g., polyimide) can be easily deposited or spun onto a substrate and patterned to form waveguides which have sufficiently low loss for optical interconnections (e.g., losses less than about 0.5 dB/cm). Polyimide, for example, is often used as the inter-metal dielectric for MCMs and provides a convenient waveguide material for such components. However, such waveguides generally have cross sectional dimensions substantially larger than the wavelength of semiconductor lasers or LEDs. Such large cross section waveguides support propagation of a multiplicity of modes (wavefront propagation angle relative to direction of waveguide). As a result, most polymer waveguides are multi-mode waveguides. As will be clear later, diffractive optical elements are sensitive to the incident angle of light beams, making such multi-mode waveguides less efficient for waveguide diffractive optics.

Single-Mode Waveguides: Non-organic materials such as SiO_2 (used as the inter-metal dielectric on ICs) can also be used as the foundation material for low loss waveguides. Cross sections comparable to the wavelength of semiconductor lasers are readily fabricated. These waveguides can be designed to propagate only a single mode, convenient for integration with diffractive elements embedded in the waveguide. For example, SiO_2-based single-mode waveguides could be fabricated directly on a silicon VLSI wafer prior to fabrication of electronic circuitry on that wafer. However, such an approach would require significant adaptation of the VLSI technologies, a significant barrier. Alternatively, assuming a sufficiently planar surface, such single mode waveguides may be fabricated above the metal layers of an IC. In the case of an MCM using a silicon substrate, single mode waveguides on the substrate are a realistic approach.

Koh et al. [13] describe an approach, illustrated in Fig. 3, to create an H-Tree using optical couplers and waveguide bends in a single-mode SiO_2 waveguide. The approach was developed for MCM-based clock distribution (considered in the next section) but is included here to illustrate both the potentials and limitations of single-mode waveguides for optical clock distribution. Figure 3(a) illustrates a planar coupler used at each node of the H-Tree whereas Fig. 3(b) illustrates a two-level waveguide approach using vertically coupled

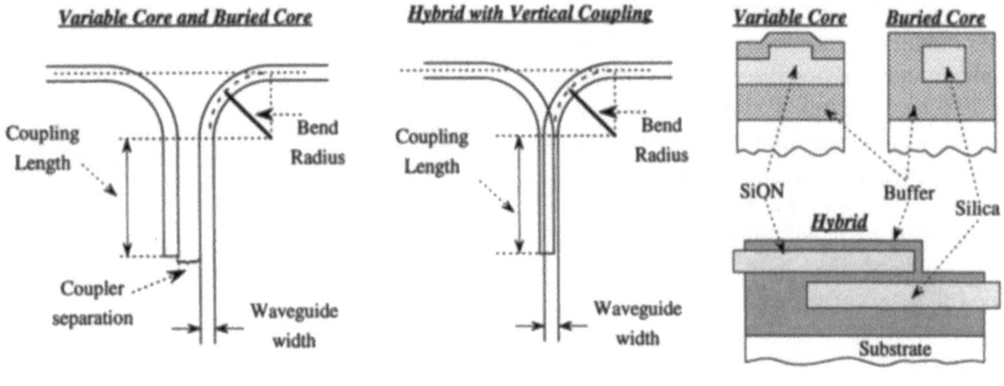

Figure 3. Single-mode waveguide implementing a H-Tree network. (a) Planar waveguide layout with coupler and bend. (b) Vertical coupler hybrid approach. (c) Cross section of SiON waveguide, buried channel silica waveguide, and hybrid approach. After [13].

waveguides (the two sections are shown slightly separated for clarity). The variable core (SiON core surrounded by an SiO_2 cladding) and buried core (silica core surrounded by an SiO_2 cladding) structures in Fig. 3(c) are used for the planar coupled waveguides whereas the hybrid structure (using both a SiON core and a silica core) in Fig. 3(c) is used for the vertically coupled waveguides.

The bend regions in Figs. 3(a) and (b), consists of a coupler through which incident optical power P_0 is equally divided (3-dB coupler) between the left and right branches followed by corners with a gradual bend (to minimize losses due to the evanescent component of the propagating optical beam). The waveguide dimensions (width w_{wg}, waveguide separation s_{cp} in the coupler, length L_{cp} in the coupling region, and radius R_{bnd} of the bend) determine the surface area for the nodes of the H-Tree. Table 1 summarizes the dimensions of the examples reported in [13].

Table 1. Waveguide parameters for coupler/bend design of [13]. The coupling loss for the hybrid design includes only the bending loss. An additional loss (<2 dB) occurs at the vertical coupler of the hybrid design.

Optical element of H-Tree	SiON waveguide	Silica waveguide	Hybrid design
Channel waveguide			
Upper buffer thickness (μm)	0.3	0.1	0.1
Thickness at channel (μm)	0.6	3.0	3.0
Thickness out of channel (μm)	0.48	NA	—
Lower buffer thickness (μm)	5.0	5.0	5
Channel width (μm)	4.0	2.5	2.5
Grating coupler			
Output coupling angle (degrees)	80	80	80
Grating period (μm)	0.99	0.99	0.99
Grating index	1.55	1.48	1.55
Grating depth (μm)	0.28	0.27	0.28
Grating length (80% efficiency)	392	17057	333
3-dB coupler			
Channel separation s_{cpl} (μm)	4.0	4.0	0.1
Bending radius R_{bnd} (μm)	3026	2000	2000
Coupling length L_{cpl} (μm)	1022	1208	70 or 210
Coupler bending loss (dB)	<0.01	<0.01	<0.01

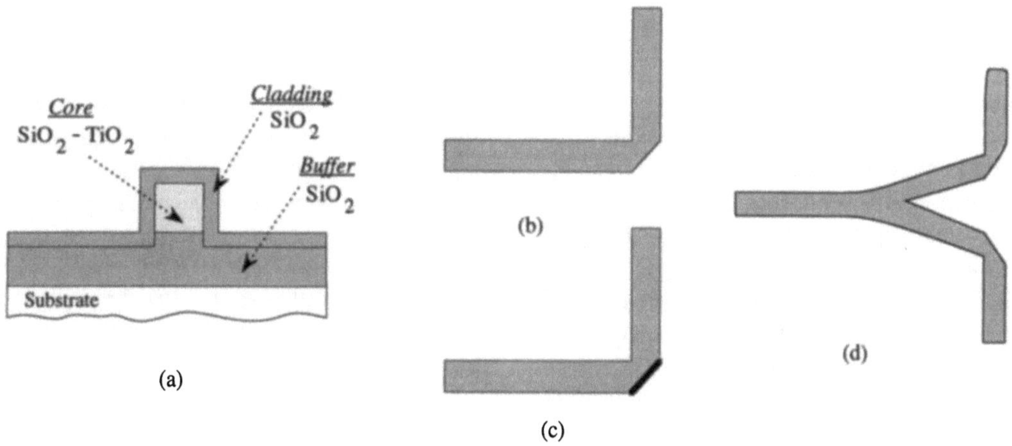

Figure 4. Right angle bends using corner mirror on waveguide. (a) Cross section of high-silica single-mode waveguide. After [14]. (b) Right angle bend with air at reflection region. (c) Right angle bend with aluminum selectively placed on reflection sidewall. (d) Illustration of a beam splitter followed by bends (less than 90°) to complete node of H-Tree.

The power $P_L(L)$ and $P_R(L)$ into the left and right branches, respectively, is related to the input power P_0 by [13]

$$\begin{aligned} P_L(L_{\mathrm{cpl}}) &= P_0 \cdot \sin^2(C_c L_{\mathrm{cpl}}) \\ P_R(L_{\mathrm{cpl}}) &= P_0 \cdot \cos^2(C_c L_{\mathrm{cpl}}) \end{aligned} \tag{1}$$

where C_c is the coupling coefficient of the coupler. For a 3 dB coupler, $C_c L_{\mathrm{cpl}} = \pi/4$. The coupling coefficient is related to the properties of the waveguide by

$$C \approx \frac{2h^2 p \exp(-p s_{\mathrm{cp}})}{k n_{\mathrm{eff}}(w_{\mathrm{wg}} + 2/p)(h^2 + p^2)} \tag{2}$$

where

$$h^2 = k^2\left(n_{\mathrm{core}}^2 - n_{\mathrm{eff}}^2\right) \tag{3a}$$

$$p^2 = k^2\left(n_{\mathrm{eff}}^2 - n_{\mathrm{buffer}}^2\right). \tag{3b}$$

In (3b), $k = 2\pi/\lambda$ (λ the wavelength of the optical signal), n_{eff} is the effective refractive index. n_{core} in the refractive index of the core, and n_{buffer} is the refractive index of the buffer layer (SiO_2 in Fig. 3). For the planar coupler, Table 1 shows a relatively long coupling length (about 0.1 cm), contributing significantly to the area of the H-Tree node. The close spacing between the vertically coupled waveguides in Fig. 3(b) leads to a considerably shorter coupling distance (see Table 1).

The minimum radius R_{bd} of the bend region is limited by the allowed optical loss around the bend. For a loss of about 0.01 dB, Table 1 shows a bending radius between 0.2 and 0.3 cm, dominating the area of the H-Tree node. The total area A_{nd} is $N_{\mathrm{nd}} = 2R_{\mathrm{bd}} \cdot (R_{\mathrm{bd}} + L_{\mathrm{cp}})$, large for ICs using the values from Table 1 but reasonable for use on MCMs (for which the study in [13] was performed).

The relatively large bending radius in the example above results from the relatively small difference in the refractive index of the core and cladding of the SiON waveguide, a limit which may be encountered if the waveguide is buried under overlaying layers of metal and insulator. If the waveguide is on the surface of the IC or MCM, then right angle bends can be obtained by drawing on the larger index change between the waveguide and air. For example, Himeno et al. [14] investigated sharp bends (up to 90°) using the reflecting mirror structure illustrated in Fig. 4. For a reflecting corner (air interface at mirror—Fig. 4(b)), the bending loss was found to be in the range 0.9–2.5 dB per corner up to 90° but increased sharply at bend angles slightly greater the 90°. Metalized corners (Fig. 4(c)) showed a bending loss less than 3 dB, higher than for the air-terminated mirror but extending to quite large bend angles (e.g., up to 140°). The 90° bend would significantly reduce the area of the H Tree node (compared to the results in [13], though a sharper bend would be obtained in that case if increased loss was allowed). High losses at the bend are problematic in the H-Tree network since there are potentially many bends between the root and leaf nodes of the tree. The mirrored bends in [14] could be combined with a "Y-junction" beam splitter, as shown in Fig. 4(d). However, the gradual bend required for a low loss splitter would impose a lower limit on the area of the node.

At the terminal points of the H-Tree network in Fig. 3(a), a grating coupler is etched in the waveguide, redirecting the beam toward the underlying surface. Suitable detectors and receiver circuitry completes the conversion of the optical clock to an electronic clock signal. Table 1 provides the grating length for the various waveguides used in [13], showing the considerable variation in length required for different waveguide structures.

Larger cross section, multi-mode waveguides can also be used to route the optical clock signal in an H-Tree configuration. The bends themselves would probably be right-angle splitters (e.g., as in Fig. 4(d)) but with a substantially shorter Y-splitter length. Diffractive gratings to couple light into the IC would be less efficient and the multi-mode waveguides may favor a mirror reflection at terminal points of the network. The lower performance of this multi-mode waveguide approach may be compensated by the more convenient waveguide fabrication technology, depositing and patterning polymers such as polyimide on the surface of a completed IC or MCM substrate.

2.4. *Planar Diffractive Units for Optical Clock Distribution*

Rather than placing optical structures for planar routing of optical beams directly on the IC (or MCM substrate), a separate optical planar distribution structure can be used, as illustrated in Fig. 2(b) (using optical waveguides with gratings) and Fig. 2(d) (using optical waveguides, a beam splitter, and coupling gratings). Figure 5 illustrates two adaptions of the beam splitter approach. Both examples support equal length optical paths to multiple destinations, similar to the H-Tree discussed earlier (with Fig. 5(b) directly illustrating a four-leaf H-Tree using two optical planes).

In the experimental version of the structure in Fig. 5(a) reported in [15], the waveguide was a low-loss, ion-exchanged glass planar waveguide, with the planar holograms formed in a layer of dichromated gelatin deposited on the glass place. 1-to-3 interconnection fanout was demonstrated at $\lambda = 633$ nm at selected angles between 30° and 60°, using an coupler interaction length of 700 μm. Like other diffractive optical systems, wavelength shifts modify the expected behavior of the optical elements. Lin et al. [16] consider the effect of such wavelength dispersion on the input grating coupler, the waveguide hologram, the output grating coupler, and together the overall end-to-end connection.

Figure 5(b) illustrates a "planar micro-optic system" described by Walker et al. [17]. This example is designed to split an incident beam into four beams, the four beams exiting from the opposite surface as the entering beam. By following the three-dimensional paths, the optical power distribution is seen to be a 3-D H-Tree structure. Such a folded, 3-D approach reduces the vertical height of the overall unit, collapsing several diffractive planes into a pair of planes. The experimental example was constructed on a 3 mm thick, silver-coated glass plate. Beam splitters were formed using binary gratings while the beam deflectors used 4-level gratings, both designed for 850 nm light. Performance was evaluated for 1 and 2 μm grating periods.

Figure 6(a) illustrates an approach developed by Kubota and Taneda [18] which, like the general approach in Fig. 2(b), uses a wide waveguide, covering the full area of the IC (or MCM). Grating couplers are formed in openings in the reflective layer covering the waveguide. Optical power is input through a prism element (shown at the left hand side of the wide waveguide) which injects light at an incident angle Θ_{in} (measured relative to the normal to the waveguide direction), with Θ_{in} greater than the critical angle for total internal reflection. On striking the phase grating (Fig. 6(b)), part of the light is coupled out of the waveguide. Letting t_{wgd} be the thickness of the waveguide, p_{grt} the period of the grating, n_{core} the index or refraction of the core, n_{top} the index of refraction of the medium above the grating, and $k_{opt} = 2\pi/\lambda$ (λ the wavelength of the optical signal), the exit angle Θ_{out}

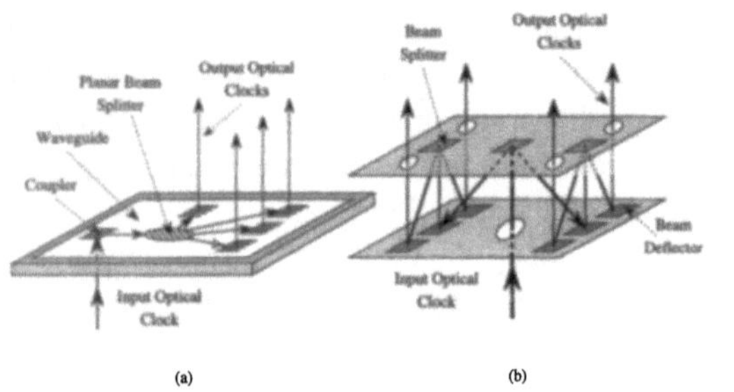

Figure 5. Example of planar diffractive optics for clock distribution. (a) Approach using waveguide holograms (after [15]). (b) Approach using free space optics (after [17]).

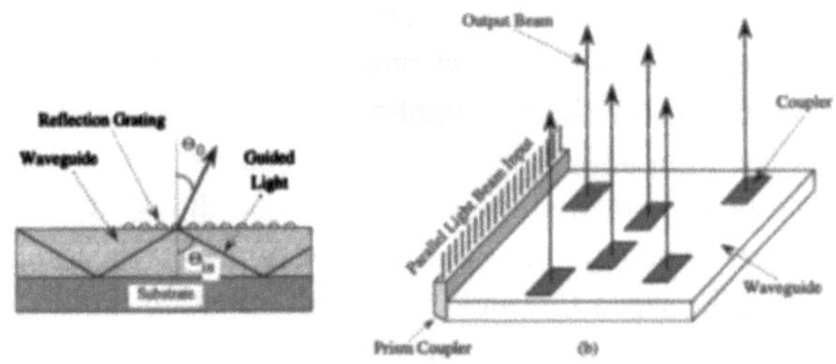

Figure 6. Parallel beam generation using a wide waveguide and grating couplers. (a) Basic coupler/waveguide structure (after [18]). (b) Beam generator with input beam coupled to waveguide using a prism and couplers placed arbitrarily on the waveguide (after [18]).

and Θ_{in} are related by [18]

$$k_{opt}n_{top}\sin(\Theta_{out}) = k_{opt}n_{core}\sin(\Theta_{in}) - 2\pi m p_{grt} \tag{4}$$

with m is an integer. Given Θ_{in}, λ, and the waveguide materials, the only variable parameter in (4) is p_{grt}, which is selected in this example to provide output beams normal to the surface of the plate.

Holograms have been studied to generate a regular array (linear or 2-D array) of optical beams from a single beam. To achieve efficient generation of high contrast spots, multi-level holograms (i.e., the surface contours are defined by surface relief quantized to several levels) rather than binary holograms (surface relief quantized to two levels, on and off) are normally used. Such multi-level relief structures provide a closer approximation to the continuously varying surface relief of an "analog" hologram by coding the surface relief into a binary coded approximation of that continually varying surface relief. The resulting binary coded surface relief is fabricated through a sequence of etching and masking steps. For example, starting with the smallest etch depth, successively deeper etches (increasing by a factor of 2) are made, with the total etch depth at any point on the surface the sum of the individual etch depths. Constraints on the fabrication of high efficiency, multi-level holograms for generation of arrays of optical beams is discussed, for example, in [19, 20]. High efficiency is important to avoid excessive background light accompanying the optical beams.

2.5. *Holographic Distribution of Optical Beams to IC Clock Nodes*

There is a very rich literature on the potential application of holograms for optical interconnections at the IC and MCM levels. Most of the investigations have considered providing a multiplicity of optical point-to-point data interconnections, with the optical signal both originating and terminating on the circuit module. Holographic distribution of optical clocks is a specialized case, using a single source but requiring the splitting of a single optical beam into several beams.

Basic Holographic Optical Data Interconnects. Three basic hologram schemes for transmitting light from a source on the IC to a detector on the IC are illustrated in Fig. 7. The reflection hologram in Fig. 7(a) gathers the diverging beam from the optical source, reflecting it back to a detector while focusing the beam

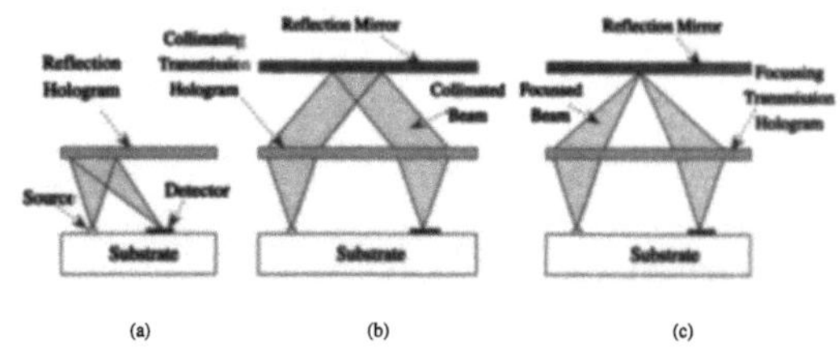

Figure 7. Basic holographic optical distribution approaches. (a) Single reflection hologram. (b) Two-pass approach using reflection mirror and collimated beams between hologram and mirror. (c) Two-pass approach using reflection mirror and optical beam focussed on mirror.

on the detector. The two-pass approaches in Figs. 7(a) and (b) use the hologram to first collect the light from the source and to focus the light beam returning from the mirror on the detector. Figures 7(b) and (c) differ in the use of a collimated beam or a focusing beam between the hologram and the mirror. A variety of hologram structures can be used. For distribution of light from several sources, each source's output directed to one or more locations, a multifacet hologram is generally used. In this case, the area of the hologram is divided into regions (facets), each facet illuminated by a single laser source.

The hologram planes in Fig. 7 must be about 1 cm from the surface of the IC due to the $f/\#$ of the hologram facet and the divergence angle of light from a semiconductor laser. The $f/\#$ of a lens is the ratio of its focal length f_{lens} to its diameter (aperture) a_{lens} (i.e., $f/\# = f_{lens}/a_{lens}$). Semiconductor lasers generally produces an output beam with divergence angles of about 30° (8° for VCSELs—Vertical Cavity Surface Emitting Lasers). To collect all the light from the source, the aperture of the hologram facet redirecting the light is related to the divergence angle Θ_{div} of the laser beam by

$$\tan(\Theta_{div}) = \frac{a_{lens}}{2 \cdot f_{lens}} = \frac{1}{2 \cdot f/\#}. \tag{5}$$

A lens with an $f/\# \approx 1$ therefore collects light originating at a distance equal to its focal length with a divergence angle of about 26.5°, reasonably well matched to the output characteristics of representative semiconductor lasers.

Alignment and Speed Issues for Holographic Interconnections. Alignment requirements for such holographic approaches have been widely discussed (e.g., [21–23]). Patra et al. [21] provide a recent theoretical treatment of the effects of mechanical misalignments (lateral shift in plane of hologram, longitudinal shift in separation between IC and hologram, and tilt

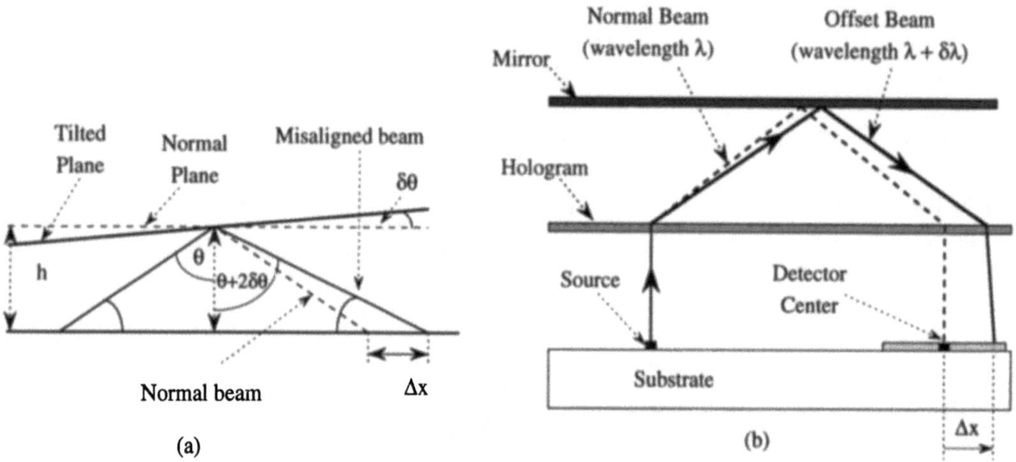

Figure 8. Examples of beam misalignment. (a) Tilt of mirror for two-pass approaches in Figs. 7(b) and (c). (b) Wavelength shift $\delta\lambda$ in source wavelength, leading to changed diffraction angle at hologram and beam shift Δx at detector.

of the hologram relative to the IC), misalignments due to thermal effects (e.g., coefficient of thermal expansion acting on physical distances in the hologram), and wavelength shifts of the optical source from the wavelength for which the hologram was designed. In [23], such alignment errors were evaluated to determine the relative advantages of the two approaches shown in Figs. 7(b) and (c). Representative misalignment mechanisms and beam area issues, including those illustrated in Fig. 8, are as follows.

- A small tilt angle $\delta\theta$ of the mirror (Fig. 8(a)), shifts the position of the collimated beam at the collecting lens by [23]

$$\Delta x = \frac{2h \cdot \delta\theta}{\cos^2(\theta) - 2\delta\theta \cdot \cos(\theta)\sin(\theta)} \tag{6}$$

  This lateral shift can become substantial for large nominal angles θ approaching $\pi/2$, as needed to connect to a detector well separated from the source.
- In Fig. 8(b), the diffractive grating (with grating period p_{grt}) has a nominal reflection angle θ_{grt} for the design wavelength λ. However, the reflection angle varies with wavelength λ. Bradley et al. [24] and others have discussed the resulting shift in the position of an optical beam. Using the formulation of [24], $d\theta_{\text{grt}}/d\lambda = 1/[p_{\text{grt}} \cdot \cos(\theta_{\text{grt}})]$, leading to a horizontal displacement Δx of the beam (after a total propagation distance L_{prop}) due to a wavelength shift $\Delta\lambda$ given by

$$\Delta x = \left(\frac{d\theta_{\text{grt}}}{d\lambda}\right) \cdot \Delta\lambda \cdot L_{\text{prop}} \tag{7}$$

  For $p_{\text{grt}} = 1\ \mu\text{m}$, $L_{\text{prop}} = 1$ cm, and $\theta_{\text{grt}} < 45°$, each 1 Å shift in wavelength displaces the beam position by about 1 μm.

  There may also be a significant temperature dependent shift in beam position. For example, temperature changes can lead to changes in the grating periods due to thermal expansion effects, leading to changes in the focal length of a diffractive lens and changes in the diffraction angle of a linear grating. Jahns et al. [25] considered such temperature changes in an optical interconnection scheme and, for the materials use in the study, the effect of a 100 K temperature change was very small (e.g., lateral shift and defocusing well below 1 μm). Behrmann and Bowen [26] describe the general influence of temperature on diffractive lenses. In addition, the wavelength of a semiconductor laser shifts by about 1 Å per degree (Celcius). In an environment with temperatures changing substantially (e.g., 100°C), the temperature-dependent wavelength shift can become a serious practical limit. Achromatic holograms may reduce this limitation [27].
- A collimated beam's width $w_{\text{bm}}(z)$ increases with increasing propagation distance z due to diffraction. Assuming a Gaussian beam with initial width $w_{\text{bm}}(0)$, [23]

$$w_{\text{bm}}(z) = w_{\text{bm}}(0)\left\{1 + \frac{\lambda^2 z^2}{\pi^2 n^2 w_{\text{bm}}(0)^4}\right\}^{1/2} \tag{8}$$

  where n is the index of refraction of the propagating medium. Assuming a 100 μm diameter hologram facet and a 1 μm wavelength, the beam width doubles at a distance $z = 85$ mm when propagating in air.

- Due to diffraction effects, light emitted from an aperture of dimension a in a medium with refractive index n diverges at an angle $\phi = \lambda/(n \cdot a)$. In the case of a perfectly collimated beam incident on a collector lens (acting as a source with aperture a defined by the facet diameter), the diffraction limited focused spot size s_{spot} at a detector a distance d from the aperture is

$$s_{spot} = \frac{\lambda \cdot d}{n \cdot a}. \tag{9}$$

Timing Errors in Optical Interconnections. Delay, jitter, and timing errors across an end-to-end optical interconnection for electronic systems includes contributions from a variety of sources. In the case of optical clock distribution, fixed delays introduced by each of the active elements and by propagation delay merely impose an upper limit on the rate at which clock pulses can be delivered to the system.

Timing jitter related to the electrical clock generated from the received optical clock signal includes jitter associated with electrical noise in the laser driver, with electrical noise in the laser bias, with the turn-on jitter of the laser, with the detector's opto-electric conversion timing jitter, with electrical noise in the detector, and with electrical noise in the receiver circuitry. Generally, timing jitter in semiconductor lasers is rather small (see, for example, [28, 29]). Timing jitter associated with the overall source end is not a source of timing skew, since all end points receive their clock from a common optical clock source. However, jitter associated with the source end produces a statistical distribution in the arrival time of a given clock pulse, with the potential of causing occasional clock pulses to arrive before electrical signals have settled to steady values if sufficient timing margin is not provided. On the other hand, timing jitter at the detector/receiver end does produce a dynamic clock skew effect (i.e., the arrival times of a single clock pulse differs for that specific pulse among the regenerated electrical clock signals).

Crosstalk can appear in optical signals as they propagate between source and receiver, though no interactions occur between beams intersecting in free space. Crosstalk can result from coupling between waveguides and from higher-mode effects in a diffractive element for one signal coupling light into a diffractive element or detector of another signal. In the case of optical clocks, such crosstalk is another source of timing jitter, qualitatively similar to electrical noise at the detector end in impacting clock integrity.

If the response time of a detector/receiver depends on the optical power received, then variations in the simultaneously received optical power at each of the detectors produces timing skew among the regenerated electrical clocks from those detectors.

A general review of time skew, timing jitter, and turn-on delay for holographic interconnections is provided in [30]. However, the comments above suggest that the precision timing often presumed for optical clock distribution may be compromised in real-world systems.

3. Optical Clock Distribution to ICs within an MCM

3.1. General MCM Packaging Approaches and Related Optical Interconnection Approaches

The previous section reviewed techniques for intra-IC optical interconnections. Many of those techniques are also applicable to inter-IC interconnections within an MCM, where distances are longer and alignment among ICs on the MCM is substantially less precise than among devices on an IC. Discussions in Section 2 of approaches also applicable for MCMs are not repeated here. MCMs are between ICs and PCBs in terms of the distance between components (transistors and ICs, respectively) and the precision of component alignment. For this reason, the new techniques discussed in this section for optical clocks to ICs in an MCM are often adaptable for clock distribution to packaged ICs on a PCB. However, the larger area of the MCM limits application of reflection or diffraction/reflection holograms such as illustrated in Fig. 7.

MCMs provide an efficient technology for modules containing not only unpackaged VLSI ICs but also other ICs, including III-V components such as semiconductor lasers and III-V detector/receiver circuits. The placement of lasers within the MCM allows generation of the optical clock within the MCM package, rather than requiring input of an external optical clock. For advanced, MCM-level systems, communications between different MCMs may be increasingly asynchronous data/packet transfers, avoiding the need to distribute a common clock to all the MCMs of a system. Compared to PCB assemblies using packaged ICs, the MCM simplifies the delivery of optical clocks to the ICs (since no package intervenes). For

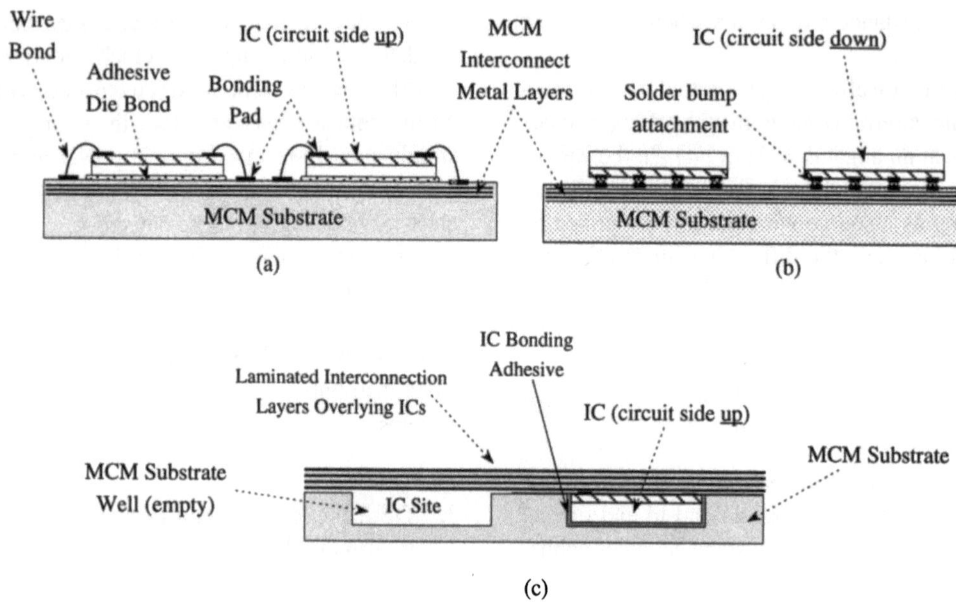

Figure 9. Basic IC mounting in MCM substrates. (a) "Chips last" epoxy bonding of IC (circuit side up) to MCM with wire bonding (or TAB bonding) to MCM substrate. (b) "Chips last" flip chip mounting (circuit side down) of ICs using solder bump connections to MCM substrate. (c) "chips first" approach with ICs placed (circuit side up) in recessed wells of MCM substrate and MCM interconnections placed on metallization layers running over ICs.

such reasons, the MCM level of packaging is considered here a quite different environment for optical clocks than either the IC or PCB environment. Cinato and Young provide a broad overview of optical interconnection approaches suggested for use within MCM modules [31]. A view of inter-MCM optical interconnections, arguing that this is the lowest level for optical data interconnections, is given in [32, 33].

Figure 9 illustrates the three primary approaches for placing and interconnecting unpackaged ICs on a multichip module. The approaches with the circuit side up (Figs. 9(a) and (c)) generally require that the optical interconnections be placed above the ICs. The circuit-side down approach in Fig. 9(b) generally requires that the optical signals be routed on the MCM substrate, again assuming detectors on the silicon ICs. These conditions are relaxed by using "through-wafer" optical interconnects [34, 35], discussed in Section 4.

In the case of wire bonding (Fig. 9(a)), the alignment accuracy is modest (based on the accuracy of eutectic bonding of the IC to the substrate). The chips last approach in Fig. 9(c) also has only modest alignment accuracy (chips must fit without difficulty into the wells of the substrate), with the general requirement that interconnect wires from the MCM substrate to the IC be custom drawn for each MCM (accounting for the IC misalignment). The approach in Fig. 9(b) provides good IC-to-MCM alignment since the solder bumps provide a self-alignment mechanism (e.g., [36, 37]). In particular, forces exerted by the many solder bumps tend to move the overall IC into an equilibrium position in which the solder bumps are vertical, even though the starting position (before the solder is melted) may be offset.

3.2. *Waveguide Distribution of Clock*

Optical waveguides can be used for clock distribution to the ICs within an MCM in much the same manner as discussed in the previous section. Early work on polymer waveguides on MCMs was reported by Selvaraj et al. [38], in that case applied to wafer-scale integrated circuits. The results of [13] discussed earlier as an example of possible H-Trees on ICs was actually addressing such H-Trees on MCM substrates (where the area of the nodes of the tree is less constraining).

Figure 10 illustrates representative approaches for addition of waveguide to MCM modules. In Fig. 10(a), the optical plate containing the waveguides is shown as being part of the top cover of the MCM package, suitable for circuit-side-up MCM modules. By providing a mechanical alignment feature within the MCM package to precisely position the MCM substrate, alignment relative to the optical "cover plate" can be improved. Alternatively, the optical "cover plate" can perhaps be

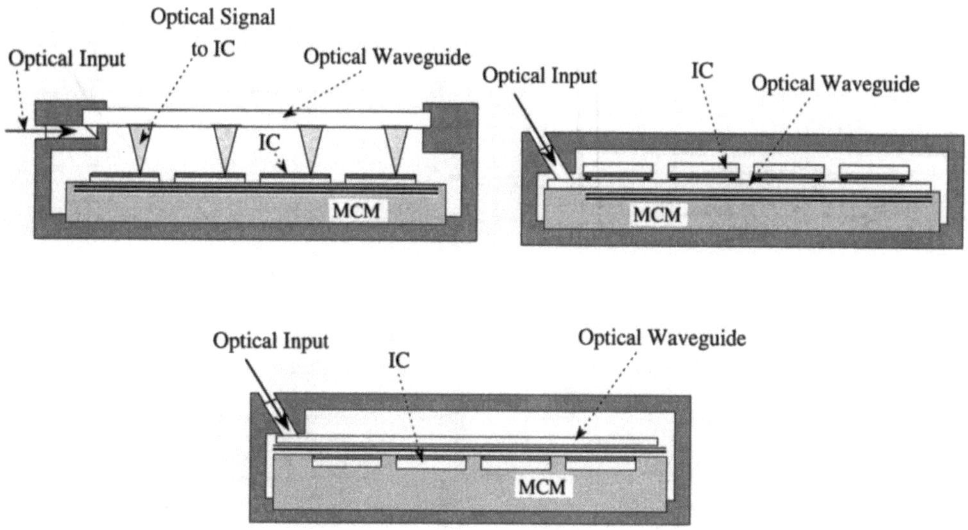

(c)

Figure 10. Examples of possible optical waveguide distribution of clock signals for approaches in Fig. 9. (a) Waveguide plate in top cover of MCM package for wire-bonded IC approach of Fig. 9(a). (b) Waveguide distribution on top of planarized MCM substrate running under solder bumped ICs in approach of Fig. 9(b). (c) Waveguide distribution in laminate layers overlaying ICs in chips first approach of Fig. 9(c).

actively aligned to the MCM substrate during final assembly of the packaged MCM component.

In Fig. 10(b), the optical waveguides are fabricated directly on a chips-last, MCM substrate, with the waveguides aligned to the IC bonding pads on the MCM substrate. The waveguides might be polymer waveguides residing on top of the electrical connections of the MCM substrate, though surface topology may present a limitation. In the case of silicon MCM substrates (or some other MCM substrate materials), SiO_2-based waveguides can be fabricated on the silicon substrate prior to fabrication (deposited and etched layers or laminated layers) of the MCM interconnection layers.

Figure 10(c) illustrates optical waveguides formed in one of the laminate layers placed on the MCM substrate and over the ICs. The HDI (High Density Interconnect) MCM technology discussed in [39], a chips-first MCM technology extensively studied by General Electric Company, is a good example of the use of laminates overlaying the ICs (in recessed wells) with a planar surface well matched to the needs of waveguides.

In the case of single-mode waveguides, diffractive beam splitters, bends, gratings or mirrors can be used much as described in Section 2. However, MCMs may favor multi-mode waveguides, with their larger cross sectional areas more compatible with the larger features sizes seen in MCM interconnections (e.g., electrical line thicknesses of about 10 μm and line widths of 25 $\rightarrow$ 60 μm) and the correspondingly larger surface topographies (for the chips last approaches).

3.3. Planar Diffractive Optical Modules for Optical Signal Distribution Among ICs

As discussed in Section 6, the reflection and diffraction/reflection hologram approaches shown in Fig. 7 confront increasingly severe tolerance requirements as the reflection angle increases to route a signal to a larger distance from the source. The angle can be reduced by increasing the height of the hologram from the substrate, but this creates a large volume component from a large, planar area component such as an MCM.

To provide a compact MCM with holographic distribution of optical signals, the optical beam path of the approach in 7 can be folded, as illustrated in Fig. 11. Unfolding of the end-to-end optical path (maintaining a single, reflection point near the center of the path) in Fig. 11 illustrates its qualitative equivalence to a reflective/diffractive approach with a large separation between the mirror and the hologram. Alignment issues (e.g., [23, 24, 40]) for such folded systems are similar to those encountered in the non-folded approach.

The folded, planar diffractive optical element in Fig. 11(a) has been widely discussed (e.g., [25, 41–44]). In Figs. 11(b) and (c), the holograms are placed

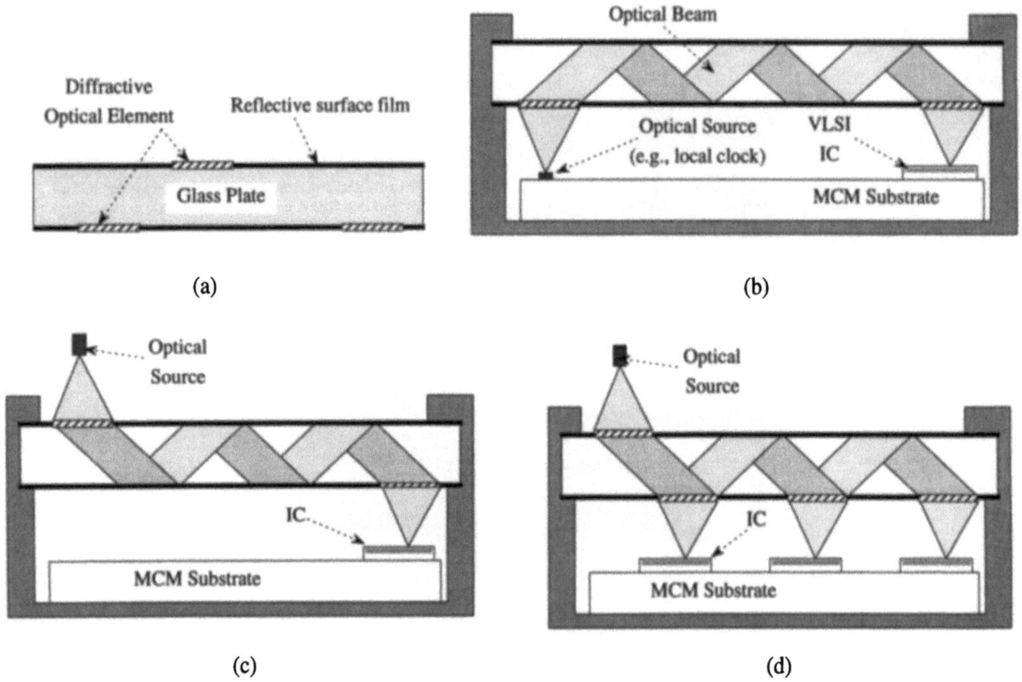

Figure 11. Folded geometry diffractive optics. (a) Thick optically transparent plate (e.g., glass) with reflective film (typically metal) with openings in the metal holding diffractive optical components (gratings, couplers, etc.). (b) Illustration of folded geometry approach for chip-to-chip interconnection within an MCM (including use of a locally generated clock). (c) Similar to (b) but using an externally provided optical signal, rather than an internally generated optical signal. (d) 1-to-many optical connection, as might be used to distribute an optical clock to several ICs in an MCM.

on the lower surface of the plate, which is placed about 1 cm above the MCM substrate for the same reasons discussed earlier regarding 7. In Figs. 11(c) and (d), a hologram is also placed on the top surface of the plate to receive optical clocks from outside the MCM. The hologram at the source end of the optical path collimates the optical beam and outputs that beam at an angle giving totally internal reflection within the plate (which typically has a mirrored surface), leading to the bouncing beam illustrated. At each point where the beam encounters a surface (top or bottom) of the plate, micro-optical elements can be placed allowing a variety of manipulations of the beam. In the example shown in Figs. 11(b) and (c), the beam is simply reflected between the two surfaces, terminating on a second hologram which focuses the beam onto the detector. Figure 11(d) illustrates multiple detector sites, as would occur for optical clock distribution. At each reflection point on the lower surface, the optical clock can be "tapped" to provide an optical clock to a detector lying below that point. The separation between tap points can be varied by changing the reflection angle of the light beam within the plate, so long as the angle meets the requirement for total internal reflection.

4. Optical Interconnections for Compact, Multi-MCM Modules

Electronic systems almost exclusively use planar packaging for the various levels from the IC itself to the rack level of system packaging. Low-level, 3-D packaging has recently received increasing attention. Examples include stacked ICs, with individual ICs thinned and then physically assembled into a compact stack[1]. Such 3-D ICs structures exploit fine dimension wiring along the edges of the stack, such as shown in Fig. 12(a). Compact MCM-based systems may also benefit from 3-D stacking, in this case a stack of MCMs such as illustrated in Fig. 12(b). 3-D MCM stacks, for example, have been investigated using the chips first approach which has advantages for stacking due to the flatter surface and more protected ICs, compared to the chips last approaches. Figure 12(c) illustrates another compact multi-MCM structure (using unpackaged MCMs in a backplane, rather than the 3-D, configuration) with the thin MCM modules mounted vertically on a miniature "backplane."

Although Fig. 12 illustrates electrical interconnections among elements of the "stack", optical

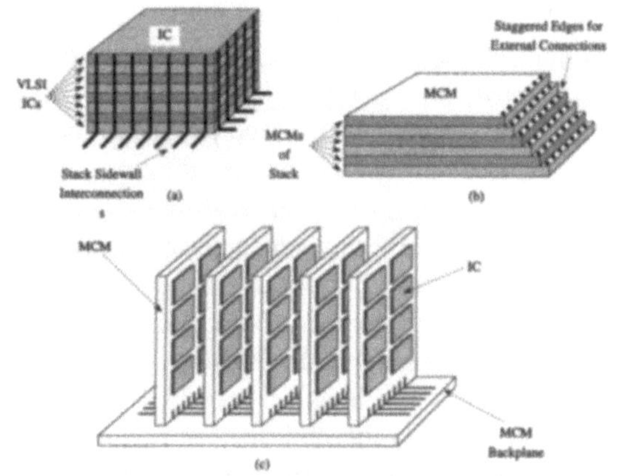

Figure 12. Three-dimensional, low level packaging approaches. (a) Stacked ICs (e.g., memory ICs). (b) stacked MCMs. (c) vertically oriented MCMs on MCM backplane "board".

interconnections provide some interesting opportunities. Figure 13 illustrates representative approaches which might be used for optical connections among layers of the 3-D stacked MCMs in Fig. 12(b). Figure 13(a) shows optical fiber ports into an optical distribution place between pairs of MCM substrates, though such optical fiber connectors will be difficult to implement efficiently. In Fig. 13(b), a sidewall optical plate is shown, receiving optical signals through an optical input port and redirecting the optical signals to optical distribution planes between pairs of MCMs. Figure 13(c) illustrates routing of external optical signals through via holes through the MCMs, with optical power intercepted by optical distribution planes serving adjacent MCMs. In Figs. 13(a) through (c), only one optical plane is needed for each pair of MCM

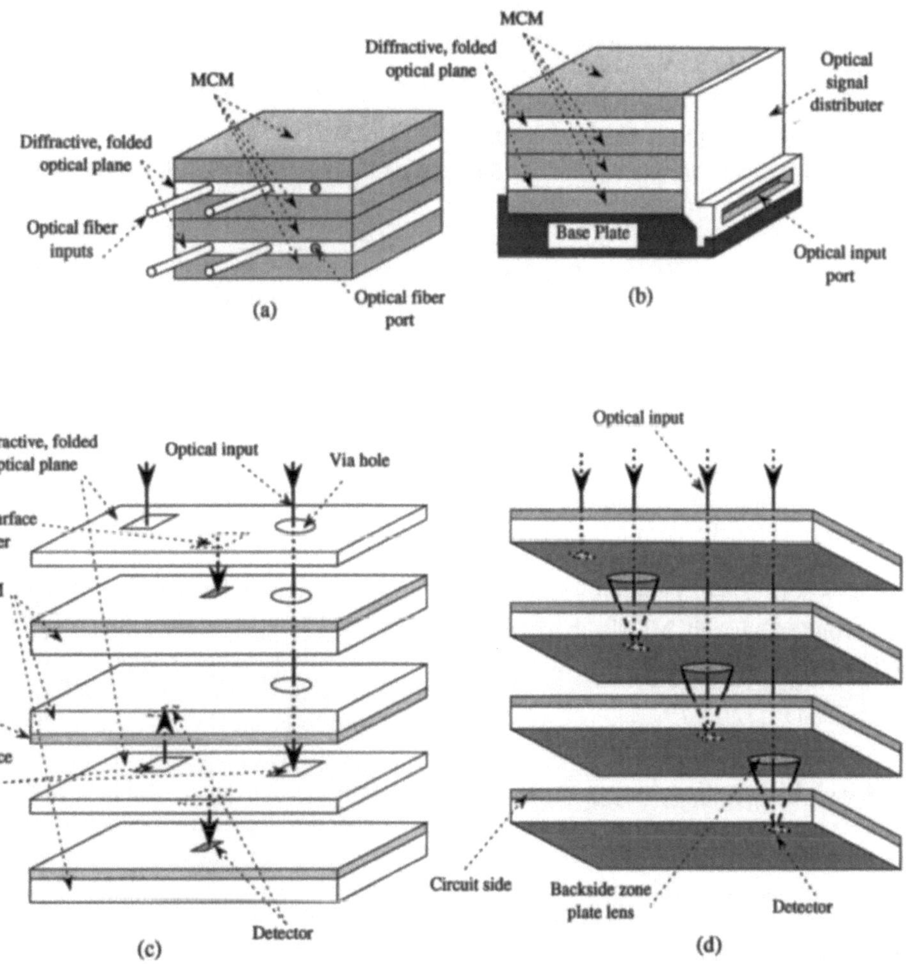

Figure 13. General approaches for optical interconnections to stacked MCMs. (a) Optical fiber connections to waveguides through sidewall of MCM. (b) Free space optical connection to waveguides through sidewall of MCM. (c) Vertical optical interconnections through optical "vias" in MCM stack. (d) Through wafer optical interconnections for MCM stack.

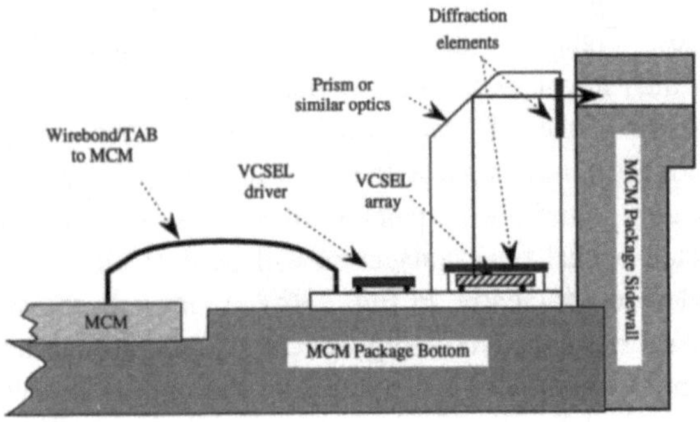

Figure 14. Direct free space optical connections between adjacent MCMs. A preassembled optical module is mounted on the package and includes optoelectronic conversions, diffractive optical elements to interface the package sidewall ports to the source/detector arrays, with wire bonding used to connect the optical elements to the fully electronic MCM substrate. (After [45].)

substrates. Figure 13(d) illustrates a through-wafer optical interconnection approach first reported in [34] for stacked WSI circuits. In this example, long wavelength light passes directly through the MCM substrates (assumed silicon) and is focussed using diffractive lenses etched directly on the backside of the MCM (WSI) substrates (e.g., SiN Fresnel zone plate lenses in [34]).

Figure 14 illustrates an approach suggested in [45], mounting a preassembled optical module within an MCM package to provide parallel optical free space connections through optical ports in the MCM package. To eliminate the need for precision mechanical alignment of the MCM substrate to the optical elements, wire (or TAB) bonding connects the optoelectronics of the optical I/O module to the fully electrical MCM substrate. This approach is only conceptual but was developed to illustrate that precision alignment of electronic components (or the need to adapt of standard electronic components) for insertion of optical interconnections can be avoided. In this example, the optical signals connect adjacent MCMs on a plane by positioning those MCMs very close to one another (an approach suggested in [32]).

5. Optical Clock Distribution Among PCBs

Several investigators (e.g., [27, 46–53]) have addressed optical interconnections between and on PCBs, addressing the difficulty of achieving very high speed electrical signal transfers at these packaging levels.

Optical waveguides for intra-PCB interconnections have been investigated by a few groups (e.g., [48, 54]). To handle the larger surface topography of PCBs (relative to MCMs and ICs) and the increased errors in alignment, large cross section, polymer waveguides were routinely used in the studies. Separate packaged components generally were assumed for the optoelectronic devices and the silicon IC circuitry. Though several interesting approaches were investigated, the emergence of MCMs has significantly changed the conditions for integration of optical interconnections on PCBs (e.g., easy placement of optoelectronics within the same package holding the MCM electronics).

Optical interconnections between PCBs moves the discussion closer to the optical fiber ribbon connections noted briefly in the introduction for high-level system interconnections. Despite the expected commercial development of such optical fiber ribbon connections, there has been considerable investigation of the extension of techniques discussed in earlier sections to the inter-PCB case. The discussion below is restricted to such extensions.

5.1. *Optical Buses for Vertically Oriented PCBs*

Figure 15 illustrates a planar diffractive optical backplane for use with PCBs (e.g., [40, 46, 47, 50, 55, 56]). These examples extend the application of planar diffractive optical elements from the use with planar mounted components discussed earlier to vertically mounted components. Figure 15(a) illustrates a direct, point-to-point optical link between two boards. Figure 15(b) provides a bidirectional broadcast capability, better reflecting the functionality of a conventional bus. In both examples in Fig. 15, the optical paths shown typically reflect parallel data paths (e.g., each optical path shown can be, in fact, a parallel array of optical beams). High density parallel optical beams have been extensively studied for several applications, including optical computing, and the backplane bus considered here is a particularly appropriate application of such approaches. Such optical backplanes (and the direct board-to-board free space connections discussed later) are capable of substantially higher numbers of parallel beams than the optical fiber ribbon cables presently under development for commercial products.

Beech and Ghosh [40] address the issue of misalignments for point-to-point optical interconnections between pairs of optical backplane connected boards such as shown in Fig. 15(a), with results similar to those already discussed for related folded diffractive optics.

Natarajan et al. [56] describes the more interesting case of a bidirectional optical bus shown in Fig. 15(b),

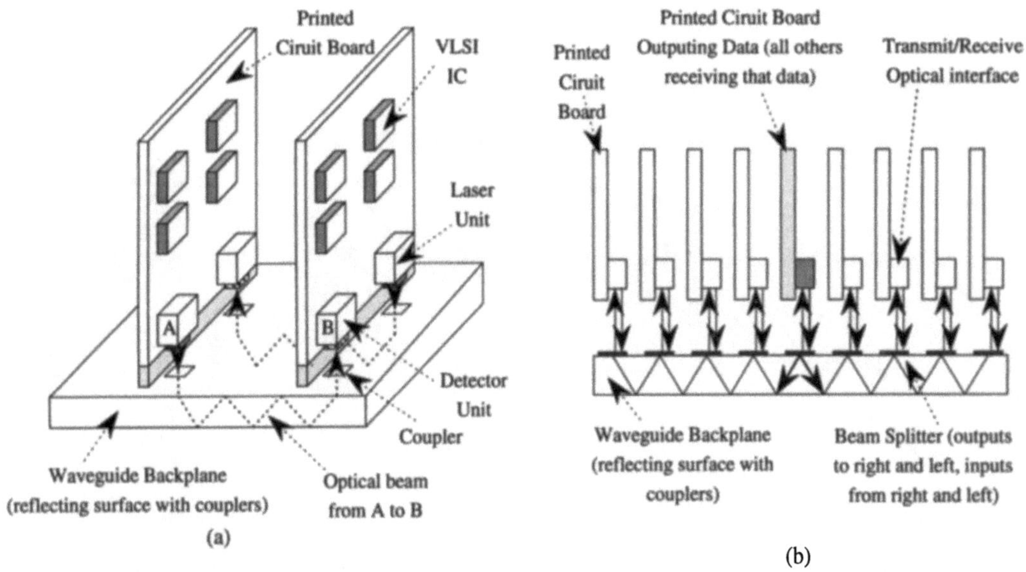

Figure 15. Optical backplane approach. (a) PCB backplane with direct point-to-point connections between pairs of PCBs (e.g., as in [40]). (b) PCB backplane with bidirectional bus (after [56]).

with each PCB reading optical data placed on the bus by any PCB. The bidirectionality is achieved through use of a holographic beam splitter on the top plate of the optical distribution plate, the hologram splitting the output optical beam from the PCB into two beams (one propagating to the right and the other propagating to the left, thereby reaching the connection sites of all other PCBs). For light propagating along the plate, the hologram also extracts a portion of the light (received from either the right or left of the hologram) and directs that portion to a detector on the PCB. As optical power is extracted from the propagating optical beam in the plate, the remaining optical power propagating to subsequent sites is reduced. For a backplane supporting several PCBs, the decrease in optical power can be considerable if the fraction α_{out} of power extracted from the beam is significant. For an initial optical power P_0, such extraction leads to a remaining optical power P_n after N_r reflections of $P_n = (1-\alpha_{out})_r^N \cdot P_0$ with the optical power $P_{in}(N)$ provided to the $N_r + 1$st PCB being $P_{in}(N_r + 1) = \alpha_{out} \cdot (1 - \alpha_{out})^{N_r} \cdot P_0$. For $N_r = r$, and $\alpha_{out} = 0.7$ ($\alpha_{out} = 0.9$), the ratio of the optical power into the 1st PCB to the optical power into the last PCB is 0.04 (0.39). For efficient optical signal transfers, it is advantageous to choose α_{out} to be sufficiently small that most of the optical power entered into the backplane is extracted to detectors on the PCBs, though the disparity in the amount of power received increases as α_{out} increases. The prototype described in natarajan.95 used a bouncing angle of 45° and demonstrated 1.2 Gb/s data transfers per optical "wire" at 1.3 μm wavelength.

An approach similar to that in Fig. 15b was implemented by Yamanaka et al. [47]. In this example, the sensitivity of the diffraction angle of the grating beam splitters to wavelength variation was reduced by passing the light through two identical grating beam splitters (important since the multiple beams must be accurately focussed on the individual detectors of the receiver array). Using a 2 mm coupling lens to the PCB and assuming typical board spacings of about 2 cm, it is suggested in [47] that such a lens can handle 100 parallel connections. The experimental unit used pnpn vertical-to-surface transmission electrophotonic (VSTEP) devices [49], providing the combination of light emission, light detection, thresholding, and latching.

5.2. *Direct Optical Interconnections between Vertically Oriented PCBs*

Figure 16 illustrates direct free space interconnections between PCBs. In both cases shown, the length of the direct optical interconnection between the two boards can be substantially less (e.g., about 2.5 cm) than the length of interconnections routed from the outer edge of one board to the backplane, across the backplane, and then to the outer edge of the receiver board (e.g., a total distance of about 25 cm for 10 cm wide PCBs).

Figure 16(a) illustrates the approach investigated at MIT/Lincoln Labs [51–53]. In this approach [53], graded index (GRIN) lenses are used, one collimating

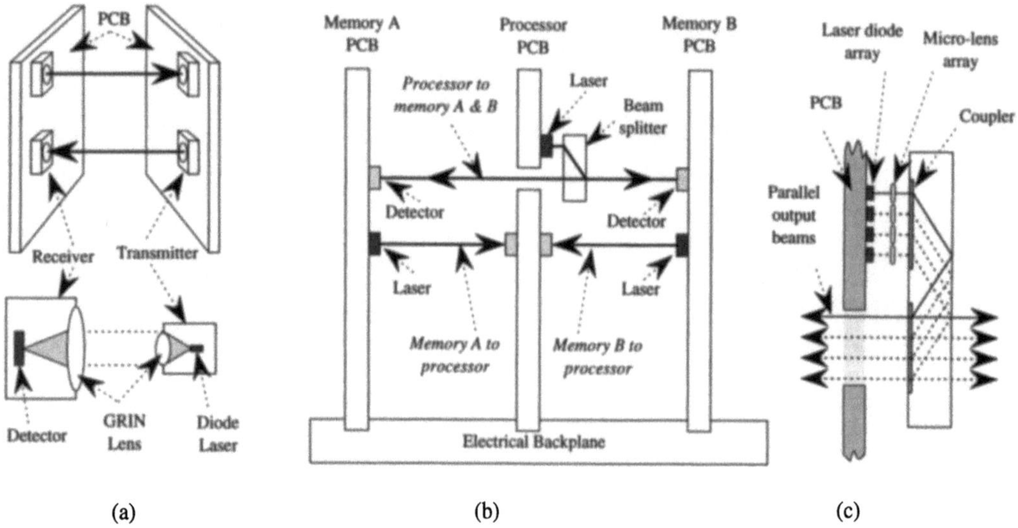

Figure 16. Direct free space optical interconnections between vertically oriented PCBs (a) Direct point-to-point, unidirectional connection between adjacent PCBs (after [53]). (b) Direct connections between a PCB and multiple other PCBs (after [57]). (c) Parallel beam generator and splitter on processor board in (b) (after [57]).

the beam from the source laser on one board and the other focusing the beam on the detector of the second board. In such arrangements, the lens on the source board can be aligned with good precision to the source laser and the lens on the receiver board can similarly be aligned with good precision to the detector. As a result, if the free space beam from the transmitter board strikes the lens of the receiver board, the transmitted light will be efficiently collected by the detector. In this manner, the sensitivity of the optical efficiency (end-to-end) can be made less sensitive to lateral misalignments of the two boards (that alignment tolerance being less well controlled than alignment of components on a circuit board) by using the collimated beam between the two boards and by using an oversized receiver lens. Angular misorientations of the source board relative to the transmitter board must be small to ensure that the transmitted beam is directed toward the receiver lens (that angular misalignment having to decrease as the board separation increases). In [53], precision stops associated with the card cage holding the boards assist in establishing the required lateral alignment of the two boards (important since a small angular tilt of the board will lead to substantial displacements of points at the outer edge of the board for their nominal positions).

In the prototype [53] of the example in Fig. 16(a), the boards were separated by 3 cm, the transmitter's GRIN lens had a 1 mm diameter, the receiver's GRIN lens had a 2 mm diameter, and the focal length of the GRIN lenses was 1.7 mm. Lateral misalignments as great as ± 0.7 mm produced less than 20% loss of light at the detector. Similarly, only 20% light loss was produced by angular misalignments (of the two PCB planes) by $\pm 2°$.

Figure 16(b) illustrates a more aggressive example, in this case using short optical interconnections between a central processor board and two memory boards to support fast memory access between the processor board and memory boards [57]. Rather than a single optical beam as in the example above, this example used parallel optical beams to implement a parallel data bus (each optical "path" in Fig. 16(b) is, in fact, an array of beams). The transmit end at the processor board communicates the same data to both memory boards, requiring that the processor board's output optical beam be split into two beams (see optoelectronic source module of processor board shown in Fig. 16(c)), one propagating to the left side memory board and the other propagating to the right side memory board. Several practical issues related to this general approach are addressed in [57], providing a good reference for the general technique of direct, inter-PCB optical interconnects. Assuming a 2.5 cm distance between boards in Fig. 16(b), the propagation delay of light between the two boards is about 83 psec, substantially shorter than would be encountered for an electrical signal traversing the two boards and connected to the backplane. However, as noted in [57], the total delay imposed by the optoelectronic elements greatly exceeds this propagation delay. The various electrical and optoelectronic components interposed between the electronic devices which would transmit and receive the signal in the absence of the optical interconnection include the line driver between the digital logic and the laser driver, the laser driver circuitry, the laser itself (with a given turn-on delay), the optical detector (e.g., photodiode), the optical receiver circuitry, and the line driver from the receiver to the logic circuitry. In the example implemented in [57], the laser driver delay (about 100 psec), the laser diode turn-on delay (about 50 psec), and the receiver (high performance design) delay (about 500 psec) combine to greatly exceed the propagation delay.

Alignment issues are also addressed in [57]. In this prototype, the focal length of the collimator and collector lens is 1.5 mm, the board separation is 2.5 cm, and the optical wavelength is 850 nm. The collimator len's radius was 165 μm. Under perfect alignment conditions, the radius of the the optical beam at the collector lens is 40 μm, leading to a beam radius of 10 μm at the detector. A longitudinal misalignment of 500 μm increased the beam width at the collector lens by only about 2 μm. Chromatic dispersion of ± 5 nm increased the beam radius at the collector lens by about 20 μm. The total beam radius (in the absence of lateral and angular misalignments) is therefore 62 μm. With this nominal beam size, the size of the collector lens sets the bounds on lateral and angular misalignments. For example, a collector lens with radius 250 μm would allow $\pm(250 - 62) = \pm 188$ μm, allowing lateral misalignments of ± 188 μm or angular misalignments (tolerance at lens divided by distance between boards) of $\pm 0.43°$. Such strict requirements for angular alignment are typically encountered, with increasing length of the beam path magnifying the effect of a given angular misalignment.

Design of optical interconnects such as shown in Fig. 16 is complex, with several competing design constraints. Computer-aided design tools promise to reduce the design complexity, with an early example of such a tool provided in [58].

The examples above have used distinct elements for the optical source (detector) and the lens arrays associated with the source (detector). However, lenses have been directly integrated on the optoelectronic devices directly. For example, Dhoedt et al. [59] describe the fabrication of the lens array directly on the back side of the GaAs substrate in which the optical sources (in this case LEDs) are fabricated. The optical wavelength is restricted to those for which the substrate is transparent (in this example, 925 nm optical beans were used). Figure 17(a) illustrates the PCB interconnection architecture investigated in [59]. Figure 17(b) illustrates the

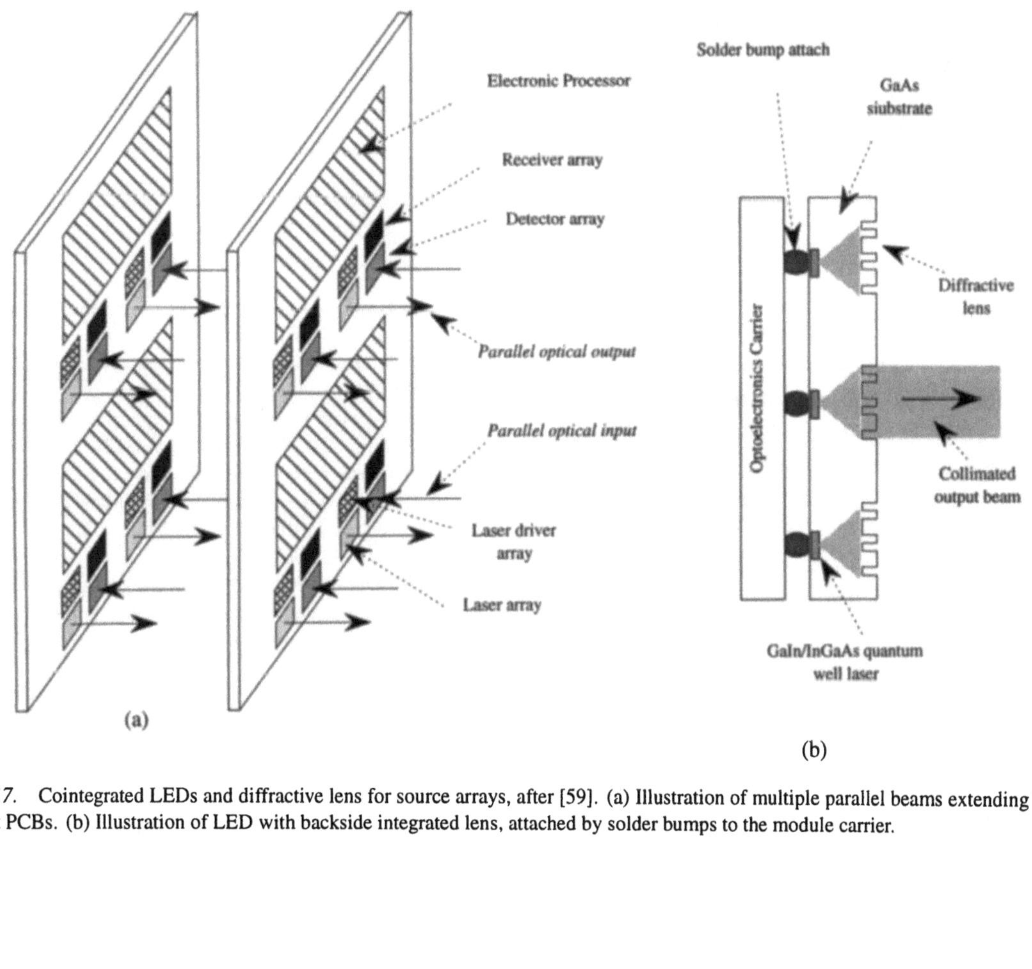

Figure 17. Cointegrated LEDs and diffractive lens for source arrays, after [59]. (a) Illustration of multiple parallel beams extending between adjacent PCBs. (b) Illustration of LED with backside integrated lens, attached by solder bumps to the module carrier.

overall structure of the source array module, with the InAs/InGaAs quantum well LEDs toward the solder bump side and Fresnel lenses fabricated on the back of the substrate. Arrays of such devices allow compact source array and detector array units to be mounted at various points on the PCB, with parallel optical data beams providing direct point-to-point connections for each pair of source/detector sites. Design rules for the binary lens are given in [59]. In this prototype, solder bump attachment to the optical carrier's substrate provides close alignment of the arrays to the carrier, as discussed earlier.

6. Summary

An overview of the many approaches and studies of optical interconnections, with a focus here on optical clock distribution using optical interconnections, has been provided, starting with the intra-IC level of connection and extending to the inter-PCB level of connection. A common thrust in the various approaches studied is the importance of diffractive optical elements, waveguide structures, and free-space paths as elements with which the overall connection can be created. Miniaturization of the optical interconnection elements is a continuing priority in such applications. The techniques presented in this review are largely drawn from approaches which provide such miniaturization, consistent with the decreasing size of high performance electronic systems as the VLSI electronics technologies advance. Another priority for optical interconnections will be manufacturable optical interconnection modules and their practical insertion into real electronic systems. Several of the examples described here (e.g., the folded optical path approach) are attractive in moving toward mass producible optical interconnections. Despite the progress to date, practical insertion of optical interconnections into low levels of electronics has been elusive to date. This is to a large extent a result of the high cost of providing the initial optical interconnections, before a manufacturing environment and equipment have been developed for mass production. However, another barrier, and perhaps a more profound barrier, lies in the interface between the optical interconnections and the standard VLSI circuits which are mass produced in vast numbers. Any adaptions of the manufacturing environment for such VLSI circuits to handle optical interconnections as a specialty item would face considerable barriers to being supported. For such reasons, despite the considerable research and exploratory development and despite the increasing need for a cost effective alternative for critical electrical interconnections, there remains considerable research and exploratory development ahead.

Note

1. Both Irvine Sensors, Corp and Texas Instruments, Corp., for example, have developed such 3-D chip stacks, particularly for the example of high-density memory modules.

References

1. Y. Ota, R.C. Miller, S.R. Forrest, D.A. Kaplan, C.W. Seabury, R.B. Huntington, J.G. Johnson, and J.R. Potopowicz, "Twelve-channel individually addressable InGaAs/InP p-i-n photodiode and InGaAsP/InP LED arrays in a compact package," *IEEE J. Lightwave Technol.*, Vol. 5, No. 4, pp. 1118–1122, 1987.
2. Y.M. Wong et al., "Technology development of a high-density 32 channel 16 Gb/s optical data link for optical interconnections for the optoelectronic technology consortium (OETC)," *IEEE J. Lightwave Technol.*, Vol. 13, No. 6, pp. 995–1016, 1995.
3. H. Karstensen, C. Hanke, M. Honsberg, J.-R. Kropp, J. Wieland, M. Blaser, P. Weger, and J. Popp, "Parallel optical interconnection for uncoded data transmission with 1 Gb/sec-per-channel capacity, high dynamic range, and low power consumption," *IEEE J. Lightwave Technol.*, Vol. 13, No. 6, pp. 1017–1030, 1995.
4. Special issue: *Diffractive Optics: Design, Fabrication, and Applications*, Appl. Optics, Vol. 32, No. 14, 1993.
5. Special issue: *Micro-Optics, Optical Eng.*, Vol. 33, No. 11, pp. 3504–3669, 1994.
6. S.K. Tewksbury, "Interconnections within microelectronic systems," in *Microelectronic System Interconnections: Performance and Modeling*, S.K. Tewksbury (Ed.), IEEE Press, Piscataway, pp. 1–49, 1994.
7. S.K. Tewksbury (Ed.), *Microelectronic System Interconnections: Performance and Modeling*, IEEE Press, Piscataway, NJ, 1994.
8. J.W. Goodman, F.I. Leonberger, S.-Y. Kung, and R.A. Athale, "Optical interconnections for VLSI systems," *Proc. IEEE*, Vol. 72, pp. 850–866, 1984.
9. Special issue: *Optical Interconnects*, Appl. Opt., Vol. 29, pp. 1067–1177, 1990.
10. D.B. Clymer and J.W. Goodman, "Optical clock distribution to silicon chips," *Opt. Eng.*, Vol. 25, pp. 1103–1108, 1986.
11. L.A. Bergman, W.H. Wu, A.R. Johnston, R. Nixon, S.C. Esener, C.C. Guest, P. Yu, T.J. Brabik, M. Feldman, and S.H. Lee, "Holographic optical interconnects for VLSI," *Opt. Eng.*, Vol. 25, pp. 1109–1118, 1996.
12. R.K. Kostuk, L. Wang, and Y.-T. Huang, "Optical clock distribution with holographic optical elements," in *Real-Time Signal Processing XI*, J.P. Letellier (Ed.), *Proc. SPIE*, Vol. 977, pp. 24–36, 1988.
13. S. Koh, H.W. Carter, and J.T. Boyd, "Synchronous global clock distribution on multichip modules using optical waveguides," *Optical Eng.*, Vol. 33, No. 5, pp. 1587–1595, 1994.
14. A. Himeno, H. Terui, and M. Kobayashi, "Loss measurement and analysis of high-silica reflection bending waveguides," *IEEE J. Lightwave Technol.*, Vol. 6, No. 1, pp. 41–46, 1988.

15. F. Lin, E.M. Strzelecki, and T. Jannson, "Optical multiplanar VLSI interconnects based on multiplexed waveguide holograms," *Appl. Optics*, Vol. 29, No. 8, pp. 1126–1133, 1990.
16. F. Lin, C. Nugyen, J. Zhu, and B.M. Hendrickson, "Dispersion effects in a single-mode holographic waveguide interconnect system," *Appl. Optics*, Vol. 31, No. 32, pp. 6831–6835, 1992.
17. S.J. Walker, J. Jahns, L. Li, W.M. Mansfield, P. Mulgrew, D.M. Tennant, C.W. Roberts, L.C. West, and N.K. Ailawadi, "Design and fabrication of high-efficiency beam splitters and beam deflectors for integrated planar micro-optic systems," *Appl. Optics*, Vol. 32, No. 14, pp. 2494–2501, 1993.
18. T. Kubota and M. Takeda, "Array illuminator using grating couplers," *Optics Letts.*, Vol. 14, No. 12, pp. 651–652, 1989.
19. J.M. Miller, M.R. Tagizadeh, J. Turunen, and N. Ross, "Multilevel-grating array generators: Fabrication error analysis and experiments," *Appl. Optics*, Vol. 32, No. 14, pp. 2519–2525, 1993.
20. E. Sidick, A. Knoesen, and J.N. Mait, "Design and rigorous analysis of high-efficiency array generators," *Appl. Optics*, Vol. 32, No. 14, pp. 2599–2605, 1993.
21. S.K. Patra, J. Ma, V.H. Ozguz, and S.H. Lee, "Alignment issues in packaging for free-space optical interconnects," *Optical Eng.*, Vol. 33, No. 5, pp. 1561–1570, 1994.
22. J. Schwider, W. Stork, N. Streibl, and R. Völkel, "Possibilities and limitations of space-variant holographic optical elements for switching networks and general interconnects," *Appl. Optics*, Vol. 31, No. 35, pp. 7403–7410, 1992.
23. K.-H. Brenner and F. Sauer, "Diffractive-reflective optical interconnects," *Appl. Optics*, Vol. 27, No. 20, pp. 4251–4254, 1988.
24. E. Bradley, P.K.L. Yu, and A.R. Johnston, "System issues relating to laser diode requirements for VLSI holographic optical interconnections," *Optical Eng.*, Vol. 28, No. 3, pp. 201–211, 1989.
25. J. Jahns, Y.H. Lee, C.A. Burrus, Jr., and J. Jewell, "Optical interconnects using top-surface-emitting microlasers and planar optics," *Appl. Optics*, Vol. 31, No. 5, pp. 592–597, 1992.
26. G.P. Behrmann and J.P. Bowen, "Influence of temperature on diffractive lens performance," *Appl. Optics*, Vol. 32, No. 14, pp. 2483–2489, 1993.
27. J. Schwider, "Achromatic design of holographic optical interconnects," *Optical Eng.*, Vol. 35, No. 3, pp. 826–831, 1996.
28. T.M. Shen, "Timing jitter in semiconductor lasers under pseudorandom word modulation," *IEEE J. Lightwave Technol.*, Vol. 7, pp. 1394–1399, 1989.
29. A. Weber, W. Ronghan, E. Bottcher, M. Schell, and D. Bimberg, "Measurement and simulation of the turn-on delay time jitter in gain-switched semiconductor lasers," *IEEE J. Quantum Electronics*, Vol. 28, pp. 441–446, 1992.
30. V.N. Morozov and W. Thomas Cathey, "Practical speed limits of free-space global holographic interconnects: Time skew, jitter and turn-on delay," *Appl. Optics*, Vol. 33, No. 8, pp. 1380–1390, 1994.
31. P. Cinato and K.C. Young, Jr., "Optical interconnections within multichip modules," *Optical Eng.*, Vol. 32, No. 4, pp. 852–860, 1993.
32. S.K. Tewksbury, *Wafer Level System Integration: Implementation Issues*, Kluwer Academic Publishers, Boston, 1989.
33. S.K. Tewksbury and L.A. Hornak, "Multichip modules: A platform for optical interconnections within microelectronic systems," *Int. J. Optoelectronics, Devices, and Technologies*, MITA Press, Japan, Vol. 9, No. 1, pp. 55–80, 1994.
34. L.A. Hornak and S.K. Tewksbury, "On the feasibility of through-wafer optical interconnects for hybrid wafer-scale integrated architectures," *IEEE Trans. Elect. Dev.*, Vol. 34, No. 7, pp. 1557–1563, 1987.
35. D.S. Wills, W.S. Lacy, C. Camperi-Ginestet, B. Buchanan, H.H. Cat, S. Wildinson, M. Lee, N.M. Jokerst, and M.A. Brooke, "A three-dimensional high-throughput architecture using through-wafer optical interconnect," *IEEE J. Lightwave Technol.*, Vol. 13, No. 6, pp. 1085–1092, 1995.
36. M.J. Wale et al., "A new self-aligned technique for the assembly of integrated optical devices with optical fiber and electronic interfaces," *Proc. ICIC '89*, paper ThA19-7, p. 368, 1989.
37. M.J. Goodwin, A.J. Mosely, M.G. Kearly, R.C. Morris, D.J. Groves-Kirkby, J. Thompson, R.C. Goodfellow, and I. Bennion, "Optoelectronic component arrays for optical interconnection of circuits and systems," *IEEE J. Lightwave Technol.*, Vol. 9, No. 12, pp. 1639–1645, 1991.
38. R. Selvaraj, H.T. Lin, and J.F. McDonald, "Integrated optical waveguides in polyimide for wafer scale integration," *IEEE J. Lightwave Technol.*, Vol. 6, pp. 1034–1037, 1988.
39. J.C. Lyke, R. Wojnarowski, G.A. Forman, E. Bernard, R. Saia, and B. Gorowitz, "Three dimensional patterned overlay high density interconnect (HDI) technology," *Journal of Microelectronic Systems Integration*, Vol. 1, No. 2, pp. 99–141, 1993.
40. R.S. Beech and A.K. Ghosh, "Optimization of alignability in integrated planar-optical interconnect packages," *Appl. Optics*, Vol. 32, No. 29, pp. 5741–5749, 1993.
41. J. Jahns and R.A. Brumback, "Integrated-optical split-and-shift module based on planar optics," *Opt. Commun.*, Vol. 76, pp. 318–323, 1990.
42. F. Sauer, "Fabrication of diffractive-reflective optical interconnects for infrared operation based on total internal reflection," *Appl. Optics*, Vol. 28, pp. 386–388, 1989.
43. J. Jahns and A. Huang, "Planar integration of free-space optical components," *Appl. Optics*, Vol. 28, pp. 1602–1605, 1989.
44. H.J. Haumann, H. Kobolla, F. Sauer, J. Schmidt, J. Schwider, W. Stork, N. Streibl, and R. Völkel, "Optoelectronic interconnections based on a light-guiding plate with holographic coupling elements," *Opt. Eng.*, Vol. 30, pp. 1620–1623, 1991.
45. L.A. Hornak, S.K. Tewksbury, J.C. Barr, W.D. Cox, and K.S. Brown, "Optical interconnections and cryoelectronics: Complimentary enabling technologies for emerging mainstream systems," *SPIE Photonics West Symposium, Optical Interconnections III Conference*, San Jose, CA, Feb. 1995.
46. J.W. Parker, "Optical interconnection for advanced processor systems: A review of the ESPRIT II OLIVES program," *IEEE J. Lightwave Technol.*, Vol. 9, pp. 1764–1773, 1991.
47. Y. Yamanaka, K. Yoshihara, I. Ogura, T. Numai, K. Kasahara, and Y. Ono, "Free-space optical bus using cascaded vertical-to-surface transmission electrophotonic devices," *Appl. Optics*, Vol. 31, No. 23, pp. 4676–4681, 1992.
48. A. Guha, J. Briston, C. Sullivan, and A. Husain, "Optical interconnects for massively parallel architectures," *Appl. Optics*, Vol. 29, pp. 1077–1093, 1990.
49. K. Kasahara, Y. Tahiro, N. Hamao, M. Sugimoto, and T. Yanese, "Double heterostructure optoelectronic switch as a dynamic memory with low-power consumption," *Appl. Phys. Lett.*, Vol. 52, pp. 679–681, 1988.
50. K. Rastani and W.M. Hubbard, "Alignment and fabrication tolerances of planar gratings for board-to-board optical interconnects," *Appl. Optics*, Vol. 31, pp. 4863–4870, 1992.

51. "High-speed optical interconnects for digital systems," *Lincoln Labs. Journal*, Vol. 4, pp. 31–43, 1991.
52. D.Z. Tsang, "Free-space board-to-board optical interconnections" in *Optical Enhancements to Computing Technology*, J.A. Heff (Ed.), SPIE Vol. 1563, pp. 66–71, 1991.
53. D.Z. Tsang and T.J. Goblick, "Free-space optical interconnection technology in parallel processing systems," *Optical Eng.*, Vol. 33, No. 5, pp. 1524–1531, 1994.
54. C.T. Sullivan, "Optical waveguide circuits for printed wire-board interconnections," *Proc. SPIE, Optoelectronic Materials, Devices and Packaging*, Vol. 994, p. 92, 1988.
55. R.C. Kim, E. Chen, and F. Lin, "An optical holographic backplane interconnect system," *IEEE J. Lightwave Technol.*, Vol. 9, pp. 1650–1656, 1991.
56. S. Natarajan, C. Zhao, and R.T. Chen, "Bi-directional optical backplane bus for general purpose multiprocessor board-to-board optoelectronic interconnects," *IEEE J. Lightwave Technol.*, Vol. 13, No. 6, pp. 1031–1040, 1995.
57. R.K. Kostuk, J.-H. Yeh, and M. Fink, "Distributed optical data bus for board-level interconnections," *Appl. Optics*, Vol. 32, No. 26, pp. 5010–5021, 1993.
58. R.K. Kostuk, "Simulation of board-level free-space optical interconnects for electronic processing," *Appl. Optics*, Vol. 31, No. 14, pp. 2438–2445, 1992.
59. B. Dheodt, P. De Dobbalaer, J. Blondelle, P. Van Daile, P. Demester, and R. Baets, "Monolithic integration of diffractive lenses with LED arrays for board-to-board free space optical interconnect," *IEEE J. Lightwave Technol.*, Vol. 13, No. 6, pp. 1065–1073, 1995.
60. R.S. Beech and A.K. Ghosh, "Optimization of alignability in integrated planar-optical interconnect packages," *Appl. Optics*, Vol. 32, No. 29, pp. 5741–5749, 1993.

Stuart K. Tewksbury received BS and PhD degrees from the University of Rochester, both in physics. From 1969 through 1990, he was with the research division of AT&T Bell Laboratories where his research included digital signal processing, low temperature electronics, advanced packaging, optical interconnections, and parallel computation engines. On retirement from AT&T Bell Laboratories, he joined the Dept. of Electrical and Computer Engineering at West Virginia University where he is a full professor. In addition to extending his earlier research interests at WVU, he is exploring advanced image processing and parallel DSP image processors.
skt@msrc.wvu.edu

Lawrence A. Hornak received his B.S. in Physics from the state University of New York at Binghamton in 1982, his M.E. from Stevens Inst. of Technology in 1986 and his Ph.D. in Electrical Engineering from Rutgers in 1991. In 1982, he joined AT&T Bell Laboratories, Holmdel, NJ where until mid-1991, he was a member of technical staff engaged in various research areas including robotic sensors, high—T_c superconducting interconnections, and optical interconnections. In 1991, Dr. Hornak joined the Department of Electrical and Computer Engineering at West Virginia University where he is currently an Associate Professor and Interim Chair. At WVU, Dr. Hornak has continued research exploring the mapping of high performance technologies into advanced wafer-level Si and MCM-based systems.
lah@msrc.wvu.edu

Journal of VLSI Signal Processing 16, 247–276 (1997)

Timing of Multi-Gigahertz Rapid Single Flux Quantum Digital Circuits

KRIS GAJ, EBY G. FRIEDMAN AND MARC J. FELDMAN
Department of Electrical Engineering, University of Rochester, Rochester, New York 14627

Received November 24, 1996; Revised December 18, 1996

Abstract. Rapid Single Flux Quantum (RSFQ) logic is a digital circuit technology based on superconductors that has emerged as a possible alternative to advanced semiconductor technologies for large scale ultra-high speed, very low power digital applications. Timing of RSFQ circuits at frequencies of tens to hundreds of gigahertz is a challenging and still unresolved problem. Despite the many fundamental differences between RSFQ and semiconductor logic at the device and at the circuit level, timing of large scale digital circuits in both technologies is principally governed by the same rules and constraints. Therefore, RSFQ offers a new perspective on the timing of ultra-high speed digital circuits.

This paper is intended as a comprehensive review of RSFQ timing, from the viewpoint of the principles, concepts, and language developed for semiconductor VLSI. It includes RSFQ clocking schemes, both synchronous and asynchronous, which have been adapted from semiconductor design methodologies as well as those developed specifically for RSFQ logic. The primary features of these synchronization schemes, including timing equations, are presented and compared.

In many circuit topologies of current medium to large scale RSFQ circuits, single-phase synchronous clocking outperforms asynchronous schemes in speed, device/area overhead, and simplicity of the design procedure. Synchronous clocking of RSFQ circuits at multigigahertz frequencies requires the application of non-standard design techniques such as pipelined clocking and intentional non-zero clock skew. Even with these techniques, there exist difficulties which arise from the deleterious effects of process variations on circuit yield and performance. As a result, alternative synchronization techniques, including but not limited to asynchronous timing, should be considered for certain circuit topologies. A synchronous two-phase clocking scheme for RSFQ circuits of arbitrary complexity is introduced, which for critical circuit topologies offers advantages over previous synchronous and asynchronous schemes.

1. Introduction

The recent achievements of superconductive circuits using *Rapid Single Flux Quantum* (*RSFQ*) logic make this technology a possible candidate to first cross the boundary of 100 GHz clock frequency in a large scale digital circuit. The success of RSFQ circuits is in part due to the unique convention used to represent digital information. Rather than using steady voltage levels, RSFQ circuits use quantized voltage pulses to transmit binary logic state information. This logic scheme has necessarily led to new timing concepts and techniques in order to coordinate the operation of the gates and sub-circuits at multigigahertz frequencies. Nevertheless, the similarities to semiconductor voltage-state timing are strong, and the two technologies can be discussed in the same language.

This paper is written with the intention that both semiconductor and superconductor communities will benefit from the mutual exchange of ideas on the timing of high speed large scale digital circuits. RSFQ designers inherit a broad range of techniques and methods developed over many years by VLSI semiconductor circuit designers. The capability of RSFQ technology offers the semiconductor community an opportunity to be made aware about existing pitfalls in the design and implementation of clocking schemes at multi-gigahertz clock frequencies, and to benefit from innovative timing schemes that have been proved to work correctly at frequencies as yet unavailable in semiconductor technologies.

1.1. Advantages of RSFQ Logic

The basic concepts and recent progress in RSFQ logic are reviewed in [1–4]. The most significant advantages are high speed, low power, and the potential for large scale integration. Today, relatively complex circuits consisting of roughly 100 clocked gates have been designed and tested at frequencies about 10 GHz by several groups [3, 5–9]. The simplest digital circuit has been demonstrated at 370 GHz [10]. The on-chip power dissipation is negligible, below 1 μW per gate, so that ultra-high device density may eventually be realized. Additional advantages are that RSFQ circuits require only a dc power supply, can employ either an external or an internal clock source, have a negligible bit error rate [11], and the fabrication technology is fairly simple. The primary disadvantages include the necessity of helium cooling, and a relatively underdeveloped fabrication infrastructure. If one recognizes that the standard feature size in today's still primitive superconductive technology is about ten times larger compared to a state-of-the art CMOS process, it is impressive that RSFQ circuits still offer two orders of magnitude speed-up in clock frequency and three orders of magnitude smaller power dissipation [4].

With these features, RSFQ can be established with a relatively modest effort as a technology of choice for high performance digital signal processing [3], wideband communication [12–14], precise high frequency instrumentation [15], and numerous scientific applications [16, 17]. In the longer term, RSFQ may also provide the speed and power characteristics required by general purpose petaflop-scale computing (petaflop = 10^{15} floating point operations per second), which is likely to remain beyond the reach of the fastest semiconductor technologies [18, 19].

The primary immediate application of RSFQ logic is digital signal processing. The current state of RSFQ technology favors the design of circuits with a regular topology, limited control circuitry, a small number of distinct cells, and limited interconnections. The analysis of timing in RSFQ circuits presented in this article focuses on but is not limited to this type of architecture, which is well suited for most DSP functions.

1.2. Introduction to RSFQ Timing

Correct timing is essential to fully exploit the high speed capability of individual RSFQ gates, and to translate this advantage into a corresponding speed-up in the performance of medium to large scale RSFQ circuits. Research in this area has only just started and has been only applied to moderate 100-gate circuits to date. Yet even for this medium scale complexity, the design of effective timing schemes in the multi-gigahertz frequency range is a challenging problem.

Timing methodologies for semiconductor VLSI circuits have been well-established and systematized [20–25]. One approach to superconductor circuit design is to rely on the application of such rules and techniques drawn from the semiconductor literature. More prevalent, however, the RSFQ clocking circuitry is developed specifically for RSFQ logic [3, 26, 27]. In this paper these two approaches are intertwined, and the similarities and the differences between semiconductor and superconductor designs are highlighted.

The emerging novel methodologies for designing the clocking circuitry in RSFQ circuits diverge from and challenge two well established rules used in the design of digital semiconductor circuits. First, the idea of *equipotential clocking*, in which the entire clock distribution network is considered to be a surface that must be brought to a specific state (voltage level) every half clock period. The analog of equipotential clocking for RSFQ circuits requires that only one SFQ pulse is present in the clock path from the clock source to the input of any synchronous RSFQ gate. This is inefficient for RSFQ circuits, in which several consecutive clock pulses can coexist within a path of the clock distribution network. Actually, equipotential clocking is inefficient for the design of ultrafast digital circuits in semiconductor technology as well, and can be easily replaced by the less restrictive *pipelined clocking* as suggested in the literature [28, 29].

Second, the ubiquitous *zero-skew clocking* is not a natural choice for RSFQ circuits. Clocking schemes that offer better performance *or* improved tolerance to process induced timing parameter variations have been proposed and analyzed [3, 26, 27]. These schemes utilize intentional clock skew to trade circuit performance with circuit robustness. Techniques that offer a significant improvement in performance over zero-skew clocking without affecting circuit yield have been developed and applied to RSFQ circuits [27, 30]. Similar schemes have been proposed earlier for semiconductor logic [31–33], but these approaches have not as yet been widely accepted. The primary reasons are conservative design conventions used within industry, complex design procedures [32, 34, 35], relatively small performance improvements (up to 40%), and difficulties in implementing well-controlled delay lines within semiconductor-based clock distribution networks [22, 35]. The success of RSFQ logic may lead to reconsidering the applicability of these techniques to ultrafast semiconductor circuits.

In addition, the emergence of multi-gigahertz RSFQ logic provides a new perspective on several early continuing controversies concerning the design of high speed digital circuits. A central dilemma is the choice between synchronous and asynchronous clocking [29, 36]. *RSFQ logic is well suited for both types of clocking.* Asynchronous event driven schemes such as *dual-rail logic* or *micropipelines* [37] appear to be easier and more natural to implement in RSFQ circuits than in semiconductor-based logic [1, 26, 38, 39]. The same applies to a *bit-level pipeline* synchronous architecture [40]. *Wave pipelining* used to increase the performance of pipelined semiconductor-based circuits [41, 42] can be used with RSFQ logic. It has also been shown that RSFQ is specifically suitable for the *Residue Number System* (*RNS*) representation of numbers [43, 44]. Operations using this representation are extremely efficient and easier to perform in RSFQ than in semiconductor-based logic but conversion difficulties and multiple frequency clocking will likely limit the use of RNS in mainstream applications.

Most medium-scale RSFQ circuits developed to date are fully synchronous circuits with one phase clocking. This trend is likely to continue, unless the problems with scalability of *multidimensional arrays* and large parameter variations require the application of *asynchronous* or *hybrid, globally asynchronous locally synchronous* schemes. In this paper a new two-phase clocking scheme is introduced which offers advantages in robustness, performance, and design simplicity over the ubiquitous single-phase clocking. However, it is far from clear that these advantages are sufficient for any multiple-phase clocking scheme to justify the device/area overhead inherent in these schemes.

2. RSFQ Logic vs. Semiconductor Logic

In this section the similarities and the differences between RSFQ and semiconductor logic elements are discussed. The most important and fundamental difference between the two technologies appear at the device level, as described in Subsection 2.1. The device level differences affect the gate design, and the basic suite of RSFQ gates differs substantially from those familiar in semiconductor logic design, as seen in Subsection 2.2. For example, several RSFQ gates with no direct analog in semiconductor-based logic appear to be the most natural components of RSFQ circuits for DSP applications [3]. All of these differences between RSFQ and semiconductor logic naturally influence the choice of timing schemes, as discussed in this paper; it will nevertheless become clear that the higher the level of abstraction the less significant the differences become.

2.1. Differences at the Device Level

Device and circuit level differences between RSFQ logic and semiconductor-based logic are summarized in Table 1. The primary difference is the use of a two-terminal *Josephson junction* as the basic active

Table 1. RSFQ vs. semiconductor voltage-stage technologies.

Characteristics	RSFQ	Semiconductor logic families
Basic active component	*Josephson junction* (2-terminal)	*Transistor* (3-terminal)
Basic passive component	*Inductance*	*Capacitance*
Information transmitted as	Quantized voltage *pulse*	Voltage *level*
Information stored as	*Current* in the *inductance* loop	*Charge* at the *capacitance*
Basic logic gates	*Synchronous*	*Asynchronous* (combinational)
Gate fanout	1	>1
Parasitic component	Parasitic inductance	Parasitic capacitance and resistance
Passive interconnects	Microstrip lines (only for long connections)	Metal RC lines (only for short connections)
Active interconnects	Josephson transmission lines + splitters	Metal RC lines with buffers

component of superconductor-based circuits, as compared to the three-terminal transistor in semiconductor-based circuits. Josephson junctions support the transmission, storage, and processing of information in RSFQ logic [1].

Magnetic field is quantized in a superconductor. It is natural to convey information in superconducting circuits in the form of quantized voltage pulses, each corresponding to the transmission of a basic quantum of the magnetic field called a *single flux quantum* (SFQ). The area of an SFQ pulse, the voltage integrated over time, is equal to

$$\int V(t)dt = \Phi_0 = h/2e = 2.07\ \mathrm{mV}\cdot\mathrm{ps}, \quad (1)$$

where h is a Planck's constant and e is the electron charge unit. The shape of an SFQ pulse is shown in Fig. 1(a). The pulse width is in the range of several picoseconds and the pulse height is sub-millivolts for a current niobium-trilayer superconductive fabrication technology [2, 4]. Note that this form of information

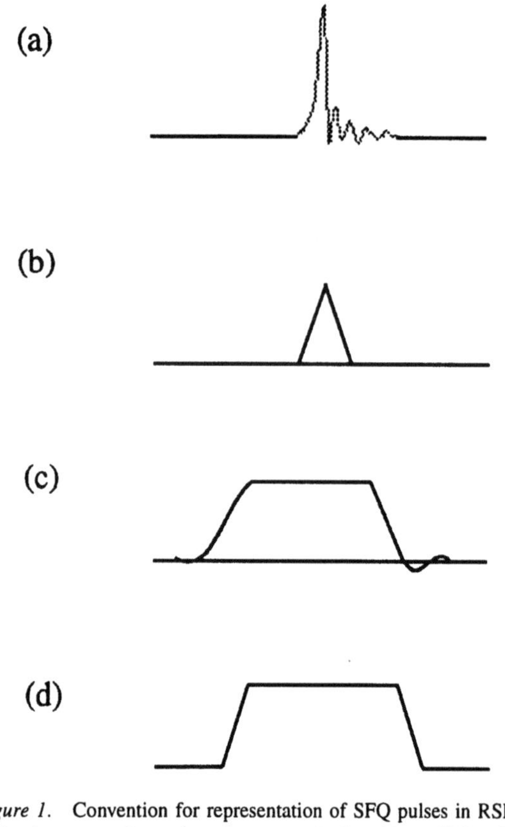

Figure 1. Convention for representation of SFQ pulses in RSFQ and voltage waveforms in voltage-state logic. (a) an SFQ pulse, (b) simplified graphical representation of an SFQ pulse, (c) voltage waveform, (d) simplified graphical representation of a voltage waveform.

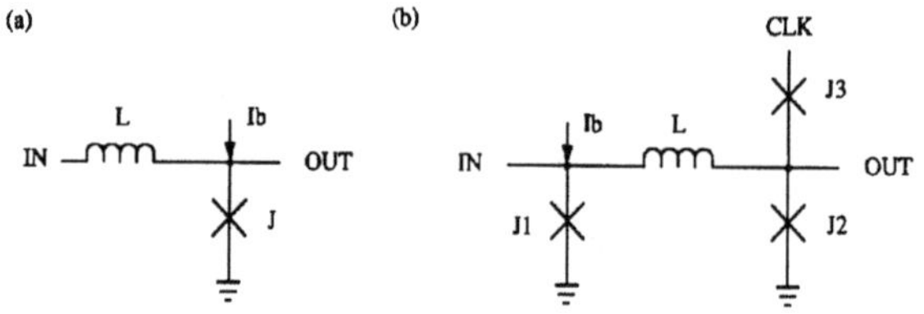

Figure 2. Circuit-level schematic of (a) single stage of a Josephson transmission line (JTL), (b) inductive storage loop including comparator. Notation: Jn—junction, L—inductor, Ib—bias current source.

is measured in fundamental physical constants and is intrinsically digital.

In this paper an SFQ pulse is graphically represented by the symbol shown in Fig. 1(b). Associated with every pulse is a single unique moment in time corresponding to the position of the peak of the pulse voltage. This convention follows the example of the simplified graphical representation of the voltage waveform commonly used in semiconductor digital circuit design, as shown in Figs. 1(c) and (d).

The basic active transmission component of SFQ circuits is called the *Josephson transmission line* (JTL). Single JTL stages (shown in Fig. 2(a)) are connected in series to transmit SFQ pulses without loss over an arbitrary distance. The delay of a single stage is several picoseconds, depending in part on the bias current, and so JTLs provide well-controlled and mutually correlated delays for the design of the clock distribution network. JTLs comprise most of the interconnections in medium to large scale RSFQ circuits, appearing both in the data paths between RSFQ gates and in the clock distribution network.

The basic storage component has the form of an *inductive storage loop* composed of two junctions (J1 and J2) and an inductor (L), as shown in Fig. 2(b). The presence of current in the loop corresponds to the logic state "1". The absence of current corresponds to the logic state "0". The current circulates around the loop without loss, until the state of the loop is evaluated. This evaluation is performed using a *Josephson comparator* which is composed of two serially connected junctions (junctions J3 and J2 in Fig. 2(b)). If the loop contains a logical "1," a pulse at the clock input generates a pulse at the output; if the loop contains a logical "0," no output pulse is generated.

The circuit shown in Fig. 2(b) (a storage loop with a comparator) constitutes the core of the simplest RSFQ clocked gate called a *Destructive Read-Out* cell or *DRO*. The behavior of a DRO for typical input stimuli is shown in Fig. 3(a). Note from Fig. 3(a) and (b) that

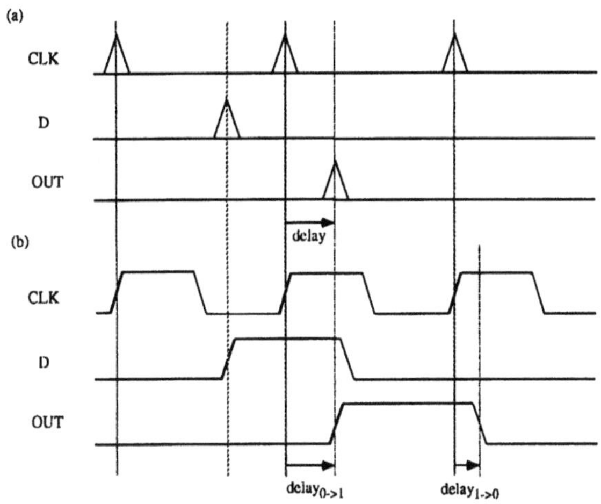

Figure 3. Comparison between the operation of (a) an RSFQ destructive read-out (DRO) cell, and (b) a semiconductor positive edge-triggered *D* flip-flop.

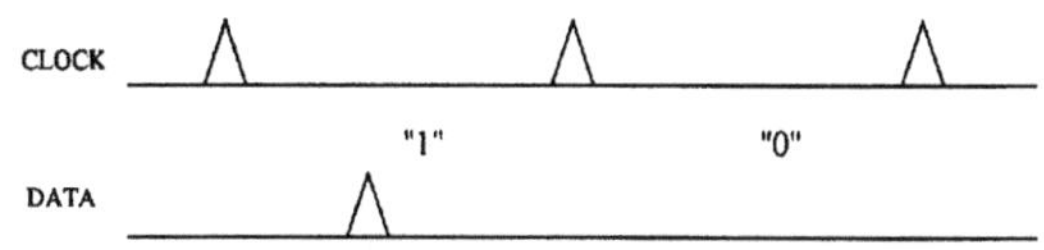

Figure 4. Basic RSFQ convention for representation of logic states.

a DRO is the RSFQ analog of the semiconductor edge-triggered *D* flip-flop. The event that changes the state of the *D* flip-flop is the rising edge of a voltage waveform; the corresponding event that changes the state of a DRO is the SFQ pulse.

Basic RSFQ logic gates (e.g., AND, OR, XOR) are composed of a combination of overlapping and interconnected inductive storage loops supplemented with JTL stages and other simple combinations of junctions and inductances [1, 2]. As a result, these gates always contain a clock input used to evaluate the contents of one or several inductive storage loops, and to release the output pulse. Therefore, *most basic RSFQ logic gates are synchronous* as compared to asynchronous combinational semiconductor gates. It is seen that the logic function of an RSFQ gate is inseparable from its storage capability.

The output logic state of an RSFQ gate is clearly determined: an output pulse (or no pulse) following the clock pulse signifies the output logic state "1" (or "0"). However, the RSFQ Basic Convention [1, 45] is required to specify the input logic state of an RSFQ gate: The appearance of a pulse at the data input of the gate in a window determined by two consecutive clock pulses corresponds to a logical "1," the absence of a pulse at the data input in the same window corresponds to a logical "0," as shown in Fig. 4. This convention distinguishes RSFQ from all semiconductor logic families and from other superconductive logic families.

Another important difference among RSFQ and other logic families is the fanout is always equal to one for all RSFQ cells, as compared with fanouts of greater than one for semiconductor logic gates and buffers. Whenever a connection to more than one input is required, a special cell called a *splitter* is used [1]. A splitter repeats at its two outputs the sequence of pulses from its input. As for the JTL, the splitter introduces a significant input- to-output delay that may affect the timing of the circuit. Splitters are inevitable components of RSFQ clock distribution networks.

Another unique feature of this superconductive technology is that an *SFQ pulse can be transferred over large distances with a speed approaching the speed of light,* using passive superconductive *microstrip lines* [1, 2, 46]. This feature was used only recently in the design of an RSFQ clock distribution network [7].

RSFQ is intrinsically a low power technology, but there is an important distinction compared to low power CMOS. In CMOS, the energy is dissipated mainly in the form of a dynamic power during voltage transitions in the circuit nodes. Therefore, the power consumption can be minimized by eliminating redundant activity in the circuit nodes even at the cost of increasing the number of transistors in the circuit. In RSFQ, the energy is consumed primarily in the form of a static power dissipated by current sources providing the bias current to the junctions. Thus, *power consumption is directly proportional to the number of junctions in the circuit.*

2.2. *Function and Complexity of Basic RSFQ Gates*

The logic function of an RSFQ circuit of any complexity can be easily described using a *Mealy state transition diagram* [1], known well from semiconductor digital circuit design. As most RSFQ gates are clocked, these gates contain an internal memory and at least two distinct internal states. A state transition diagram for the DRO cell is shown in Fig. 5(a), together with a symbol of the gate. The nodes of the Mealy diagram correspond to the two distinct logic states of the DRO storage loop. The arrows show transitions that appear as a result of input pulses (including clock pulses). Output data pulses are associated with transitions between states, and for synchronous cells appear as a result of pulses that arrive at the clock input.

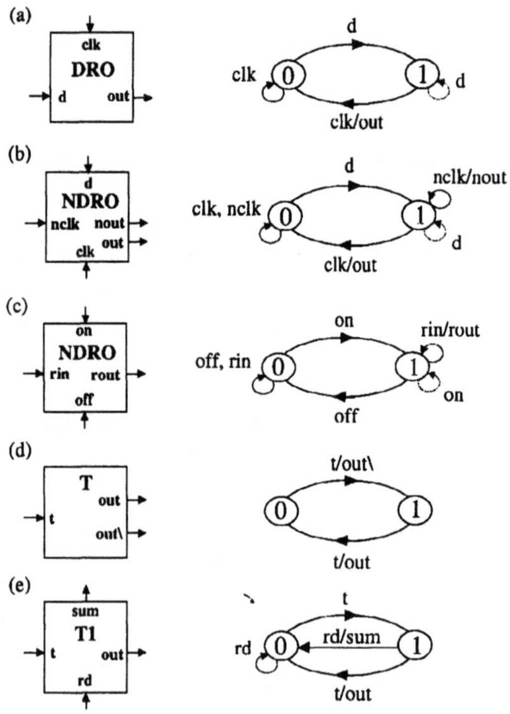

Figure 5. Symbols and Mealy state transition diagrams for basic RSFQ gates: (a) DRO, (b), (c) NDRO, (d) T flip-flop, (e) T1 flip-flop.

The function of most elementary RSFQ gates may be described by analogy to the function of their semiconductor counterparts, as shown in Table 2. Note, however, that this analogy must be correctly understood. The behavior of the two circuits is similar but not identical: the rising edge of a voltage waveform in a semiconductor circuit corresponds to an SFQ pulse in the RSFQ counterpart, as shown in Fig. 3.

Table 2. Semiconductor counterparts of RSFQ gates.

RSFQ gate	Semiconductor counterpart
Clocked	
DRO	*D* flip-flop
NOT	NOT + *D* flip-flop
AND	AND + *D* flip-flop
OR	OR + *D* flip-flop
XOR	XOR + *D* flip-flop
Non-clocked without memory	
Splitter	Buffer with fanout two
Confluence buffer	Event OR
Non-clocked with memory	
NDRO	Transmission gate
Coincidence junction	Muller C-element

Table 3. Complexity of RSFQ gates and CMOS counterparts.

RSFQ gate	# of JJs	CMOS gate	# of transistors
DRO	4	*D* flip-flop	12
NOT	4	NOT + *D* flip-flop	2 + 12
AND	14	AND + *D* flip-flop	6 + 12
OR	8	OR + *D* flip-flop	6 + 12
XOR	7	XOR + *D* flip-flop	6 + 12
T-flip-flop	5	—	—
T1-flip-flop	8	—	—
Confluence buffer	5	OR	6
Splitter	3	Buffer with fanout two	4
NDRO	9	Transmission gate	2

Review articles on RSFQ [1, 2, 47] describe state transition diagrams, circuit level schematics, and device parameters for the majority of basic RSFQ cells. The existing suite of basic RSFQ gates does not include such elementary semiconductor gates as NAND, NOR, and XNOR [48]. This difference occurs since inversion is more difficult to obtain in RSFQ than in voltage stage logic [2]. Also, the relative complexity of various cells differs substantially between the two technologies, as shown in Table 3 [1, 48]. These differences require new design methodologies, including a different set of elementary gates. These differences also make the automated logic synthesis of large RSFQ circuits particularly challenging.

Apart from clocked gates, a basic set of RSFQ cells also includes several non-clocked (asynchronous) cells that are used to build larger synchronous or asynchronous RSFQ circuits. Non-clocked cells without memory include the splitter cell (described above) and the *confluence buffer*. The confluence buffer operates as an asynchronous OR: it passes all pulses from either of its inputs to the output with appropriate delay [1]. The standard implementation of this gate has a significant drawback; it does not allow two input pulses to appear too close in time to each other. If the distance between pulses at the two inputs of the confluence buffer is smaller than the *minimum separation time*, only one pulse will appear at the output.

Most frequently used non-clocked RSFQ gates with internal memory are the NDRO cell [1, 2], T flip-flop [1], and T1 flip-flop [49]. Symbols and state transition diagrams describing each of these cells are shown in Figs. 5(b)–(d) and (e).

The *NDRO* cell can be treated as a simple extension of the DRO cell [1] (Fig. 5(b)). Apart from operating

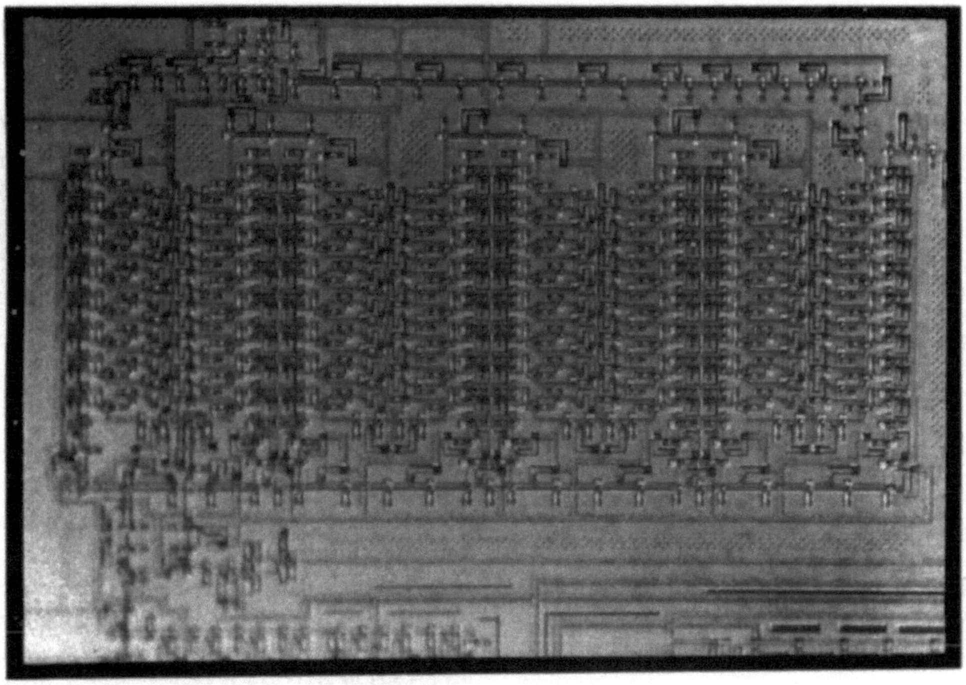

Figure 6. Photograph of the RSFQ circular shift register.

similar to a DRO, it has an additional function associated with an extra non-destructive clock input (nclk) and non-destructive clock output (nout). The non-destructive clock reads the contents of the storage loop to a non-destructive output without changing the internal state of the cell.

Another interpretation of the function of the NDRO cell is given in Fig. 5(c). In this case, the NDRO does not have a previous destructive-read output, and the remaining inputs and outputs have been renamed to better describe this new function. The cell behaves like a CMOS transmission gate [48]. Pulses at inputs ON and OFF permit the gate to transmit, and not transmit, respectively. In the transmitting mode, every pulse from the input RIN propagates with a delay to the output ROUT. In the non-transmitting mode, no pulse appears at the output ROUT regardless of the pulses at the RIN input.

A *T flip-flop* is a modulo two counter that reverses its logical state each time a pulse appears at the T input (Fig. 5(d)). A pulse is generated at its primary output every two input pulses. A T1 flip-flop is an extension of the T flip-flop that permits destructive read-out of an internal state of a T flip-flop to a separate output SUM (Fig. 5(e)).

Other more complex RSFQ cells with sophisticated logic functions have been reported in the literature. These include: a demultiplexer [47, 49, 50], B flip-flop [51], full-adder [1, 49, 52], adder-accumulator [8], carry-save adder [53, 54], and a majority AND gate [52]. Most of these cells cannot be decomposed into simpler RSFQ cells. Special cells with complementary inputs and outputs have been designed to be used with asynchronous dual rail logic [55–58] as described in Section 5. The photograph of a medium size RSFQ circuit—RSFQ circular shift register [59] is shown in Fig. 6.

In most cases, cells specifically designed for RSFQ logic are superior to functionally equivalent cells generated from semiconductor circuit design principles. As an example, in Figs. 7(a) and (b) two equivalent implementations of a half-adder in RSFQ logic are shown. From Table 3, it is seen that the RSFQ-specific implementation results in a circuit with fewer junctions, 20 versus 30 in this case, and thus also a smaller area. A more significant difference between the two implementations, however, is evident when one extends the function of the half-adder to that of a *full adder* or *adder accumulator*. The traditional half-adder in Fig. 7(a) cannot be easily changed; any extension would involve adding several new gates and multiplying the complexity of the circuit. For the RSFQ-specific implementation either modification is small and straightforward (although mutually exclusive, as

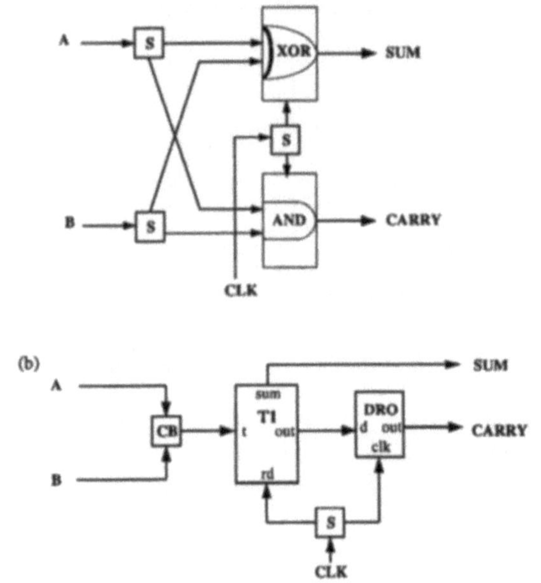

Figure 7. Two implementations of a half-adder in RSFQ logic; (a) based on elementary logic gates; (b) based on gates specific to RSFQ.

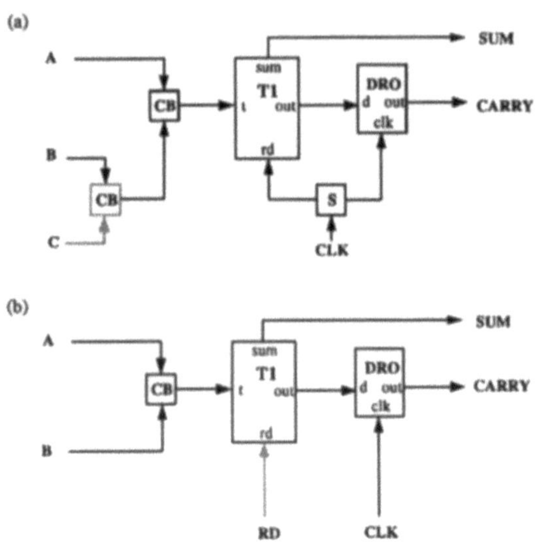

Figure 8. Modification of the half adder to (a) full-adder, (b) adder-accumulator.

a "full adder-accumulator" is not a valid gate). The full adder is obtained by adding a single confluence buffer at the data input as illustrated in Fig. 8(a); the function of an adder-accumulator is created by deleting the splitter at the clock input, and separating the clock (CLK) and read (RD) inputs, as shown in Fig. 8(b).

The best approach for choosing which basic gates should comprise a circuit may depend upon the function of the circuit, for instance digital signal processing vs. general purpose computing and operational unit vs. control unit. For example, the control units in an RSFQ microprocessor can be based on a set of asynchronous *event-driven* (*data-driven*) gates with events represented in the form of SFQ pulses [60–62]. Similarly, the synchronization scheme also influences the most effective suite of basic gates. For instance, asynchronous gates with complementary inputs and outputs are well suited for an asynchronous data-driven synchronization scheme, while basic synchronous RSFQ logic gates (AND, OR, XOR, NOT) are well suited for synchronous bit-level pipelining.

3. Single-Phase Synchronous Clocking

Single-phase synchronous clocking is the form of clocking most frequently used in semiconductor circuit design. Its primary advantages include high performance, design simplicity, small device and area overhead, and good testability. Several authors regarded this kind of clocking as inadequate for ultrafast RSFQ circuits [26, 63]. The main argument used against synchronous clocking is the deteriorating effects of *clock skew* and *phase delay* on circuit robustness and performance. Despite these theoretical limitations, single-phase synchronous clocking has been successfully used in almost all medium to large scale RSFQ circuits developed to date [3, 5–9].

In this section, it is shown that most of the limitations of single-phase synchronous clocking can be easily overcome by applying an appropriate design procedure. In Subsection 3.1, it is shown that using *pipelined* (*flow*) clocking instead of equipotential clocking eliminates the deteriorating effect of *phase delay* (the propagation delay from the clock source to the most remote cell in the clock distribution network) on the circuit performance. In Subsection 3.2, the limitations imposed by the external and internal clock sources are analyzed. In Subsection 3.3, techniques to minimize the effect of *clock skew* on the circuit performance without decreasing the circuit yield and reliability are presented. This discussion is continued in Subsection 3.4 and Subsection 3.5 by analyzing several synchronous clocking schemes with different topologies for the clock distribution network and different values of the interconnect delays. In Subsection 3.6, these clocking schemes are applied to particular circuit—a linear unidirectional pipelined array comprised of N heterogeneous RSFQ cells. A graphical model of the circuit behavior is provided, and the performance of all clocking schemes is compared in terms of circuit throughput and latency. The analysis presented in this section is extended in Section 4 by taking into account the effects of fabrication process variations.

3.1. *Equipotential vs. Pipelined Clocking*

Two basic modes of clocking apply to any general *semiconductor* clock distribution network:

Equipotential clocking [20, 25] assumes that a voltage state (voltage level) at the primary clock input does not change until the previous state has propagated through the longest path in the clock distribution network. This limitation has historically been negligible, as the *phase delay*, i.e., the worst case propagation delay in the clock path, was typically much smaller than the limitation imposed on the clock period by the most critical data path between two registers in the circuit. For high speed large scale semiconductor circuits, however, this is no longer true; the limitation imposed by the propagation delay of the clock distribution network becomes a dominant factor which limits the maximum clock frequency in this type of clocking environment [29].

As described in the literature [28, 29], the requirement of equipotential clocking can be substantially relaxed. In clock distribution networks composed of metal interconnections separated by buffers, it is sufficient that the voltage state in a given node in the network does not change until the previous state has propagated past the nearest buffer. A method of clocking that complies with this much less restrictive rule is called *pipelined clocking* [28]. In *pipelined clocking*, several consecutive clock transitions corresponding to several clock cycles may travel simultaneously along the longest path in the clock distribution network.

In RSFQ logic, even for medium size circuits, the propagation delay through the clock distribution network is often several times larger than the worst case data path delay. Two factors contribute to this. First, the clock distribution network is typically composed of JTLs and splitters, each with a delay comparable to the delay of a single RSFQ gate. Multiple JTL stages must be used to cover the physical distance between the clock inputs of neighboring cells. Second, the data path between two clocked RSFQ gates does not contain any combinational logic. Therefore, equipotential clocking is not considered to be a viable solution for medium to large scale RSFQ circuits. Instead, pipelined clocking, referred in RSFQ literature as *flow clocking* [3], is used in all medium to large scale RSFQ circuits developed to date [3, 5–9]. In flow (pipelined) clocking, several consecutive clock pulses travel simultaneously through the clock distribution network. In the clock distribution network composed of JTLs and splitters, the only limit on the distance between clock pulses originates from the width of the clock pulse [2] and the effects of the interactions between consecutive pulses [64]. Both limitations are negligible compared to the limitations imposed by the critical data path in the circuit.

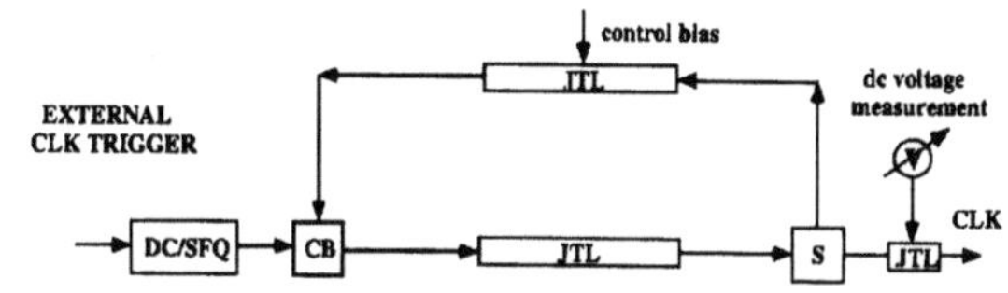

Figure 9. RSFQ clock ring used as an internal clock source. Notation: S—splitter, CB—confluence buffer.

3.2. *Clock Sources*

Additional practical limitations on the maximum clock frequency of RSFQ circuits derive from the characteristics of the available clock sources. When an external clock generator is used, the high-frequency sinusoidal signal must be converted to a string of SFQ pulses using a *DC/SFQ converter* [1, 2]. The maximum frequency of the clock is constrained by the maximum input frequency of the converter. An alternative solution is the use of an on-chip clock generator. An internal clock source can be composed of a JTL ring with a confluence buffer used to introduce the initial pulse to the ring, and a splitter used to read the data from the ring [9, 47, 65], as shown in Fig. 9. The minimum clock period of the ring is limited by the sum of the delays of the splitter and the confluence buffer to less than 100 GHz with current fabrication technology. The other form of on-chip high frequency clock, an overbiased Josephson junction [1], can generate much higher frequencies but it has limitations arising from its relatively large jitter.

3.3. *Synchronization of a Pair of Clocked Cells*

A variety of clocking schemes (single-, two-, and multiple-phase) and associated storage elements are used in semiconductor logic design [22]. *Single-phase clocking* typically requires the use of either *edge-triggered D flip-flops* or *D latches*. In Section 2, Fig. 3, it was shown that the RSFQ basic storage element, DRO, is the analog of the positive edge-triggered *D* flip-flop. The authors are unaware of any analog of a semiconductor *D* latch in RSFQ logic.

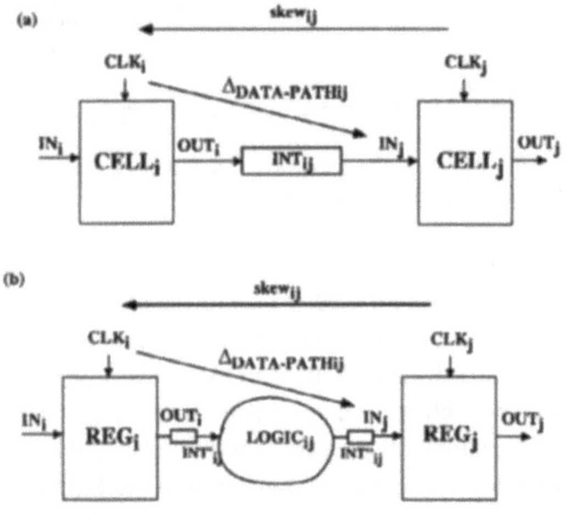

Figure 10. Data path between two sequentially adjacent cells in (a) RSFQ logic; (b) semiconductor logic. Notation: INT_{ij}—interconnection between cells i and j, REG—register composed of D-flip-flops, LOGIC_{ij}—combinational logic, skew_{ij}—clock skew between cells i and j.

Two storage (clocked) cells that exchange data between each other are called *sequentially adjacent. Conditions for the correct exchange of data between a pair of sequentially adjacent RSFQ cells are identical to the conditions for communicating between two semiconductor positive-edge triggered D flip-flops.* These conditions are demonstrated below.

Schematics of generalized *synchronous data paths* for RSFQ and for semiconductor circuits using D flip-flops are shown in Figs. 10(a) and (b). These schematics are almost identical, apart from two important differences. First, in semiconductor circuits, the actual logic function of the circuit is performed by a combinational path (labeled LOGIC_{ij} in Fig. 10(b)) between the two D flip-flop storage components (labeled REG_i and REG_j in Fig. 10(b)). In RSFQ circuits, the logic function is performed by the cells at the beginning and at the end of the data path (labeled CELL_i and CELL_j in Fig. 10(a)). The logic function of an RSFQ gate is inseparable from the storage capability. Interconnections between cells INT_{ij} are typically composed of a few JTL stages and do not perform any logic function. Second, storage cells at the beginning and at the end of the data paths in semiconductor circuits are typically identical for all data paths within the entire system, and are characterized using a single set of timing parameters (hold time, setup time, and the clock-to-output delay of a D flip-flop). In RSFQ circuits, cells at the beginning and at the end of the data paths are not identical, and change from one data path to the next. The hold and setup times of various RSFQ cells differ substantially.

For both technologies, an important parameter describing the data path is the *clock skew* [20, 21]. Clock skew (denoted $skew_{ij}$) is defined as the difference between the arrival time of the clock signal (SFQ pulse in RSFQ, rising edge of the clock waveform in voltage-state logic) at the clock inputs of the cells at the beginning and at the end of data path (t_{CLK_i} and t_{CLK_j}, respectively). The clock skew between cells i and j is

$$skew_{ij} = t_{\text{CLK}i} - t_{\text{CLK}j}. \tag{2}$$

Similarly, the *data path delay* (denoted $\Delta_{\text{DATA-PATH}_{ij}}$) is defined as the interval between the moment when the clock arrives at the *clock* input of the first cell (t_{CLK_i}), and the moment when the data appears at the *data* input of the second cell (t_{IN_j}):

$$\Delta_{\text{DATA-PATH}_{ij}} = t_{\text{IN}_j} - t_{\text{CLK}_i}. \tag{3}$$

Waveforms corresponding to the correct exchange of data between two sequentially adjacent cells in the presence of clock skew are shown in Figs. 11(a) and (b) for voltage state logic and for RSFQ, respectively. From these waveforms, two inequalities that fully describe the timing constraints of the data path between two

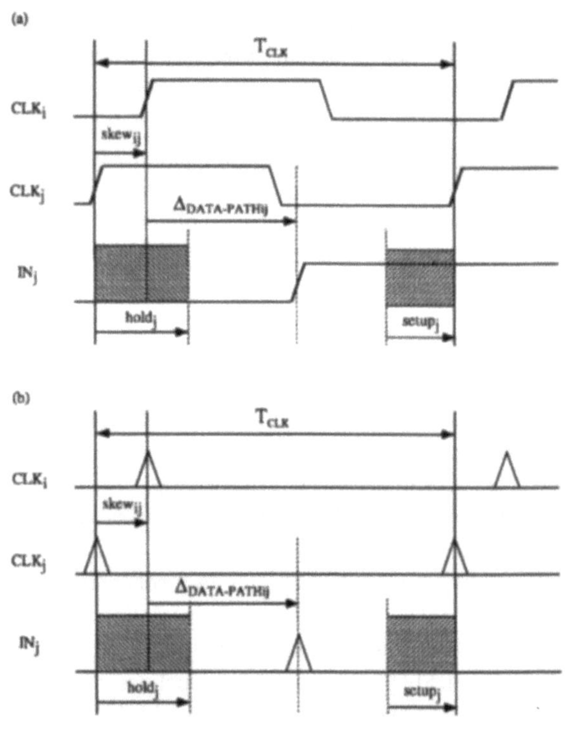

Figure 11. Timing diagram describing the exchange of data between two sequentially adjacent storage cells in (a) semiconductor voltage-state logic, (b) RSFQ logic.

adjacent cells can be derived:

$$skew_{ij} + \Delta_{\text{DATA-PATH}_{ij}} \geq hold_j, \quad (4)$$

$$T_{\text{CLK}} \geq skew_{ij} + \Delta_{\text{DATA-PATH}_{ij}} + setup_j. \quad (5)$$

These inequalities are identical for RSFQ and voltage state logic. The formulas for the data path delay differ between RSFQ and semiconductor technologies. For RSFQ,

$$\Delta_{\text{DATA-PATH}_{ij}} = \Delta_{\text{CELL}_i} + \Delta_{\text{INT}_{ij}}, \quad (6)$$

for a voltage state logic

$$\Delta_{\text{DATA-PATH}_{ij}} = \Delta_{\text{REG}} + \Delta_{\text{LOGIC}_{ij}} + \Delta_{\text{INT}_{ij}}, \quad (7)$$

where Δ_X denotes the delay introduced by the component X.

Using (4) and (5), the dependence between the clock skew and the minimum clock period in the circuit can be determined. Clock skew can be both positive and negative [25]. Positive clock skew increases the minimum clock period [see (5)], but at the same time prevents the possibility of race errors (the propagation of the data through several data paths within one clock period) that occurs when (4) is not satisfied. Negative clock skew decreases the minimum clock period, but makes a violation of the hold time constraint, and thus race errors, more likely.

The operating region of the circuit composed of two sequentially adjacent cells as a function of the clock period and the clock skew between the cells is shown in Fig. 12. The following conclusions can be drawn:

a) Changing the nominal value of the clock skew changes the minimum clock period. The minimum clock period is linearly dependent on the clock skew. *There exist values of clock skew for which the circuit does not work for any* (even an extremely small) *clock frequency.*
b) The minimum clock period is equal to

$$T_{\text{MIN}} = hold_j + setup_j, \quad (8)$$

and is obtained for a clock skew equal to

$$skew_0 = -\Delta_{\text{DATA-PATH}_{ij}} + hold_j. \quad (9)$$

Note that although the hold and setup time may be individually negative, the sum of the hold and setup time is always positive. In contrast, the optimal value of the clock skew, $skew_0$, although typically negative, can be positive for some configurations of RSFQ cells.
c) It can be seen that *zero clock skew is in no respect advantageous compared to other values of clock skew.* It is only a point on a continuum of allowed values of clock skew.

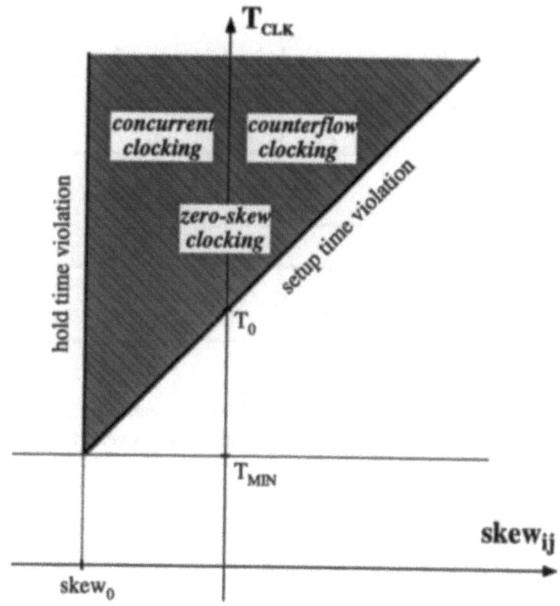

Figure 12. Operating region of a circuit composed of two sequentially adjacent cells as a function of the clock period and the clock skew.

In circuits with a closed data loop [59, 66], the sum of the local clock skews around the loop must be equal to zero. This characteristic however does not imply that all local clock skews must be equal to zero. Local skews may be different in order to minimize the clock period imposed by the most critical data path in the loop (this design procedure is referred to in the literature as "cycle stealing" or as exploiting "useful clock skew" [22, 25]).

Similarly, in many cases the module is a part of a larger (e.g., multi-chip) circuit. If communication between modules is synchronous, the requirement to maintain zero clock skew among all of the inputs and outputs of the module may be imposed [35]. This however does not apply to the current state of RSFQ technology, where the complexity of circuits within a single chip is limited, and the projected inter-chip communication is asynchronous.

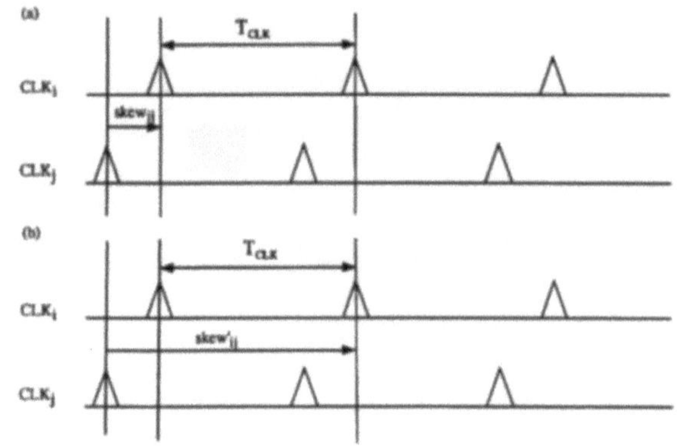

Figure 13. $skew_{ij}$ in (a) and $skew'_{ij}$ in (b) are indistinguishable from the point of view of the circuit operation at clock period T_{CLK}.

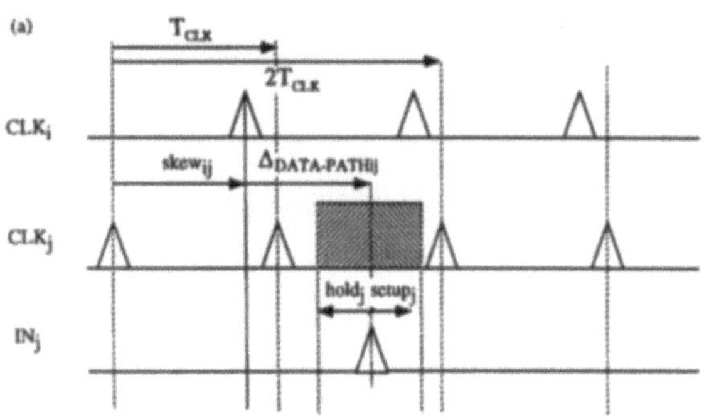

Figure 14. Timing illustration for a circuit operating with a non-conventional value of the clock skew: Counterflow clocking with $k = 1$.

These results can be generalized by the simple observation that for a fixed clock period, values of clock skew that differ by an integer multiple of the clock period are indistinguishable from the point of view of maintaining correct circuit operation, as illustrated in Fig. 13. With this observation, conditions (4) and (5) can be rewritten as follows:

$$skew_{ij} - kT_{CLK} + \Delta_{DATA\text{-}PATH_{ij}} \geq hold_j, \quad (10)$$

$$T_{CLK} \geq skew_{ij} - kT_{CLK} + \Delta_{DATA\text{-}PATH_{ij}} + setup_j. \quad (11)$$

$k = 0$ corresponds to the circuit operating in a standard manner, as shown in Fig. 11(b). The operation of the circuit for the case of $k = 1$ is shown in Fig. 14. The clocking scheme corresponding to $k = -1$ is described in Section 3.4.2. In Fig. 15, the generalized operating region of the circuit composed of two adjacent RSFQ cells as a function of the clock skew and the clock period is shown. Historically, only the operating region corresponding to $k = 0$ has been used, almost exclusively.

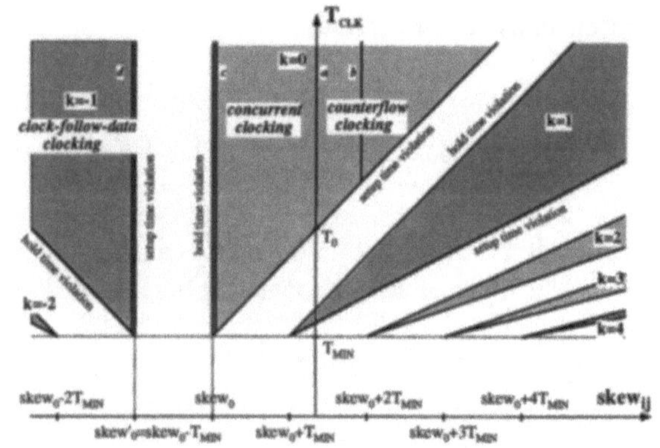

Figure 15. Complete operating space of the data path between two sequentially adjacent cells as a function of the clock frequency and the clock skew. Lines a, b, c, d, correspond to the range of allowed clock frequencies for the circuit with the clock skew fixed to the optimum value for a given clocking scheme (without taking parameter variations into account). a—zero-skew clocking, b—counterflow clocking, c—concurrent clocking, d—clock-follow-data clocking.

3.4. Basic Single-Phase Clocking Schemes

3.4.1. Standard Clocking Modes. The most popular clocking scheme used in *semiconductor* circuit design is single-phase zero-skew equipotential clocking [20, 22]. A clock distribution network used to implement this clocking scheme for a two-dimensional systolic array has the form of an H-tree network consisting of metal lines separated by large-fanout buffers [67, 68], as shown in Fig. 16(a). Buffers within the clock distribution network decrease the time of the clock propagation through the longest path in the network and substantially decrease the requirements on the fanout of the clock source [22]. Nominally, the symmetry of an H-tree clock distribution network assures the simultaneous arrival of the clock signal to the inputs of all the cells in the array. However, in a real circuit there will be timing parameter variations in both the passive and active components of the network, and so the actual clock skew between any two sequentially adjacent cells is randomly distributed around zero [69]. The worst case value of this clock skew depends on the size of the array and on the distribution of the local parameters. This problem is addressed in detail in Section 4.

Zero-skew clocking is relatively easy to implement in RSFQ circuits. In Fig. 16(b), an RSFQ H-tree network composed of JTLs and splitters suited for a square structured systolic array is shown. With some

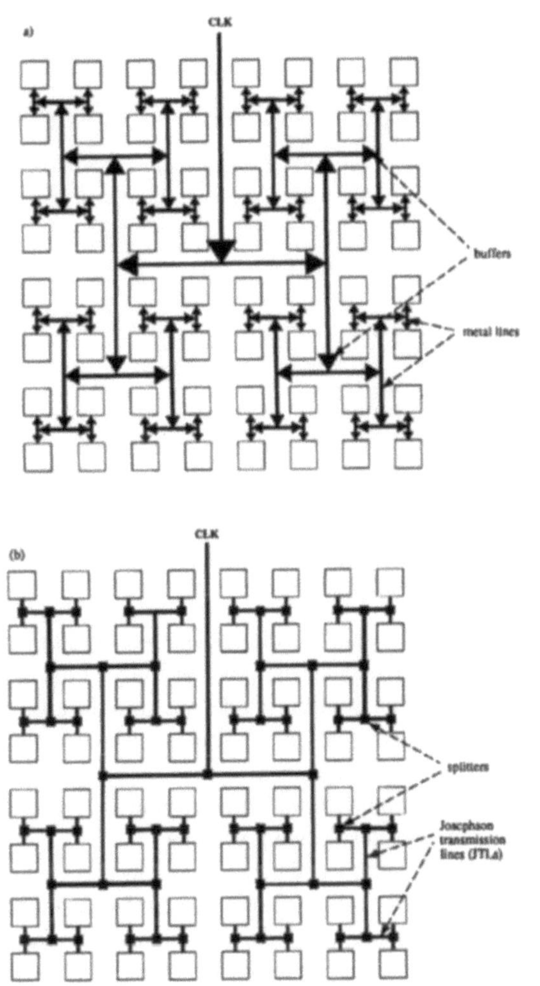

Figure 16. (a) H-tree zero-clock-skew clock distribution network in semiconductor logic. (b) H-tree zero-clock-skew clock distribution network in RSFQ logic.

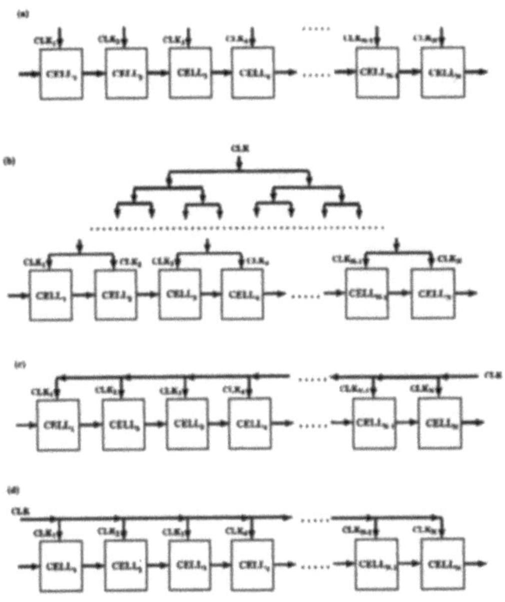

Figure 17. Clocking in the one-dimensional array. (a) General structure of the array; (b) binary-tree zero-skew clocking; (c) straight-line counterflow clocking; (d) straight-line concurrent clocking.

overhead, similar networks can be build for less symmetric circuit structures. However, as shown in the previous section, zero clock skew is in no respect advantageous to other values of the clock skew. Usually, the *optimum* clock skew for a pair of sequentially adjacent cells, $skew_0$ [defined by (9)], is substantially less than zero. Less commonly, for some configurations of RSFQ cells, $skew_0$ may be positive. In this case a circuit with zero clock skew will not operate correctly for any clock frequency. Note that this situation cannot occur in semiconductor circuits for which the hold time of the edge-triggered *D* flip-flop is typically equal to zero (and is certainly less than the delay of the *D* flip-flop) [48], and thus $skew_0$ is always negative.

A general linear pipelined array is shown in Fig. 17(a). When zero-skew clocking is applied the clock distribution network has the form of a binary tree as shown in Fig. 17(b). This has two disadvantages. First, the binary tree is composed of a large number of splitters and JTLs. Second, the skew between the clock source and the clock signals arriving at all of the cells in the array is large, and this may affect the synchronization between the array and the other circuits connected to its inputs and outputs.

An alternative to clocking a one-dimensional systolic array with a binary tree structure is *straight-line clocking* [70], in which the clock path is distributed in parallel to the data path of the array. Two types of straight-line clocking can be distinguished. In *counterflow clocking* [1], the clock flows in the direction *opposite* to the data, as shown in Fig. 17(c). In *concurrent clocking* (also referred to as con-flow [3] or concurrent-flow clocking [1, 27]), the clock and the data flow in the *same* direction, as shown in Fig. 17(d).

For straight line clocking the magnitude of the clock skew is equal to the propagation delay through the clock path between two adjacent cells. In RSFQ circuits, this delay is equal to the delay of a single splitter plus the delay of an interconnecting JTL. The *sign* of the clock skew depends upon the relative direction of the clock and data signals, which is opposite for counterflow vs. concurrent clocking.

For counterflow clocking, clock skew is positive. As shown by Eq. (4) and Fig. 12, a violation of the hold time is less likely than for zero-skew clocking. This characteristic means that counterflow clocking is a robust design strategy—the circuit timing should always be correct at a frequency low enough to satisfy the setup time constraint, even if there are large timing parameter variations. The disadvantage of counterflow clocking is that the minimum clock period of the circuit is larger than for zero-skew clocking by the magnitude of the delay in the clock path.

For counterflow clocked circuits, as shown in (5), the clock skew and hence the propagation delay in the clock path should generally be minimized. This is advantageous because the hold time constraint (4) is typically satisfied even for zero clock skew. Thus counterflow circuits are designed using the minimum number of JTL stages necessary to cover the physical distance between the clock inputs of adjacent cells. A common strategy is to scale the physical dimensions of the JTL (without changing the values of the device parameters) to permit covering the maximum physical distance with the minimum number of JTL stages, and thus with the minimum delay. The correct operating points of the circuit for a fixed clock skew and for clock periods greater or equal to the minimum clock period are indicated by the line *b* in the diagram in Fig. 15.

For concurrent clocking, clock skew is negative. The data released by the clock from the first cell of the data path travels simultaneously with the clock signal in the direction of the second cell. The clock arrives at the second cell earlier than the data. The clock releases the result of the cell operation computed during the last clock cycle, preparing the cell for the arrival of the new data.

Concurrent clocking guarantees greater maximum clock frequency than counterflow or zero-skew clocking. The clock skew in concurrent clocking may be set to the optimum nominal value corresponding to the minimum clock period by choosing an appropriate delay (number of stages) of the interconnect JTL line. The minimum clock period T_{MIN} is given by (8). This limitation is imposed only by the internal speed of the gates, and not by the clock distribution network as in previous schemes. The optimum clock skew is given by (9). Operating points for the optimal clock skew and for clock periods greater or equal to the minimum clock period form the line *c* in the diagram in Fig. 15. The data pulse arrives at the input of the second cell in the worst case data path at the *beginning* of the clock period at the boundary of the *hold* time violation as shown in Fig. 18(a).

In the presence of timing parameter variations affecting both the clock skew and the position of the hold time

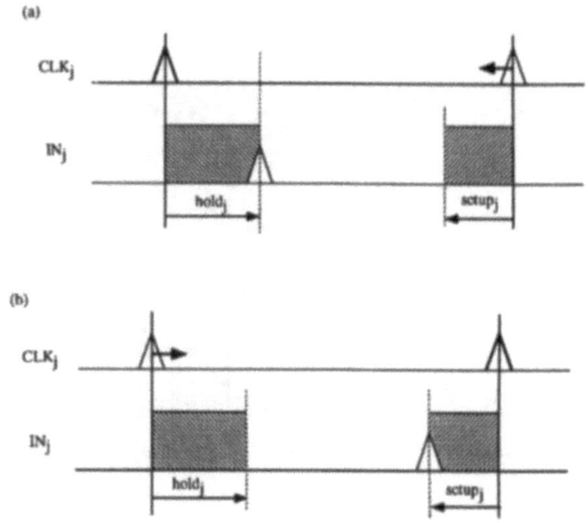

Figure 18. The position of the data pulse within the clock period for the optimal value of the clock skew in (a) concurrent clocking; (b) clock-follow-data clocking.

boundary, the circuit is vulnerable to the hold time violation, which may appear independently of the clock frequency. This is unacceptable, and thus the absolute value of the nominal clock skew must be decreased, as described in detail in Section 4. This leads to a smaller than optimum performance gain and requires a relatively complex design procedure.

Both counterflow and concurrent flow clocking can be generalized to the case of a two dimensional array. The corresponding clock distribution networks have a *corner-based* (*comb*) topology (as in Fig. 27, below).

3.4.2. Clock-Follow-Data Clocking. If the magnitude of the clock skew (the delay in the clock path) is increased in a clock distribution network with the straight-line concurrent clocking topology (Fig. 17(d)), a distinct clocking mode results. In this mode the data signal released by the clock from the first cell of the data path arrives at the second cell earlier than the clock. We call this scheme *clock-follow-data* clocking. [As the topology of the clock distribution network and the sign of the clock skew is the same as in concurrent clocking, clock-follow-data clocking has been previously referred in the literature as con-flow with data traveling faster [3] or simply concurrent-flow clocking [1]. We introduce a separate name for this mode to clearly distinguish it from the typical concurrent clocking scheme].

The operating region of the circuit in the clock-follow-data is described by (10) and (11) with $k = -1$ and is shown in Fig. 15. The typical operation of the circuit is shown in Fig. 19.

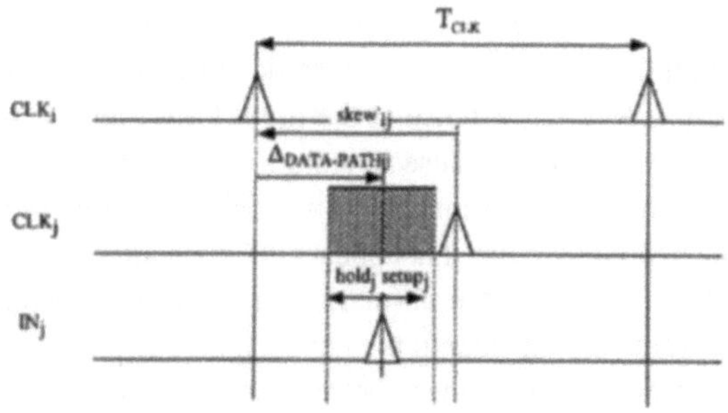

Figure 19. Timing diagram describing the exchange of data between two sequentially adjacent cells in clock-follow-data clocking.

In clock-follow-data clocking a *single* clock pulse carries the data through the whole array of N clocked cells in a time which is independent of the clock period. In concurrent clocking, $N - 1$ clock periods are necessary to carry the data through an array comprised of N cells.

The clock skew in clock-follow-data clocking may be set to the optimum value, corresponding to the minimum clock period, by choosing an appropriate number of interconnect JTL stages. The minimum clock period T_{MIN} is the same as for the concurrent clocking mode, and is given by (8). The optimum clock skew in clock-follow-data clocking, $skew_0'$, differs from the optimum clock skew for concurrent clocking, $skew_0$, by the value of the minimum clock period (see Fig. 15), i.e.,

$$skew_0' = skew_0 - T_{\mathrm{MIN}} = -\Delta_{\mathrm{DATA\text{-}PATH}_{ij}} - setup_j. \tag{12}$$

In clock-follow-data clocking the data pulse at the input of the second cell in the worst case data path lies at the *end* of the clock period, at the boundary of the *setup* time violation, as shown in Fig. 18(b). This relation means that in the presence of timing parameter variations affecting both clock skew and the position of the setup time boundary, the circuit may exhibit a setup time violation independent of the clock frequency. Therefore, the nominal magnitude of the clock skew must be *increased* above the theoretical optimum (see Fig. 15).

3.5. Minimum Clock Period in Various Clocking Schemes

The minimum clock period of the synchronous circuit $T_{\mathrm{CLK}}^{\mathrm{MIN}}$ is equal to the maximum of limitations $T_{\mathrm{CLK}_{ij}}^{\mathrm{MIN}}$ imposed by the data paths between any pair of sequentially adjacent cells, CELL_i, CELL_j, in the circuit, i.e.,

$$T_{\mathrm{CLK}}^{\mathrm{MIN}} = \max_{ij} \left\{ T_{\mathrm{CLK}_{ij}}^{\mathrm{MIN}} \right\}. \tag{13}$$

The minimum clock period imposed by a pair of sequentially adjacent cells is equal to

$$T_{\mathrm{CLK}_{ij}}^{\mathrm{MIN}} = skew_{ij} + \Delta_{\mathrm{DATA\text{-}PATH}_{ij}} + setup_j. \tag{14}$$

Let us consider the minimum clock period for different clocking schemes:

For zero-skew clocking, the minimum clock period is

$$T_{\text{zero-skew}}^{\mathrm{MIN}} = \max_{ij} \left\{ \Delta_{\mathrm{DATA\text{-}PATH}_{ij}} + setup_j \right\}. \tag{15}$$

For counterflow clocking, the minimum clock period is equal to

$$T_{\text{counterflow}}^{\mathrm{MIN}} = \max_{ij} \left\{ \Delta_{\mathrm{CLK\text{-}PATH}_{ij}} + \Delta_{\mathrm{DATA\text{-}PATH}_{ij}} + setup_j \right\}, \tag{16}$$

where $\Delta_{\mathrm{CLK\text{-}PATH}_{ij}}$ is the delay of the clock path between cells i and j. This delay is typically the delay of one splitter and the minimum number of JTL stages necessary to cover the physical distance between the clock inputs of both cells.

For concurrent clocking and clock-follow-data clocking, with the optimal clock skew between cells given by (9) and (12), respectively, the minimum clock period is

$$T_{\text{concurrent}}^{\mathrm{MIN}} = T_{\text{clock-follow-data}}^{\mathrm{MIN}} = \max_{j}\{hold_j + setup_j\}. \tag{17}$$

From (15)–(17),

$$T_{\text{concurrent}}^{\mathrm{MIN}} = T_{\text{clock-follow-data}}^{\mathrm{MIN}} < T_{\text{zero-skew}}^{\mathrm{MIN}} < T_{\text{counterflow}}^{\mathrm{MIN}}. \tag{18}$$

3.6. Performance of the Linear Pipelined Array with Synchronous Clocking

Consider a general linear synchronous array comprised of N heterogeneous cells with distinct timing parameters, as shown in Fig. 17(a). The array processes data in a pipelined fashion. The data is fed to the input of the first cell in the array, and the corresponding result appears at the output of the Nth cell after the appropriate number of clock cycles. The performance of the pipeline is described using two parameters. The *throughput* is defined as the output rate of the circuit, i.e., the inverse of the time between two consecutive outputs. In a synchronous array, throughput is equal to clock frequency. *Latency* is defined as the total time needed to process the data from the input to the output of the circuit. In an N-cell synchronous array, the latency is defined as an interval between the moment when the clock reads the data into the first cell, and the moment when the clock releases the corresponding result from the last cell of the array.

The behavior and performance of the linear array are analyzed for different clocking schemes using space vs. time diagrams shown in Figs. 21 to 24. In these diagrams, the data flows in two directions, in space—along the vertical axis, and in time—along the horizontal axis. The flow of the data in space corresponds to the data moving from one stage of the pipeline to the next stage as a result of the clock pulse at the cell separating the two stages. Each clock pulse releases the next data. The flow of the data through the data path between two clocked cells i and j is represented by a horizontal bar (rectangle), according to the convention depicted in Fig. 20. In this convention, the time necessary for processing the data within a single stage between cells i and j (interval AD in Fig. 20) is equal to the sum of the propagation delay through the data path [as defined by (6)] and the setup time of the cell j. The shaded part of the rectangle (interval CD in Fig. 20) represents the

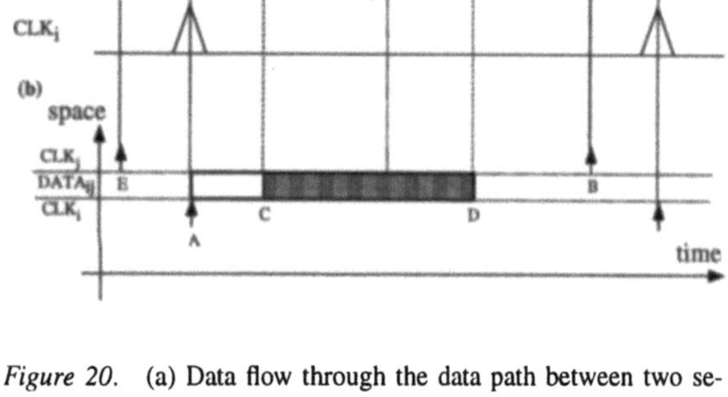

Figure 20. (a) Data flow through the data path between two sequentially adjacent clocked cells, and (b) its simplified graphical representation used in Figs. 21–24.

time interval around the position of the data pulse at the input of the cell j that is forbidden for clock pulses CLK_j. Any clock pulse at the input CLK_j appearing within this interval causes a violation of either the hold or setup time constraint, and thus a circuit malfunction. The first clock pulse that appears at CLK_j after the end of the forbidden interval (marked as the shaded rectangle) transfers the data to the next stage. The preceding clock pulse must appear before the beginning of the forbidden interval.

The operation of the pipeline for zero-skew clocking is shown in Fig. 21. The maximum clock frequency and throughput of the array is determined by the time to process the data through the slowest stage of the pipeline—data path $DATA_{23}$ between cells 2 and 3. Only one data pulse is present in the pipeline stage at any given time.

In Fig. 22, the operation of the circuit for counterflow clocking is shown. The minimum clock period of the circuit is determined by the time to process the data through the slowest stage of the pipeline *plus* the clock skew of this stage. In the most critical data path $DATA_{23}$, CLK_2 initiates processing of the data, and CLK_3 reads the result as soon as this processing is completed. In other non-critical pipeline stages, the data is ready to be transferred to the next stage long before the arrival of the clock pulse.

The operation of the pipeline for concurrent clocking is shown in Fig. 23, and for clock-follow-data clocking in Fig. 24. In both cases, the clock skew of the most critical data path $DATA_{23}$ has been chosen to be the optimal value given by (9) and (12), respectively. For all stages, the next data pulse begins propagating through the data path before the previous pulse *has been transferred* to the next pipeline stage. Additionally, for the slowest stage $DATA_{23}$ (as well as for the data path $DATA_{45}$), the next data pulse starts propagating through the data path before the previous pulse *is ready to be transferred* to the next pipeline stage. From the relation between the clock skews of the critical data path, (9), (12), and from Fig. 13, it can be seen that the timing in the circuit for the minimum clock period in concurrent and clock-follow-data clocking is indistinguishable. As a result, the maximum throughput and the minimum latency are identical in both schemes. This equality between latencies does not hold for clock periods greater than the minimum clock period.

The maximum throughput of the array is equal to the inverse of the minimum clock period. The minimum

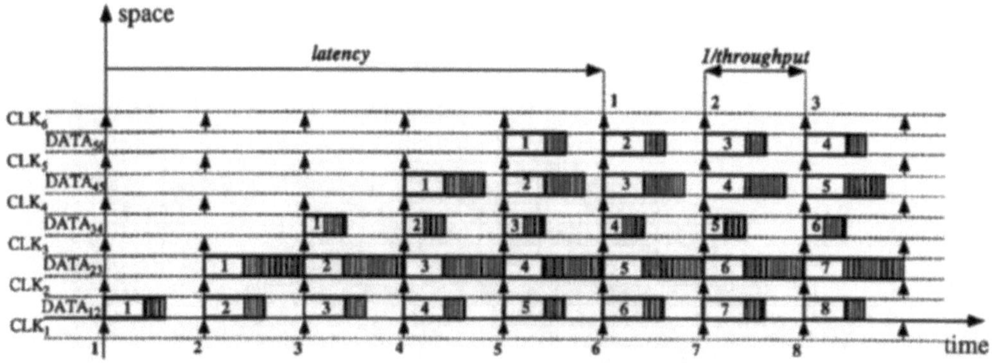

Figure 21. Space vs. time operation of the pipelined one-dimensional array with zero-skew clocking.

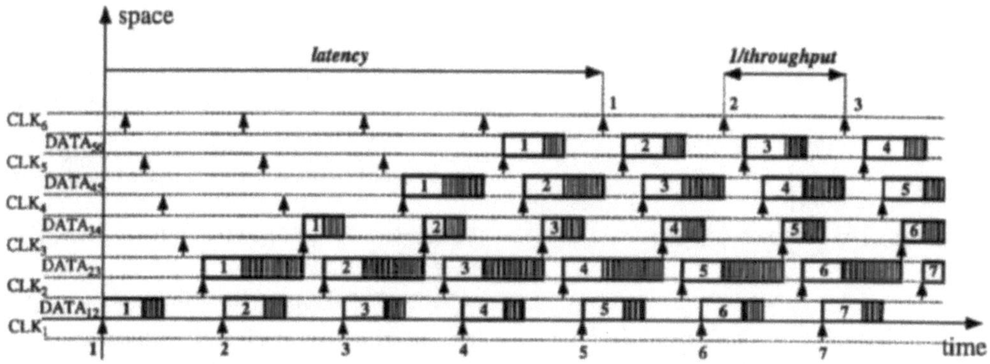

Figure 22. Space vs. time operation of the pipelined one-dimensional array with counterflow clocking.

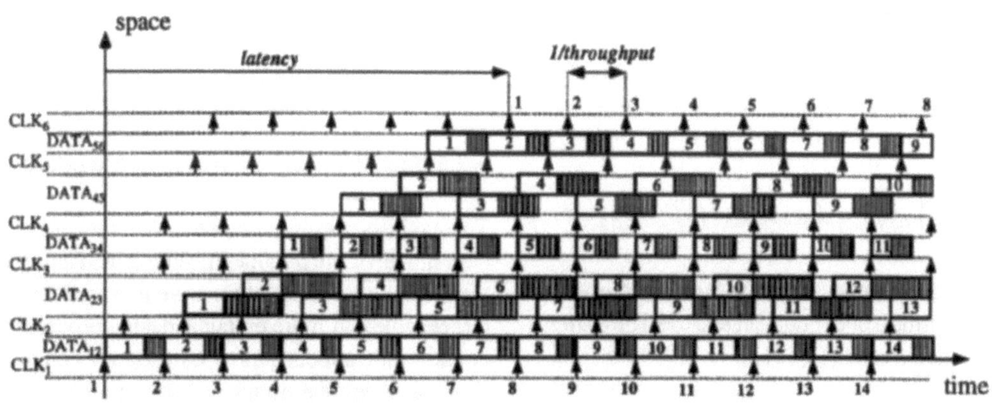

Figure 23. Space vs. time operation of the pipelined one-dimensional array with concurrent clocking.

clock period for each of the clocking schemes discussed in this section is given by (13) with $j = i + 1$. From (18), the relation between the throughputs for each of the different clocking schemes is given by

$$TH_{\text{concurrent}} = TH_{\text{clock-follow-data}} > TH_{\text{zero-skew}} > TH_{\text{counterflow}}. \tag{19}$$

The latency of the circuit for all clocking schemes is given by the following formulae:

$$L_{\text{zero-skew}}(T_{\text{CLK}}) = (N-1) \cdot T_{\text{CLK}}, \tag{20}$$

$$L_{\text{counterflow}}(T_{\text{CLK}}) = (N-1) \cdot T_{\text{CLK}} - \sum_{i=1}^{N-1} \Delta_{\text{CLK-PATH}i(i+1)}, \tag{21}$$

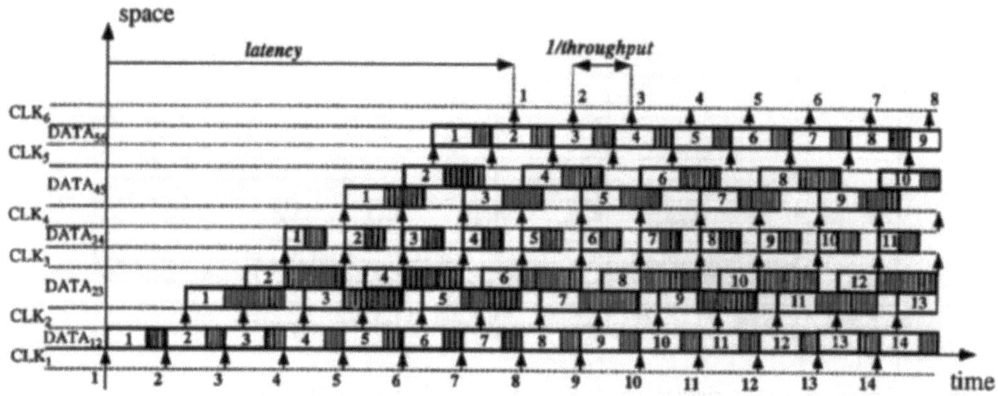

Figure 24. Space vs. time operation of the pipelined one-dimensional array with clock-follow-data clocking.

$$L_{\text{concurrent}}(T_{\text{CLK}}) = (N-1)\cdot T_{\text{CLK}} + \sum_{i=1}^{N-1} |skew_{i(i+1)}|, \quad (22)$$

$$L_{\text{clock-follow-data}}(T_{\text{CLK}}) = \sum_{i=1}^{N-1} |skew'_{i(i+1)}|. \quad (23)$$

Relations between latencies for different clocking modes are not unique. They depend on the parameters of the cells constituting the array and any physical constraints due to the layout. It is possible however to establish these relations unambiguously for most typical parameters.

If $\Delta_{\text{CLK-PATH}_{ij}}$ is constant for all cells, i.e., the physical distance between adjacent cells in the array is the same for all cells, then, from (15), (16), (20), and (21),

$$L_{\text{counterflow}}\left(T^{\text{MIN}}_{\text{counterflow}}\right) = L_{\text{zero-skew}}\left(T^{\text{MIN}}_{\text{zero-skew}}\right). \quad (24)$$

If the clock skew between cells is set to the optimum value which is distinct for concurrent and clock-follow-data clocking, then from (9), (12), (17), (22), and (23),

$$L_{\text{concurrent}}\left(T^{\text{MIN}}_{\text{concurrent}}\right) = L_{\text{clock-follow-data}}\left(T^{\text{MIN}}_{\text{clock-follow-data}}\right). \quad (25)$$

This relation holds despite the fact that in the concurrent scheme, N clock pulses are necessary to drive the data from the input of the first cell to the output of the last cell, while in the clock-follow-data scheme, a single clock pulse drives the data along the entire pipeline.

The minimum latency for concurrent clocking scheme is typically smaller than for zero-skew clocking,

$$L_{\text{concurrent}}\left(T^{\text{MIN}}_{\text{concurrent}}\right) < L_{\text{zero-skew}}\left(T^{\text{MIN}}_{\text{zero-skew}}\right). \quad (26)$$

The latency in all clocking schemes apart from the clock-follow-data scheme is a function of the clock period, and is not defined for the clock periods smaller than the minimum clock period characteristic for each scheme. For clock periods T_{CLK} permitted in all clocking modes (i.e., for T_{CLK} larger than the minimum clock period for counterflow clocking $T^{\text{MIN}}_{\text{counterflow}}$)

$$L_{\text{clock-follow-data}}(T_{\text{CLK}}) < L_{\text{counterflow}}(T_{\text{CLK}}) < L_{\text{zero-skew}}(T_{\text{CLK}}) < L_{\text{concurrent}}(T_{\text{CLK}}). \quad (27)$$

4. Effects of Timing Parameter Variations

The analysis presented in Section 3 concerns the ideal case in which the parameters characterizing devices in the circuit after fabrication are equal to their assumed target values. A more practical design process must account for the effects of process variations on the timing characteristics of a circuit. Taking parameter variations into account results in different *expected* and *worst case* maximum clock frequencies of the circuit and in different optimum values of interconnect delays in the clock distribution network. Including parameter variations in the timing analysis may also lead to the choice of a different synchronization scheme.

Specific features of present day niobium-trilayer technology used to develop medium to large scale RSFQ circuits are described in [71–73]. Two problems must be considered. First, superconducting fabrication technology is relatively immature compared to well established semiconductor technologies such as CMOS, resulting in much larger parameter variations. Secondly, because of the small volume of the integrated circuits produced by the superconducting foundries, their fabrication process is typically not well characterized.

4.1. *Global vs. Local Timing Parameter Variations*

The minimum clock period of a synchronous circuit using one of the standard clocking schemes described in Subsection 3.4.1 is

$$T_{\mathrm{CLK}}^{\mathrm{MIN}} = skew_{ij} + \Delta_{\text{DATA-PATH}_{ij}} + setup_j, \quad (28)$$

where the data path between cells i and j is the most critical data path in the circuit. In the presence of parameter variations, the timing parameters included on the right side of (28) can be modeled as random variables. The distribution of these variables is typically assumed to be normal, with mean equal to the nominal value of the timing parameter and standard deviation dependent on the deviations of the fabrication process and the effects of the internal structure of the RSFQ cell [74].

As shown in [74], the timing parameters of the basic RSFQ gates are predominantly affected by *wafer-to-wafer* variations in the *resistance per square* and the *inductance per square*. Other parameters that affect the difference between the actual and nominal values of the timing parameters are the *critical current density* (which affects the electrical characteristics of the junctions) and the global mask-to-wafer biases of the *inductor, resistor,* and *junction sizes* within the circuit.

The effects of deviations in the critical current density and global deviations in the junction size can be significantly decreased by adjusting the *global bias current* that provides the dc power supply to the integrated circuit. Both of these deviations can be approximated for a wafer (or an integrated circuit) using an auxiliary array of test structures, as described in [75]. The bias current can be changed proportionally to the actual values of the critical current density and the normalized junction area [74, 76]. Taking these adjustments into account, a relative 3σ *standard deviation in the delay of the basic RSFQ gates has been estimated to be about* 20% for an existing standard superconductive fabrication process [74]. By taking this result into account, one may estimate the *worst case* minimum clock period of a circuit to be about 20% greater than under nominal conditions.

The other more dramatic effect of parameter variations is the reduction in circuit yield. If for certain actual values of the timing parameters

$$skew_{ij} + \Delta_{\text{DATA-PATH}_{ij}} \geq hold_j \quad (29)$$

is not satisfied, then the circuit will not work properly for *any* clock frequency. This effect is greatest in the concurrent clocking mode, where the clock skew is chosen to be as close as possible to the boundary corresponding to the hold time violation (Fig. 18(a)). To the extent that the *wafer-to-wafer* variations of global parameters (such as inductance and resistance per square) change all timing parameters proportionally, (29) implies that a violation of the hold time constraint will not result from the global parameter variations. The danger cannot be completely discounted, as the timing parameters included in (29) will not necessarily change in the same proportion. However, changes in the values of these parameters tend to be correlated, which minimizes the effects of global variations on the circuit yield [74].

A more direct deleterious effect on circuit yield results from local *on-chip* variations of the individual parameters, such as the sizes of the junctions, inductors, and resistors, and *on-chip* variations of resistance per square, inductance per square and critical current density. These on-chip variations are typically not well characterized. Preliminary data imply that the local on-chip variations are several times smaller than global wafer-to-wafer parameter variations [71, 75, 77]. Deviations of the timing parameters of various components of the data path that result from local parameter variations are uncorrelated; thus a value of the optimum clock skew for concurrent clocking can be safely chosen according to

$$skew_0^{\mathrm{MIN}} = -\Delta_{\text{DATA-PATH}_{ij}}^{\mathrm{MIN}} + hold_j^{\mathrm{MAX}}, \quad (30)$$

where the minimum and maximum values are taken to account only for the effects of local parameter variations. Note from (28) that changing the *nominal* value of the clock skew given by (9) to satisfy (30) affects not only the *worst case* but also the *expected* value of the minimum clock period. A similar analysis applies to the clock-follow-data clocking approach.

As a result, in concurrent clocking and clock-follow-data clocking the local parameter variations affect both the *expected* and the *worst case* value of the minimum clock period. Global parameter variations affect primarily the *worst case* value of the minimum clock period. Both effects are smaller in concurrent clocking vs. clock-follow-data clocking because of smaller absolute delays in the clock paths between sequentially adjacent cells.

In counterflow and zero-skew clocking, global parameter variations typically do not affect the *expected* value of the minimum clock period but change substantially the *worst case* value of the minimum clock period. The effect of local parameter variations is negligible.

4.2. Local Variations within a Clock Distribution Network

Heretofore, the clock skew has been assumed to be proportional to the *difference* in delays of the clock paths from the clock source to the inputs of two sequentially adjacent cells. This model of the clock skew is referred to in the literature as a *difference model* [28]. The difference model holds well for straight-line clocking of a linear array, but is inadequate for other topologies such as a binary tree, H-tree, or a *corner clocking* structure, shown in Figs. 25–27 respectively. This is best understood by considering that the *nominal* clock skew between cells X and Y in Figs. 25 and 26 is zero, and in Fig. 27 it is determined only by the segment CC′. However, as a result of local on-chip variations in the clock distribution network the *actual* clock skew between cells X and Y will depend upon the entire crosshatched portion of the network.

Therefore, in order to discuss the clock skew caused by local on-chip variations it is necessary to introduce the more general model of the clock skew called the *summation model* [28]. In the summation model, clock skew is a function of the *sum* of the clock path delays from the nearest common node of the clock distribution network to the inputs of sequentially adjacent cells.

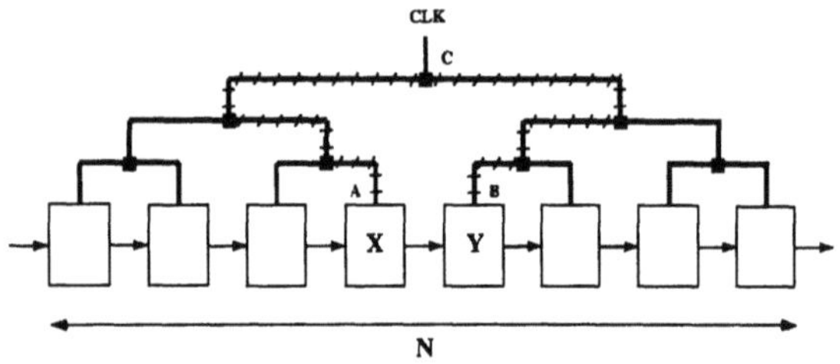

Figure 25. Binary-tree clock distribution network. Local variations in the crosshatched part of the network contribute to the clock skew between cells X and Y.

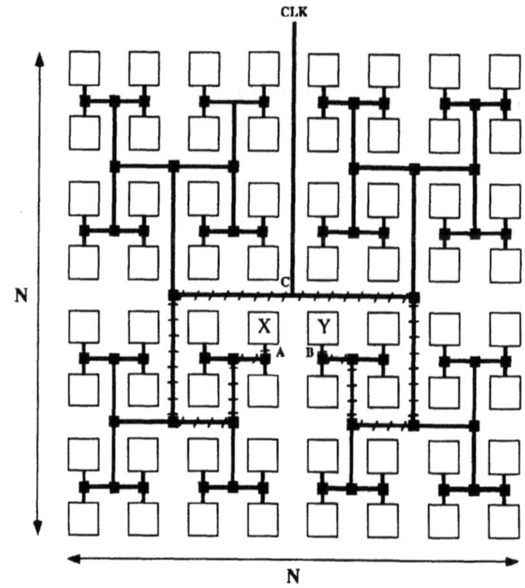

Figure 26. H-tree clock distribution network. Local variations in the crosshatched part of the network contribute to the clock skew between cells X and Y.

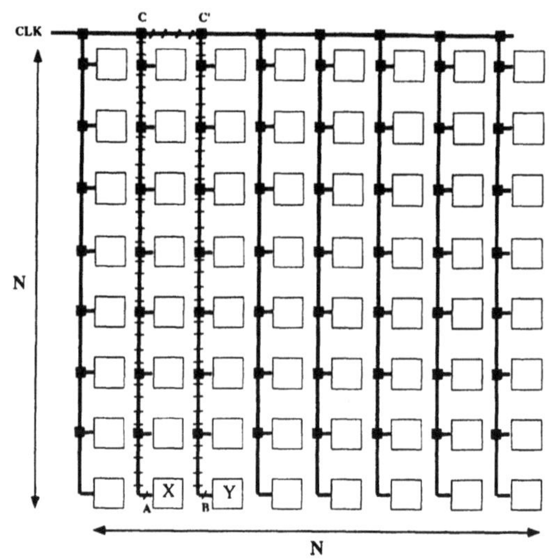

Figure 27. Corner-based clock distribution network. Local variations in the parallel paths CA, C'B contribute to the random clock skew between cells X and Y. The nominal value of the clock skew depends only on the delay CC′. The actual value of this delay changes as a result of both local and global parameter variations.

The effect of the local on-chip variations in the clock distribution network is primarily a function of a network topology, rather than the clocking scheme used within that topology:

For linear arrays, *straight-line* clocking offers an optimum solution in which the difference model of clock skew applies (see Fig. 29). As a result, this topology of the clock distribution network is perfectly scaleable and works efficiently for an arbitrary number of cells in an array. Asymmetric $M \times N$ systolic arrays with a small constant value of M scale similarly with N to linear arrays. Examples of such circuits include an N-bit serial multiplier ($N \times 3$ array) [3, 5] and an N-bit multiplier-accumulator [$(2N - 1) \times 3$ array] [8].

For the binary tree topology shown in Fig. 25 the data path most critical to local variations is likely to be between cells X and Y. Clock skew is a function of the sum of the path delays CA and CB, between each clock input node and the nearest common ancestor in the binary tree. Therefore, for relatively small N arrays, the clock skew resulting from the local variations in the clock tree increases the *worst case* minimum clock period in the circuit; for large N arrays it may additionally cause an unacceptable reduction in the circuit yield.

The effect of local parameter variations in the clock distribution network on the clock skew is particularly strong for a square array. The worst-case skew grows quickly with the increase in size of the array. Assuming that variations of the clock path delays between any two adjacent cells of the array are independent of each other, the standard deviation of the clock skew for the worst case data path grows proportionally to $\sqrt{N}$, where N is the size of the array [28] (see also [78]). Variations of the resistance per square, inductance per square, and critical current density depend strongly on the physical distance between the corresponding paths of the clock distribution network. For example, the variations tend to be larger in the H-tree network (see Fig. 26), than in the corner-clocked network (see Fig. 27).

Therefore, large-size fully synchronous two-dimensional systolic arrays are difficult to build. Since the local parameter variations are not well characterized to date, it is difficult to judge whether this effect limits the practical sizes of arrays currently developed in RSFQ logic (e.g., 16×16 parallel multiplier described in [1, 2, 26]). Certainly, there exists a limit on the size of a square $N \times N$ systolic array above which synchronous clocking will lead to an unacceptable worst case performance or a very low circuit yield. Depending on the magnitude of the on-chip variations in the timing parameters and the size of the array, either a more conservative clocking scheme (e.g., counterflow vs. concurrent), or a *hybrid* synchronization scheme may be required. In the hybrid scheme presented in [28], an entire array is divided into local synchronous

Figure 28. Scheme for resynchronization of the clock signal traveling along different paths in the clock distribution network using coincidence junctions.

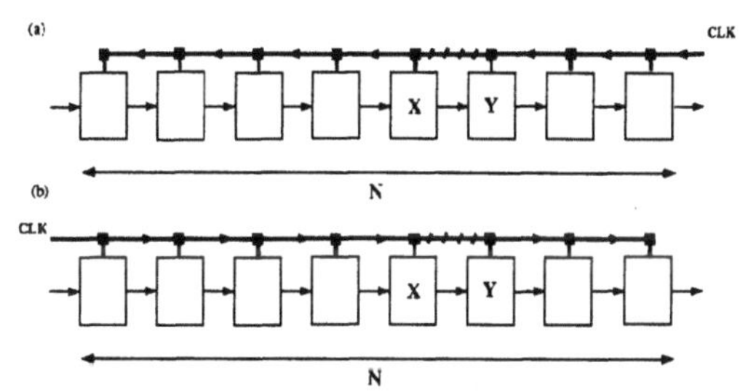

Figure 29. The portions of the straight-line clock distribution network which affect the clock skew between sequentially adjacent cells for (a) counterflow clocking, (b) concurrent clocking.

subarrays with local clocks controlled using an asynchronous handshaking protocol.

Another solution, developed specifically for RSFQ arrays, is described in [1, 26]. In this approach, shown in Fig. 28, clock signals traveling along different paths in the clock distribution network are resynchronized using coincidence junctions. A *coincidence junction* [1, 26, 47] produces an output pulse only after an input pulse has arrived at both of its inputs. Statistically, in the circuit shown in Fig. 28, the clock skew between any two neighboring cells is substantially reduced.

4.3. Optimal Choice of Interconnect Delays

A quantitative analysis of the effects of global and local variations on the performance of a circuit and thus also the optimal choice of interconnect delays is difficult to perform analytically, and usually requires computationally intensive Monte Carlo simulations. These computations can be substantially sped-up by using a behavioral simulation rather than a circuit level simulation, as described in [30, 79]. Another approach, based on an approximate worst case analysis, is presented in [27]. This approach leads to correct but not necessarily optimal solutions. The design of circuits with concurrent clocking is particularly challenging since a *nominal* value of the clock skew must be chosen considering the effects of the global and local parameter variations. An incorrect choice may lead to a large percentage of the integrated circuit not working properly for any clock frequency. Good characterization data of the fabrication process is a necessary condition for a correct quantitative analysis of the circuit performance and the design of the *optimum* clock distribution network.

5. Asynchronous Timing

In semiconductor VLSI, asynchronous timing has been for many years considered a possible alternative to synchronous clocking [20, 37]. Its main advantages include modularity, reliability and high resistance to fabrication process variations. Nevertheless, asynchronous clocking has *not* been widely accepted in semiconductor circuit design due to unsatisfactory performance in terms of area, speed, and power consumption, as well as complicated design and testing procedures [29].

Asynchronous timing requires local signaling between adjacent cells. This signaling is naturally based on the concept of *events* such as *request* and *acknowledge*. In semiconductor logic events are coded using voltage state *transitions* (rising edges in *return-to-zero* signaling, and rising and falling edges in *non-return-to-zero* signaling). Semiconductor logic elements that process voltage transitions (e.g., Muller C-element, Toggle, Select) are complex and slow compared to logic gates that process *voltage levels*. In RSFQ logic, events are coded using SFQ pulses. Asynchronous logic elements that process SFQ pulses (e.g., confluence buffer, coincidence junction) are simple and fast compared to RSFQ logic gates (such as AND, OR, XOR), and therefore RSFQ asynchronous circuits can approach the speed of synchronous circuits. Because of this asynchronous clocking appears to be easier and more natural to implement in RSFQ circuits than in semiconductor voltage-state logic. For complex RSFQ circuits, the disadvantage of larger area and power consumption required for local signaling in asynchronous circuits may be compensated by circuit modularity and the larger tolerance to fabrication process variations.

5.1. Dual-Rail Logic

The only asynchronous timing approach reported to be actually used in the design of a large scale RSFQ circuit [16] is based on *dual-rail logic*. Adapting dual-rail logic for use with RSFQ gates has been investigated in [38, 39, 55–58].

In *dual-rail logic*, each signal is transmitted using two signal lines, denoted true- and false-. The appearance of an SFQ pulse on the true-line is defined as the logical "1", and the appearance of the pulse on the false-line as the logical "0". This convention differs significantly from the Basic RSFQ Convention described in Section 2. Therefore, any RSFQ gate which should be used as the core of a dual-rail logic cell must be redesigned by adding special input and output circuitry. First, the gate is extended with a second complementary output OUT\. Each time the cell performs a logic operation, an SFQ pulse is created at one and only one of the cell outputs, OUT or OUT\. Additionally, the cell is supplemented with the input circuitry used to accept dual-rail inputs and to internally generate the clock pulse driving the core RSFQ gate.

The input circuitry for a single-input gate can have a form of a confluence buffer with two delay lines: clock line C-JTL and data line *D*-JTL as shown in Fig. 30(a). A pulse that appears at either input a or $a\backslash$ of the cell generates a pulse at the output of the confluence buffer, CB. This pulse, delayed by the JTL line C-JTL, is used to clock the RSFQ gate. The timing constraints in the circuit are described by

$$\Delta_{\text{D-JTL}} + \text{setup} \leq \Delta_{\text{CB}} + \Delta_{\text{C-JTL}}, \tag{31}$$

$$\Delta_{\text{D-JTL}} + T_{\text{IN}} \geq \Delta_{\text{CB}} + \Delta_{\text{C-JTL}} + \text{hold}, \tag{32}$$

where T_{IN} is the period of the input data signal. From (31) and (32),

$$T_{\text{IN}} \geq \Delta_{\text{CB}} + \Delta_{\text{C-JTL}} - \Delta_{\text{D-JTL}} + \text{hold}. \tag{33}$$

The minimum value of the input period is

$$T_{\text{MIN}} = \text{hold} + \text{setup}, \tag{34}$$

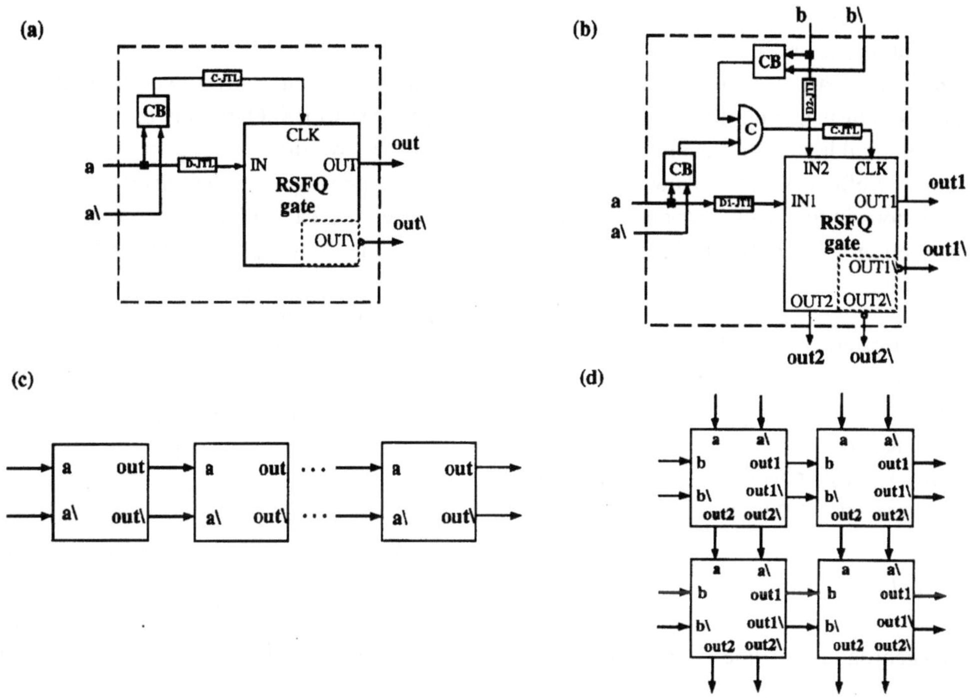

Figure 30. Internal structure of a dual-rail cell based on (a) one-input RSFQ gate, (b) two-input RSFQ gate; and the method of connecting dual-rail cells into (c) a linear array, (d) a rectangular array. Notation: CB—confluence buffer, C-JTL—clock path JTL, D-JTL—data path JTL, C—coincidence junction.

and is obtained for a choice of interconnect delays according to

$$\Delta_{\text{C-JTL}} - \Delta_{\text{D-JTL}} = \text{setup} - \Delta_{\text{CB}}. \tag{35}$$

For the optimum choice of the interconnect delays, the data pulse appears at the input of the RSFQ gate exactly a setup time before the clock pulse arrives (the same as for clock-follow-data synchronous clocking). This makes the circuit vulnerable to fabrication process variations. The actual optimum value of the interconnect delays $\Delta_{\text{C-JTL}}$ and $\Delta_{\text{D-JTL}}$ must be derived taking parameter variations into account.

Dual-rail cells designed according to these rules can be connected into a linear array with unidirectional data-flow without any additional circuitry, as shown in Fig. 30(c). Note that in this configuration no *acknowledge* signal is used, and the *request* signal does not appear explicitly but rather is integrated with the dual-rail data signals. As a result, the circuit is vulnerable to timing violations resulting from the next data appearing at the cell input before the previous data is accepted. *The maximum input rate of the signal driving the first cell of the array is limited by the maximum input rate of the slowest gate in the array.* If the interval between any two external input data pulses is smaller than the minimum input period for any cell in the array, the timing constraints are violated, leading to a circuit malfunction.

Therefore, the overall performance of this simple array in dual-rail logic in terms of the latency and the maximum throughput is comparable to the performance in synchronous clock-follow-data clocking. The device overhead and design complexity of a dual-rail logic is significantly greater.

In case of a two-input dual-rail cell, the cell input circuitry becomes even more complicated (Fig. 30(b)). The output of the confluence buffer associated with each of the dual-rail inputs feeds the input of the coincidence junction. The coincidence junction generates the clock pulse only after both input data signals have arrived. The maximum input rate of the cell can be derived using an analysis similar to that performed for one-input cells. The important difference is that the maximum data rate for each input depends not only on the internal delays in the circuit but also on the interval between the arrival of the dual-rail data signals at two different inputs of the cell. As a result, the maximum input rate for a gate becomes dependent on the circuitry surrounding the gate and the timing characteristics of the external input data sources.

A two-dimensional array composed of two-input dual-rail cells is shown in Fig. 30(d). For a square $N \times N$ array, dual rail logic offers a unique advantage by eliminating the effect of clock skew due to local parameter variations in the clock distribution network discussed in Section 4.2. However, disadvantages of the scheme include

a) a large device overhead resulting from using two confluence buffers, one coincidence junction and complementary output circuitry per every two-input gate in the circuit;
b) vulnerability to discrepancies between input rates at any two inputs in the circuit.

5.2. *Micropipelines*

The other asynchronous scheme considered for application in RSFQ one-dimensional arrays is the *micropipeline*. This scheme, known from semiconductor circuit design [37], appears to be easily adaptable to RSFQ logic [1, 26, 60, 61]. The scheme is based on the use of coincidence junctions (Muller C-elements in semiconductor logic) to generate the clock for each cell in the pipeline on the basis of the *request* signal generated by the previous cell in the pipeline, and the *acknowledge* signal generated by the next cell in the pipeline.

From the analysis presented in [37], this scheme does not offer any advantage in speed compared to a fully synchronous methodology (e.g., concurrent clocking). The design of the circuitry for generation signaling events (*acknowledge* and *request*) must take into account the effects of the local timing parameter variations. The disadvantage of the scheme lies in its large device overhead (one coincidence junction plus multiple JTL stages per each clocked cell), and the requisite complex operation.

6. Two-Phase Synchronous Clocking

Two-phase clocking is a common approach used in semiconductor circuit design in which a two-phase master-slave double latch is used as a storage component [22]. Multiple phases of the clock relax the timing constraints in the circuit, and thus increase circuit tolerance to variations in the fabrication process. The disadvantage is the area/device overhead resulting from the second clock path and more complex storage components.

In this section a novel two-phase clocking scheme applicable to RSFQ circuits of any complexity is introduced. We show that high performance, robustness, and design simplicity may justify two-phase clocking despite the area overhead inherent in this scheme.

An initial attempt to apply two-phase clocking to RSFQ circuits was reported in [80]. In this paper, two-phase clocking is used to drive a long linear shift register. The motivation is to assure that the circuit works correctly at a very low clock frequency applied during functional testing, independently of the parameter variations in the circuit. No attempt is made to optimize the performance of the circuit. Two phases of the clock are generated using complementary DC/SFQ converters, and distributed independently along the data path of the shift register. As a result, the design is vulnerable to independent local parameter variations occurring in the two parallel clock paths used to distribute each phase of the clock.

An enhanced version of this *two-phase concurrent clocking* scheme applicable to any general one- and two-dimensional arrays as well as to RSFQ circuits with a less regular topology is presented here. The performance of this scheme is analyzed, and its advantages and disadvantages are compared to single-phase concurrent clocking.

In RSFQ two-phase clocking, the phases of the clock are shifted from each other by half of the clock period, as shown in Fig. 31(a). Both phases of the clock can be generated from one signal with twice the clock frequency using a T flip-flop, as shown in Fig. 31(b). A separate T flip-flop can be associated with each clocked cell in the circuit, or can be used to generate both phases for a whole sequence of clocked cells. In the latter case,

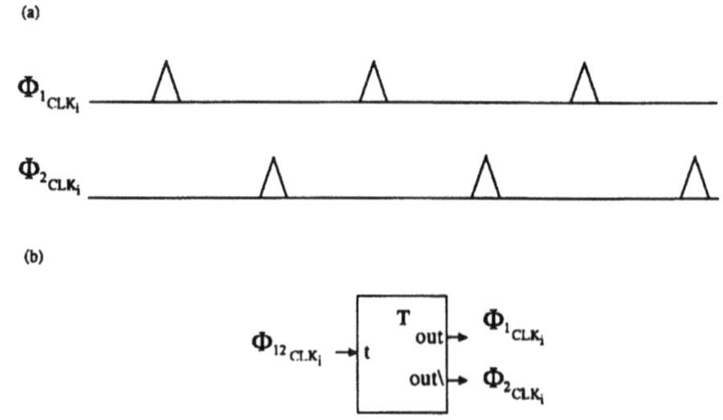

Figure 31. Two-phase clocking in RSFQ logic. (a) phases of the clock; (b) method of generating both phases from a single signal operating at twice the clock frequency.

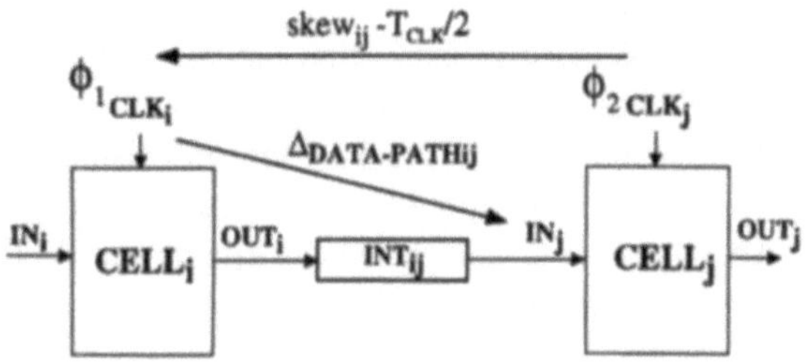

Figure 32. Data path between two physically adjacent RSFQ cells in two-phase clocking.

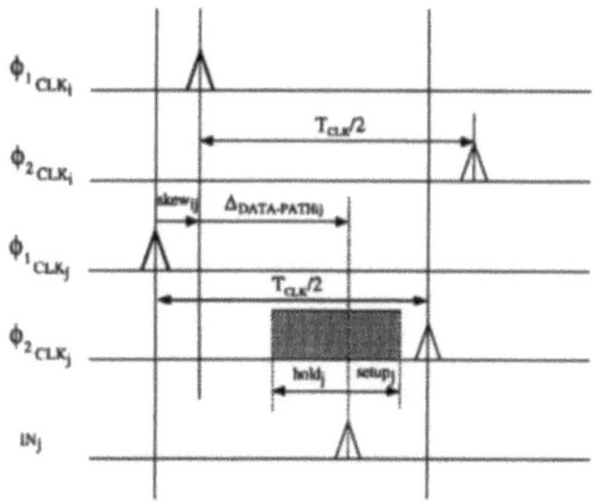

Figure 33. Exchange of data between two physically adjacent cells in two-phase clocking.

both clock phases are distributed independently at an interval between two consecutive T flip-flops.

The data path between two sequentially adjacent cells is shown in Fig. 32. Timing diagrams depicting the exchange of data between two sequentially adjacent cells are given in Fig. 33.

Conditions for the correct operation of the circuit are

$$T_{\text{CLK}}/2 \geq skew_{ij} + \Delta_{\text{DATA-PATH}_{ij}} + setup_j, \quad (36)$$

$$skew_{ij} + \Delta_{\text{DATA-PATH}_{ij}} + T_{\text{CLK}}/2 \geq hold_j. \quad (37)$$

In Fig. 34, the operating region of the circuit as a function of the clock skew and the clock period is shown. By comparing the shape of this operating region with the regions for single-phase clocking presented in Fig. 12, it can be concluded that for two-phase clocking:

a) There does not exist a region of clock skew values for which the circuit does not work for any clock frequency. For any possible value of the clock skew, there exist a minimum clock period above which the circuit works correctly for any clock period.

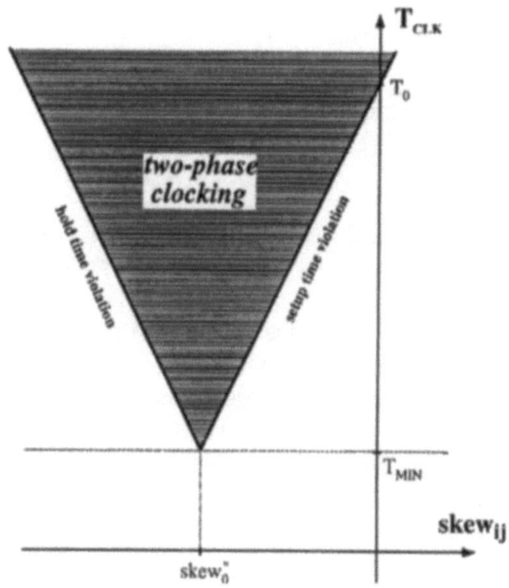

Figure 34. Operating region for two phase clocking of a circuit composed of two sequentially adjacent cells, as a function of the clock period and the clock skew.

b) The minimum clock period in the circuit is limited by

$$T_{\text{MIN}} = hold_j + setup_j, \quad (38)$$

and the optimal choice of the clock skew is

$$skew_0'' = -\Delta_{\text{DATA-PATH}_{ij}} - setup_j/2 + hold_j/2. \quad (39)$$

The optimum clock skew in two-phase clocking, $skew_0''$, is related to the optimum clock skew for single-phase concurrent clocking, $skew_0$, [given by (9)] and single-phase clock-follow-data clocking, $skew_0'$, [given by (12)] according to

$$skew_0'' = \frac{skew_0 + skew_0'}{2} \quad (40)$$

The minimum clock period, without taking parameter variations into account, is identical for all three clocking schemes. The position of the data pulse within the clock period for the optimum value of the clock skew in two-phase clocking is shown in Fig. 35.

c) The optimal value of the clock skew is identical regardless of whether parameter variations are considered. This feature simplifies considerably the design of the circuit by eliminating the need for the computationally intensive Monte Carlo simulations

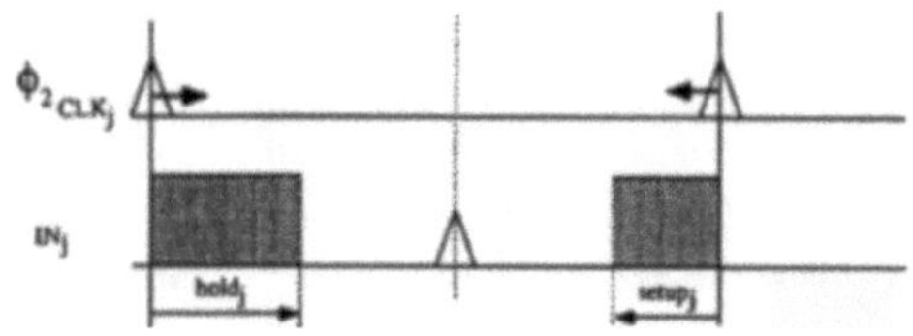

Figure 35. The position of the data pulse within the clock period in two-phase clocking for the optimum value of the clock skew.

necessary to determine the optimal clock skew in single-phase concurrent clocking.

d) The *expected* value of the minimum clock period is the same with or without taking parameter variations into account. As a result, the expected value is smaller than in single-phase concurrent clocking. The worst case value of the clock period in both schemes is comparable.

Thus the advantage of the two-phase synchronous clocking are robustness, high-performance, and the simplicity of the design procedure. The disadvantage of this approach the additional overhead circuitry required to generate and distribute the second phase of the clock.

7. Conclusions

The timing of medium to large scale RSFQ circuits follows the well-established principles and methodologies of the timing applied to high speed VLSI semiconductor-based circuits. There are significant qualitative differences which arise from

a) the lack of purely combinational logic in RSFQ circuits;
b) the low fanout of RSFQ gates;
c) a different suite of elementary gates in the two technologies.

There are other differences which primarily arise from the much greater operating frequencies in RSFQ logic. The most important of these is the inefficiency of applying equipotential clocking to multi-gigahertz large clock distribution networks. Pipelined (flow) clocking should be used instead in both RSFQ and semiconductor-based circuits.

Zero-skew clocking, which is ubiquitous in semiconductor circuits, has no particular advantage when applied to RSFQ logic. Non-zero-skew clocking schemes can be chosen either for superior performance or for extended tolerance to fabrication process variations. Although these advantages may be easier to exploit in RSFQ circuits, the same clocking schemes also apply to the design of high speed semiconductor circuits.

The choice of clocking scheme for a particular RSFQ circuit depends upon:

a) the topology of the circuit (one-dimensional vs. two-dimensional array, regular vs. irregular structure);
b) the performance requirements (throughput, latency) of the circuit;
c) global and local parameter variations in the circuit;
d) complexity of the design procedure (computationally intensive Monte Carlo analysis vs. analytical estimations);
e) the device, area, and power consumption overhead;
f) the complexity of the physical layout.

For circuits which are essentially one-dimensional, $N \times 1$ arrays and asymmetric $N \times M$ arrays with small M, the natural choices are the *straight-line* synchronous clocking schemes. *Counterflow clocking* offers the advantages of high robustness to timing parameter variations, small area, and a simple design procedure, but at the cost of reduced circuit throughput. When the highest clock frequency is of primary concern, *concurrent clocking* should be considered. An aggressive application of this scheme will reduce the expected yield of the circuit unless there is a good quantitative knowledge of the fabrication process variations. The design procedure leading to the optimum solution may require intensive Monte Carlo simulations, although suboptimal solutions can be obtained using simpler analytical methods. Concurrent clocking tends to require a larger number of JTL stages in the clock paths compared to counterflow clocking, and thus a greater overhead in circuit area and in layout complexity is expected.

This paper introduces a new clocking scheme, *two-phase clocking*, which is expected to offer better performance than concurrent clocking, better tolerance to fabrication process variations than counterflow clocking, and an extremely simple design procedure. Also in two-phase clocking, the choice of the optimum interconnects in the circuit does not require any knowledge of the timing parameter variations. Interconnect delays within the clock distribution network are similar to concurrent clocking. The only disadvantage of two-phase clocking is the area overhead resulting from the necessity to generate both clock phases for every cell of a linear $N \times 1$ array or every column of an asymmetric $N \times M$ array. A single T flip-flop (5 Josephson junctions) per cell or column of cells is sufficient for this purpose. In all of these synchronous schemes applied to $N \times 1$ arrays or asymmetric $N \times M$ arrays with small M, the maximum clock frequency is independent of N.

Asynchronous schemes such as *dual-rail clocking* and *micropipelines* can also be successfully applied to linear and asymmetric arrays, but these schemes do not offer any advantages over synchronous schemes in either performance, robustness, or design complexity. Either scheme can be adjusted (by the appropriate choice of interconnect delays) to provide either the performance equivalent of concurrent clocking or the robustness of counterflow clocking. Both schemes, however, require a significant overhead, which is comparable to or greater than required by two-phase clocking. The design for optimum performance is equally complex as for concurrent clocking and requires good knowledge of the timing parameter variations.

For a two-dimensional symmetric square $N \times N$ arrays the situation is more complicated. The additional effects of the local parameter variations in corresponding paths of the clock distribution network (a *summation model* of the clock skew) must be considered. In all of the synchronous schemes, the performance of the circuit deteriorates with an increase of the array size N by a factor proportional to at least $\sqrt{N}$. Depending on the magnitude of the on-chip variations and the topology of the clock distribution network, these effects may become critical for different sizes of N. In particular, it is possible that the constant factors may be sufficiently small for practical sizes of RSFQ arrays. For all synchronous schemes, the worst case maximum clock frequency deteriorates with increasing N. Additionally, for all single-phase clocking schemes, there exists a value of N above which the yield of the circuit begins to decrease. This value of N is smallest for concurrent clocking and largest for counterflow clocking. In two-phase clocking, increasing the array size deteriorates only the *worst case* circuit performance. Neither the *expected* performance of the circuit nor the functional circuit yield at low speed is affected by an increase of the array size N.

Asynchronous schemes scale better with increasing N. Again, the primary disadvantage of these schemes is the large circuit overhead. These schemes are also more difficult to analyze and test than synchronous schemes. As a result, the use of asynchronous timing methodologies may be limited to circuits of large N. Finally, hybrid synchronization schemes which use asynchronous strategies in tandem with simpler synchronous schemes are likely to be advantageous for large RSFQ circuits.

Acknowledgment

This work was supported in part by the Rochester University Research Initiative sponsored by the US Army Research Office.

References

1. K.K. Likharev and V.K. Semenov, "RSFQ logic/memory family: A new Josephson-junction technology for sub-terahertz-clock frequency digital systems," *IEEE Trans. Appl. Supercond.*, Vol. 1, pp. 3–28, 1991.
2. K.K. Likharev, "Rapid single-flux-quantum logic," in *The New Superconductor Electronics*, H. Weinstock and R. Ralston (Eds.), Kluwer, Dordrecht, pp. 423–452, 1993.
3. O.A. Mukhanov, P.D. Bradley, S.B. Kaplan, S.V. Rylov, and A.F. Kirichenko, "Design and operation of RSFQ circuits for digital signal processing," *Proc. 5th Int. Supercond. Electron. Conf.*, Nagoya, Japan, Sept. 1995, pp. 27–30.
4. K.K. Likharev, "Ultrafast superconcductor digital electronics: RSFQ technology roadmap," *Czechoslovak J. Phys.*, Suppl. S6, Vol. 46, 1996.
5. O.A. Mukhanov and A.F. Kirichenko, "Implementation of a FFT radix 2 butterfly using serial RSFQ multiplier-adders," *IEEE Trans. Appl. Supercond.*, Vol. 5, pp. 2461–2464, 1995.
6. J.C. Lin, V.K. Semenov, and K.K. Likharev, "Design of SFQ-counting analog-to-digital converter," *IEEE Trans. Appl. Supercond.*, Vol. 5, pp. 2252–2259, 1995.
7. V.K. Semenov, Yu. Polyakov, and D. Schneider, "Preliminary results on the analog-to-digital converter based on RSFQ logic," *CPEM96 Conf. Digest Suppl.*, Braunschweig, Germany, June 1996, pp. 15–16.
8. Q.P. Herr et al. "Design and low speed testing of a four-bit RSFQ multiplier-accumulator," *IEEE Trans. Appl. Supercond.*, Vol. 7, 1997.
9. Q.P. Herr, K. Gaj, A.M. Herr, N. Vukovic, C.A. Mancini, M.F. Bocko, and M.J. Feldman, "High speed testing of a four-bit RSFQ decimation digital filter," *IEEE Trans. Appl. Supercond.*, Vol. 7, 1997.
10. P.I. Bunyk et al., "High-speed single-flux-quantum circuit using planarized niobium-trilayer Josephson junction technology," *Appl. Phys. Lett.*, Vol. 66, pp. 646–648, 1995.
11. Q.P. Herr and M.J. Feldman, "Error rate of a superconducting circuit," *Appl. Phys. Lett.*, Vol. 69, pp. 694–695, 1996.
12. D.Y. Zinoviev and K.K. Likharev, "Feasibility study of RSFQ-based self-routing nonblocking digital switches," *IEEE Trans. Appl. Supercond.*, Vol. 7, 1997.
13. Q. Ke, B.J. Dalrymple, D.J. Durand, and J.W. Spargo, "Single flux quantum crossbar switch," *IEEE Trans. Appl. Supercond.*, Vol. 7, 1997.

14. N.B. Dubash, P.-F. Yuh, V.V. Borzenets, T. Van Duzer, and S.R. Whiteley, "SFQ data communication switch," *IEEE Trans. Appl. Supercond.*, Vol. 7, 1997.
15. O.A. Mukhanov and S.V. Rylov, "Time-to-digital converters based on RSFQ digital counters," *IEEE Trans. Appl. Supercond.*, Vol. 7, 1997.
16. A.V. Rylyakov and S.V. Polonsky, "All digital 1-bit RSFQ autocorrelator for radioastronomy applications: Design and experimental results," *IEEE Trans. Appl. Supercond.*, Vol. 7, 1997.
17. A.V. Rylyakov, "New design of single-bit all-digital RSFQ autocorrelator," *IEEE Trans. Appl. Supercond.*, Vol. 7, 1997.
18. G. Taubes, "Redefining the supercomputer," *Science*, Vol. 273, pp. 1655–1657, 1996.
19. G. Gao, K.K. Likharev, P.C. Messina, and T.L. Sterling, "Hybrid technology multithreaded architecture," in *Proc of PetaFlops Architecture Workshop*, to be published; see also the Web site http://www.cesdis.gsfc.nasa.gov/petaflops/peta.html.
20. C. Mead and L. Conway, *Introduction to VLSI Systems*, Addison-Wesley, Reading, MA, 1980.
21. M. Hatamian, "Chapter 6, Understanding clock skew in synchronous systems," in *Concurrent Computations (Algorithms, Architecture, and Technology)*, S.K. Tewksbury, B.W. Dickinson, and S.C. Schwartz (Eds.), Plenum Publishing, New York, pp. 87–96, 1988.
22. H.B. Bakoglu, *Circuits, Interconnections and Packaging for VLSI*, Addison-Wesley, 1990.
23. H.M. Teresa, *Synchronization Design for Digital Systems*, Kluwer Academic Publishers, 1991.
24. E.G. Friedman, "Clock distribution design in VLSI circuits—an overview," *Proc. IEEE Int'l Symp. Circuits Syst.*, pp. 1475–1478, May 1993.
25. E.G. Friedman (Ed.), *Clock Distribution Networks in VLSI Circuits and Systems*, IEEE Press, 1995.
26. O.A. Mukhanov, S.V. Rylov, V.K. Semenov, and S.V. Vyshenskii, "RSFQ logic arithmetic," *IEEE Trans. Magnetics*, Vol. 25, pp. 857–860, 1989.
27. K. Gaj, E.G. Friedman, M.J. Feldman, and A. Krasniewski, "A clock distribution scheme for large RSFQ circuits," *IEEE Trans. Appl. Supercond.*, Vol. 5, pp. 3320–3324, 1995.
28. A.L. Fisher and H.T. Kung, "Synchronizing large VLSI processor arrays," *IEEE Trans. Comput.*, Vol. C-34, pp. 734–740, 1985.
29. M. Afghahi and C. Svensson, "Performance of synchronous and asynchronous schemes for VLSI systems," *IEEE Trans. Comput.*, Vol. C-41, pp. 858–872, 1992.
30. K. Gaj, C.-H. Cheah, E.G. Friedman, and M.J. Feldman, "Optimal clocking design for large RSFQ circuits using Verilog HDL," (in preparation).
31. J.P. Fishburn, "Clock skew optimization," *IEEE Trans. Comput.*, Vol. 39, pp. 945–951, 1990.
32. J.L. Neves and E.G. Friedman, "Topological design of clock distribution networks based on non-zero clock skew," *Proc. 36th Midwest Symp. Circuits Syst.*, pp. 468–471, Aug. 1993.
33. J.L. Neves and E.G. Friedman, "Design methodology for synthesizing clock distribution networks exploiting nonzero localized clock skew," *IEEE Trans. VLSI Syst.*, Vol. 4, pp. 286–291, 1996.
34. J.L. Neves and E.G. Friedman, "Circuit synthesis of clock distribution networks based on non-zero clock skew," *Proc. IEEE Int'l Symp. Circuits Syst.*, pp. 4.175–4.178, June 1994.
35. J.L. Neves and E.G. Friedman, "Automated synthesis of skew-based clock distribution networks," *Int'l J. VLSI Design*, March 1997.
36. S.Y. Kung and R.J. Gal-Ezer, "Synchronous versus asynchronous computation in very large scale integrated (VLSI) array processors," *Proc. of SPIE*, Vol. 341, pp. 53–65, May 1982.
37. I.E. Sutherland, "Micropipelines," *Comm. ACM*, Vol. 32, pp. 720–738, 1989.
38. Z.J. Deng, S.R. Whiteley, and T. Van Duzer, "Data-driven self-timing of RSFQ digital integrated circuits," ext. abstract, *5th Int'l Supercond. Electr. Conf. (ISEC)*, Nagoya, Sept. 1995, pp. 189–191.
39. M. Maezawa, I. Kurosawa, Y. Kameda, and T. Nanya, "Pulse-driven dual-rail logic gate family based on rapid single-flux-quantum (RSFQ) devices for asynchronous circuits," *Proc. 2nd Int. Symposium Advanced Research in Asynchronous Circuits and Systems*, pp. 134–142, March 1996.
40. M. Hatamian and G.L. Cash, "Parallel bit-level pipelined VLSI design for high-speed signal processing," *Proc. IEEE*, Vol. 75, pp. 1192–1202, Sept. 1987.
41. D.C. Wong, G.D. Micheli, and M.J. Flynn, "Designing of high-performance digital circuits using wave pipelining: Algorithms and practical experiences," *IEEE Trans. Comp. Aid. Design Int. Circ. and Syst.*, Vol. 12, pp. 25–46, 1993.
42. D.A. Joy and M.J. Ciesielski, "Clock period minimization with wave pipelining," *IEEE Trans. Comp.-Aid. Design Int. Circ. and Syst.*, Vol. 12, pp. 461–472, 1993.
43. Q. Ke and M.J. Feldman, "Single flux quantum circuits using the residue number system," *IEEE Trans. Appl. Supercond.*, Vol. 5, pp. 2988–2991, 1995.
44. Q. Ke, "Superconducting single flux quantum circuits using the residue number system," Ph.D. Thesis, University of Rochester, 1995.
45. K.K. Likharev, O.A. Mukhanov, and V.K. Semenov, "Resistive single flux quantum logic for the Josephson-junction technology," in *SQUID'85*, Berlin, Germany, W. de Gruyter, pp. 1103–1108, 1985.
46. S.V. Polonsky, V.K. Semenov, and D.F. Schneider, "Transmission of single-flux- quantum pulses along superconducting microstrip lines," *IEEE Trans. Appl. Supercond.*, Vol. 3, pp. 2598–2600, 1993.
47. S.V. Polonsky et al., "New RSFQ circuits," *IEEE Trans. Appl. Supercond.*, Vol. 3, pp. 2566–2577, 1993.
48. N. Weste and K. Eshraghian, *Principles of CMOS VLSI Design—A Systems Perspective*, Addison-Wesley, Reading, MA, 1985.
49. S.B. Kaplan and O.A. Mukhanov, "Operation of a superconductive demultiplexer using rapid single flux quantum (RSFQ) technology," *IEEE Trans. Appl. Supercond.*, Vol. 5, pp. 2853–2856, 1995.
50. A.F. Kirichenko, V.K. Semenov, Y.K. Kwong, and V. Nandakumar, "4-bit rapid single-flux-quantum decoder," *IEEE Trans. Appl. Supercond.*, Vol. 5, pp. 2857–2860, 1995.
51. S.V. Polonsky, V.K. Semenov, and A.F. Kirichenko, "Single flux, quantum B flip-flop and its possible applications," *IEEE Trans. Appl. Supercond.*, Vol. 4, pp. 9–18, 1994.
52. S.S. Martinet and M.F. Bocko, "Simulation and optimization of binary full-adder cells in RSFQ logic," *IEEE Trans. Appl. Supercond.*, Vol. 3, pp. 2720–2723, 1993.

53. A.F. Kirichenko and O.A. Mukhanov, "Implementation of novel 'push-forward' RSFQ carry-save serial adders," *IEEE Trans. Appl. Supercond.*, Vol. 5, pp. 3010–3013, 1995.
54. S.V. Polonsky, J.C. Lin, and A.V. Rylyakov, "RSFQ arithmetic blocks for DSP applications," *IEEE Trans. Appl. Supercond.*, Vol. 5, pp. 2823–2826, 1995.
55. Z.J. Deng, N. Yoshikawa, S.R. Whiteley, and T. Van Duzer, "Data-driven self-timed RSFQ digital integrated circuit and system," *IEEE Trans. Appl. Supercond.*, Vol. 7, 1997.
56. I. Kurosawa, H. Nakagawa, M. Aoyagi, M. Maezawa, Y. Kameda, and T. Nanya, "A basic circuit for asynchronous superconductive logic using RSFQ gates," in *Extended Abstracts of 5th Int'l Supercond. Electr. Conf. (ISEC)*, Nagoya, Sept. 1995, pp. 204–206.
57. I. Kurosawa, H. Nakagawa, M. Aoyagi, M. Maezawa, Y. Kameda, and T. Nanya, "A basic circuit for asynchronous superconductive logic using RSFQ gates," *Supercond. Sci. Technol.*, Vol. 8, pp. A46–A49, 1995.
58. M. Maezawa, I. Kurosawa, M. Aoyagi, H. Nakagawa, Y. Kameda, and T. Nanya, "Rapid single-flux-quantum dual-rail logic for asynchronous circuits," *IEEE Trans. Appl. Supercond.*, Vol. 7, 1997.
59. C.A. Mancini, N. Vukovic, A.M. Herr, K. Gaj, M.F. Bocko, and M.J. Feldman, "RSFQ circular shift registers," *IEEE Trans. Appl. Supercond.*, Vol. 7, 1997.
60. P. Bunyk and A. Kidiyarova-Shevchenko, "RSFQ microprocessor: New design approaches," *IEEE Trans. Appl. Supercond.*, Vol. 7, 1997.
61. P. Bunyk and V.K. Semenov, "Design of an RSFQ microprocessor," *IEEE Trans. Appl. Supercond.*, Vol. 5, pp. 3325–3328, 1995.
62. P. Patra and D.S. Fussell, "Conservative delay-insensitive circuits," *Proc. 4th Workshop on Physics and Computation: PhysComp96*, Boston, 1996, pp. 248–259.
63. J. Fleischman and T. Van Duzer, "Computer architecture issues in superconductive microprocessors," *IEEE Trans. Appl. Supercond.*, Vol. 3, pp. 2716–2719, 1993.
64. V.K. Kaplunenko, "Fluxon interaction in an overdamped Josephson transmission line," *Appl. Phys. Lett.*, Vol. 66, pp. 3365–3367, 1995.
65. J.-C. Lin and V.K. Semenov, "Timing circuits for RSFQ digital systems," *IEEE Trans. Appl. Supercond.*, Vol. 5, pp. 3472–3477, June 1995.
66. A. Yu. Kidiyarova-Shevchenko and D. Yu. Zinoviev, "RSFQ pseudo-random generator and its possible applications," *IEEE Trans. Appl. Supercond.*, Vol. 5, pp. 2820–2822, 1995.
67. H.B. Bakoglu, J.T. Walker, and J.D. Meindl, "A symmetric clock-distribution tree and optimized high-speed interconnections for reduced clock skew in ULSI and WSI circuits," *IEEE Int'l Conf. Computer Design*, pp. 118–122, Oct. 1986.
68. M. Shoji, "Elimination of process-dependent clock skew in CMOS VLSI," *J. Solid- State Circuits*, Vol. SC-21, pp. 875–880, 1986.
69. D.C. Keezer, "Design and verification of clock distribution in VLSI," *Proc. IEEE Int'l Conf. Commun. ICC'90*, Vol. 3, pp. 317.7.1–317.7.6, April 1990.
70. M.D. Dikaiakos and K. Steiglitz, "Comparison of tree and straight-line clocking for long systolic arrays," *J. VLSI Signal Processing*, Vol. 3, pp. 1177–1180, 1991.
71. "Hypres niobium process flow and design rules," Available from Hypres, Inc., 175 Clearbrook Road, Elmsford, NY 10523.
72. "TRW topological design rule for Josephson junction technology JJ-110A," available from TRW, One Space Park, Redondo Beach, CA 90278.
73. Z. Bao, M. Bhushan, S. Han, and J.E. Lukens, "Fabrication of high quality, deep-submicron Nb/AlO_x/Nb Josephson junctions using chemical mechanical polishing," *IEEE Trans. Appl. Supercond.*, Vol. 5, pp. 2731–2734, 1995.
74. K. Gaj, Q.P. Herr, and M.J. Feldman, "Parameter variations and synchronization of RSFQ circuits," *Applied Superconductivity 1995*, Institute of Physics Conf. Series #148, Bristol, UK, 1995, pp. 1733–1736.
75. A.D. Smith, S.L. Thomasson, and C. Dang, "Reproducibility of niobium junction critical currents: Statistical analysis and data," *IEEE Trans. Appl. Supercond.*, Vol. 3, pp. 2174–2177, 1993.
76. Q.P. Herr and M.J. Feldman, "Multiparameter optimization of RSFQ circuts using the method of inscribed hyperspheres," *IEEE Trans. Appl. Supercond.*, Vol. 5, pp. 3337–3340, June 1995.
77. L.A. Abelson, K. Daly, N. Martinez, and A.D. Smith, "LTS Josephson junction critical current uniformities for LSI applications," *IEEE Trans. Appl. Supercond.*, Vol. 5, pp. 2727–2730, 1995.
78. S.D. Kugelmass and K. Steiglitz, "An upper bound of expected clock skew in synchronous systems," *IEEE Trans. Comput.*, Vol. 39, pp. 1475–1477, 1990.
79. K. Gaj, C.-H. Cheah, E.G. Friedman, and M.J. Feldman, "Functional modeling of RSFQ circuits using Verilog HDL," *IEEE Trans. Appl. Supercond.*, Vol. 7, 1997.
80. P.-F. Yuh, "Shift registers and correlators using a two-phase single flux quantum pulse clock," *IEEE Trans. Appl. Supercond.*, Vol. 3, pp. 3009–3012, 1993.

Kris Gaj received the M.S. and Ph.D. degrees in Electrical Engineering from Warsaw University of Technology, Poland, in 1988 and 1992, respectively.

He has worked in computer-network security, computer arithmetic, testing of integrated circuits, and VLSI design automation. In 1991 he was a visiting scholar at the Simon Fraser University in Vancouver, Canada, where he worked on the analysis of various BIST (build-in-self-test) techniques for VLSI digital circuits. In 1992-93 he headed a research team at the Warsaw University of Technology developing an implementation of the Internet standard for secure electronic mail (Privacy Enhanced Mail), and software for secure Electronic Data Interchange per UNO standard UN-EDIFACT. He was a founder of ENIGMA, a company that generates practical software and hardware applications from new cryptographic research.

He has been with the Department of Electrical Engineering at the University of Rochester, Rochester, NY, since 1994, where he is a postdoctoral research fellow working on logic-level design and timing analysis of high-speed superconducting circuits. He currently teaches a graduate course on cryptology and computer-network security at the University of Rochester, and supervises student research projects on high-speed implementations of cryptography, VLSI circuit design and superconducting electronics.

He is the author of a book on code-breaking.

Eby G. Friedman was born in Jersey City, New Jersey in 1957. He received the B.S. degree from Lafayette College, Easton, PA. in 1979, and the M.S. and Ph.D. degrees from the University of California, Irvine, in 1981 and 1989, respectively, all in electrical engineering.

He was with Philips Gloeilampen Fabrieken, Eindhoven, The Netherlands, in 1978 where he worked on the design of bipolar differential amplifiers. From 1979 to 1991, he was with Hughes Aircraft Company, rising to the position of manager of the Signal Processing Design and Test Department, responsible for the design and test of high performance digital and analog IC's. He has been with the Department of Electrical Engineering at the University of Rochester, Rochester, NY, since 1991, where he is an Associate Professor and Director of the High Performance VLSI/IC Design and Analysis Laboratory. His current research and teaching interests are in high performance microelectronic design and analysis with application to high speed portable processors and low power wireless comunications.

He has authored two book chapters and many papers in the fields of high speed and low power CMOS design techniques, pipelining and retiming, and the theory and application of synchronous clock distribution networks, and has edited one book, *Clock Distribution Networks in VLSI Circuits and Systems* (IEEE Press, 1995). Dr. Friedman is a Senior Member of the IEEE, a Member of the editorial board of *Analog Integrated Circuits and Signal Processing*, Chair of the VLSI Systems and Applications CAS Technical Committee, Chair of the VLSI track for ISCAS '96 and '97, and a Member of the technical program committee of a number of conferences. He was a Member of the editorial board of the *IEEE Transactions on Circuits and Systems II: Analog and Digital Signal Processing*, Chair of the Electron Devices Chapter of the IEEE Rochester Section, and a recipient of the Howard Hughes Masters and Doctoral Fellowships, an NSF Research Initiation Award, an Outstanding IEEE Chapter Chairman Award, and a University of Rochester College of Engineering Teaching Excellence Award.
friedman@ee.rochester.edu

Marc J. Feldman received the Ph.D. degree in physics from the University of California at Berkeley in 1975. He worked at Chalmers University in Sweden and at the NASA/Goddard Institute for Space Studies in New York City in the development of superconducting receivers for radio astronomy observatories. He joined the faculty of Electrical Engineering at the University of Virginia in 1985, where he developed a variety of superconducting diodes for receiver applications.

He is now Senior Scientist and Professor of Electrical Engineering at the University of Rochester. Dr. Feldman's current research activities are directed towards the development of ultra-high-speed large-scale digital circuits using superconducting single-flux-quantum logic.

MIX
Papier aus verantwortungsvollen Quellen
Paper from responsible sources
FSC® C105338

If you have any concerns about our products,
you can contact us on
ProductSafety@springernature.com

In case Publisher is established outside the EU,
the EU authorized representative is:
Springer Nature Customer Service Center GmbH
Europaplatz 3, 69115 Heidelberg, Germany

Printed by Libri Plureos GmbH
in Hamburg, Germany